D1319743

THE FORGOTTEN EXPEDITION, 1804–1805

The FORGOTTEN EXPEDITION

1804–1805

THE LOUISIANA PURCHASE JOURNALS OF DUNBAR AND HUNTER

Edited by TREY BERRY,
PAM BEASLEY, *and*
JEANNE CLEMENTS

LOUISIANA STATE UNIVERSITY PRESS
BATON ROUGE

Published with the assistance of the V. Ray Cardozier Fund

Published by Louisiana State University Press

Manufactured in the United States of America
First Printing

DESIGNER: Amanda McDonald Scallon
TYPEFACE: Whitman
TYPESETTER: The Composing Room of Michigan, Inc.
PRINTER AND BINDER: Edwards Brothers, Inc.

Library of Congress Cataloging-in-Publication Data
Dunbar, William, 1749–1810.
The forgotten expedition, 1804–1805 : the Louisiana Purchase journals of Dunbar and Hunter / edited by Trey Berry, Pam Beasley, and Jeanne Clements.
p. cm.
Includes bibliographical references and index.
ISBN 0-8071-3165-2 (cloth : alk. paper)
1. Louisiana Purchase. 2. Ouachita River Region (Ark. and La.)—Discovery and exploration. 3. Dunbar, William, 1749–1810—Diaries. 4. Hunter, George, fl. 1804–1805—Diaries. 5. Explorers—Southwest, Old—Diaries. I. Hunter, George, fl. 1804–1805. II. Berry, Trey. III. Beasley, Pam. IV. Clements, Jeanne. V. Title.
F353.D86 2006
973.4′6—dc22 2005031032

The paper in this book meets the guidelines for permanence and durability of the Committee on Production Guidelines for Book Longevity of the Council on Library Resources.∞

For Kathy

Contents

Acknowledgments

The list of people who have worked to put this book together is expansive. First, my co-editors Pam Beasley and Jeannie Clements contributed in some very vital ways: organization, proofing, locating illustrations, and basically being there whenever they were called upon. This work would not exist without their contributions. Thank you, Jeanne and Pam.

The staff of the American Philosophical Society in Philadelphia, especially Rob Cox and Valerie Lutz, provided important assistance in opening the Dunbar and Hunter documents and giving guidance to other materials that turned out to be very relevant. In the same manner, the people at the Ouachita Parish Public Library in Monroe, Louisiana, and Wendy Richter and Janice Ford at the Riley-Hickingbotham Library at Ouachita Baptist University helped in locating several of the illustrations and secondary sources used throughout the book. Jeff Rogers at the Mississippi Department of Archives and History also helped locate several pictures from his department's Dunbar Collection.

Special appreciation goes to Clyde Gousett, who surprised everyone by revealing the existence of the original "Dunbar Trip Journal." Mr. Gousett, your gift continues to fascinate everyone who reads that two-hundred-year-old description of Arkansas and Louisiana.

Ouachita Baptist University awarded a research grant that provided both the time and the funding for this book. I wish to thank the administration of Ouachita, especially President Andrew Westmoreland and Vice-President Stan Poole, for their support. My fellow teachers in the School of Social Science at Ouachita, Tom Auffenberg, Hal Bass, Erik Benson, Wayne Bowen, Kevin Brennan, Ray Granade, Mark Miller, Doug Reed, and Randall Wight, have given me tremendous support during the research and writing stages. They are both friends and colleagues. The staff at the Arkansas Natural Resources Museum in Smackover, Arkansas, supported the project in many ways; they found sources and illustrations and also provided the time for my fellow editors to contribute.

Other scholars, including Tim Knight of the Biology Department at Ouachita Baptist University, biologist Henry Robison at Southern Arkansas University, and Mickie Warwick in the Spatial Information Systems Department at the University of Arkansas at Monticello, gave graciously of their time to help me fully understand the expedition's scientific and technical aspects. Their contributions enhance the clarity of this edited work.

Finally, I wish to thank Dr. Joe Nix and Professor Lavell Cole for being the inspiration behind this project. They first introduced me to the story of the "Forgotten Expedition," and I will forever be grateful to them.

TREY BERRY

Editors' Introduction

Beginning May 13, 1804, Meriwether Lewis and William Clark led the Corps of Discovery into the mostly unknown areas of the northern Louisiana Purchase. In response to the charge from President Jefferson, they were to establish relationships with the area Indians, map the various geographic features, catalog the native flora and fauna, and attempt to discover an all-water route to the Pacific Ocean. The journey took them through thousands of miles of territory, some of it known to Europeans and Americans and much of it unknown. Since their return in 1806, the chronicling and analysis of Lewis and Clark's accomplishments have continued to fascinate scholars and millions of Americans.[1]

The Corps of Discovery, however, was not alone in the exploration of the Louisiana Purchase. President Jefferson sanctioned three other fact-finding ventures into the new territory. These included the Dunbar-Hunter trip up the Ouachita River in 1804–1805, the Red River exploration of Thomas Freeman and Peter Custis of 1806, and the Zebulon Pike expedition into the Rocky Mountains and the Southwest from 1806 to 1807.[2] All of these Jeffersonian-era expeditions provide important information about a new and vibrant U.S. frontier. Of the journals from these four journeys, only the records of Dunbar and Hunter have not (until the publication of this volume) been edited together in a thorough fashion.

At the same time that Thomas Jefferson began recruiting and organizing the Corps of Discovery for the northern expedition, he initiated correspondence concerning a southern expedition. Jefferson envisioned an exploratory venture along the Arkansas and Red Rivers to the source of each. The proposed southern journey was later called the Grand Expedition; if completed, it would have been an excursion that rivaled the explorations of Lewis and Clark. After pursuing several prominent men to lead the southern trip, President Jefferson settled on an acquaintance of his from Natchez, Mississippi, named William Dunbar. Dunbar, a

1. Moulton, *Journals of Lewis and Clark*, 175–179.

2. Pike's expedition was authorized by Jefferson only after he had already departed St. Louis. It was viewed as a reconnaissance mission for the military. General James Wilkinson, serving at St. Louis, had sent Pike on this venture. As Jefferson said, it was "to reconnoitre the country, and to know the positions of his [Wilkinson's] enemies, Spanish and Indian." Jefferson to Dearborn, October 27, 1818, Jefferson Papers.

Scottish immigrant, shared Jefferson's passion for science, his belief in the superiority of an agricultural life, and his intense intellectual curiosity about the natural world.

On March 13, 1804, Jefferson wrote to Dunbar,

> Congress will probably authorize me to explore the greater waters on the Western side of Mississippi and Missouri, to their sources in case I should propose to send . . . another party up the Arcansa to its source, thence that to its mouth . . . These several surveys will enable us to prepare a map of Louisiana . . . as you live near the point of departure of the lowest expedition and can acquire so much better the information . . . I have thought if Congress should authorize the enterprize to propose to you the unprofitable trouble of directing it.[3]

With this letter, Jefferson charged Dunbar with the task of assembling and conducting the first American scientific expedition into the lower Louisiana Purchase.

Along with Dunbar, Jefferson assigned a prominent Philadelphia chemist-apothecary, George Hunter (another Scot), to be the coadjutor and to provide important information concerning the possible resources found in this new U.S. territory. Jefferson said of Hunter, "his fort[é] is chemistry, in that practical branch of science he has probably no equal in the US. and he is understood to be qualified to take the necessary astronomical observations."[4] Together, William Dunbar and George Hunter were prepared to embark on a journey that would take them into the vast hinterland of a growing nation.

Although Dunbar and Hunter exercised rigorous planning for the proposed Grand Expedition, their trip was drastically altered, resulting in a shorter journey. The brevity of the excursion consequently provided the earliest scientific study of the Ouachita River and the area known as "the hot springs." Their descriptions of the primitive environs of Arkansas and Louisiana provide an early-nineteenth-century portrait of the region's natural systems. Although a few scholars are aware of the journey, the expedition's meaning and significance are not widely understood. It is important that the Dunbar-Hunter trip along the Red, Black, and Ouachita Rivers be placed in the proper historical perspective as a piece of the complete story of Louisiana Purchase exploration, together with Lewis and Clark's expedi-

3. Jefferson to Dunbar, March 13, 1804, Jefferson Papers; McDermott, "Western Journals of Hunter," 8–9.

4. Jefferson to Dunbar, April 15, 1804, Jefferson Papers.

tion, the Zebulon Pike exploration from the central Purchase to the Spanish-held Southwest, and the Freeman-Custis Red River expedition.

William Dunbar and George Hunter were not the first people to explore the Ouachita River. The region had long been familiar to various indigenous peoples as well as countless European traders, trappers, hunters, and explorers. Europeans such as Don Juan (Jean-Baptiste) Filhiol had plied the waters long before the territory became U.S. property.[5] French and Spanish hunters and traders established settlements such as Ecore à Fabri (at the site of present-day Camden, Arkansas) and later Fort Miró (Monroe, Louisiana) as outposts in the late 1700s. Trade between displaced Native American populations and Europeans had also been fruitful for many years prior to the U.S. acquisition of the Purchase. In fact, the daily journal entries of both Dunbar and Hunter report several encounters with individuals and groups hunting, farming, and traveling along the Black and Ouachita Rivers.[6]

What this American team attempted to accomplish, however, was different from what other inhabitants and travelers were doing. They planned to complete precise mapping of the waterways, to categorize the flora and fauna, and to scientifically test the waters along the Ouachita River. They hoped to emerge from their short expedition with the first official view and description of the southern Louisiana Purchase. Although the historical record does not often reflect their journey, they did succeed in their effort.

The Dunbar and Hunter journey was not as long as the Corps of Discovery and did not provide an equal expanse of geographic and environmental information, but this mostly forgotten expedition remains significant for several reasons. First, the results of the astronomical and directional observations of both men supplied the information to draw accurate period maps of the Ouachita and Black Rivers and the many streams that have confluences along their banks. By using the explorers' data in 1805–1806, cartographer Nicholas King prepared a detailed map of the rivers and their confluences.[7] The Dunbar and Hunter journals and maps contain some of the first records in English of names and locations of settlers, set-

5. For more on Don Juan Filhiol, see Dunbar's entries for November 6 and 22.

6. Morris S. Arnold, *Colonial Arkansas, 1686–1804: A Social and Cultural History* (Fayetteville: University of Arkansas Press, 1991), 20, 219; Whayne, *Cultural Encounters in the Early South*, 39–60, 76–87. For more on Ecore à Fabri as well Fort Miró, see Dunbar's entries for November 6 and 22. The southern portion of the Ouachita River (beginning at Jonesville, Louisiana) is known as the Black River. The Black River flows into the Red River approximately 25 miles above the Red's confluence with the Mississippi River.

7. King, "Map of the Washita River in Louisiana."

tlements, trails, and footpaths along these waterways. They found locations that were simply named for the hunters who regularly deposited their cache of pelts at the site; others were small-to-medium-sized settlements functioning with varying degrees of prosperity. Although many of the named venues do not often appear on previous or subsequent maps, the names, locations, and trails are important markers that reveal the demographic and geographic details of the Ouachita River valley shortly following the U.S. purchase of the area.

A second argument for the importance of the journey is that significant early information was gathered about the region's human populations. In their journals, the explorers described the Red, Black, and Ouachita Rivers as well-known and frequently traveled waterways. Dunbar and Hunter's written depictions recount interactions between frontier settlers and traders of European descent with the indigenous populations. As the group progressed, they detailed many encounters with European trappers, hunters, planters, settlers, and fellow river travelers plying the waters of the Ouachita and southern Mississippi River valleys. Their copious notes portray a region where Europeans and their Indian hunters harvested abundant natural resources along the rivers and in the lands beyond. The reports from both men show that the hot springs near the upper reaches of the Ouachita River had already become a frequented destination for persons seeking relief from ailments and infirmities.[8] When the explorers arrived at the springs, they discovered a crude cabin and shacks constructed by visitors to the purportedly healing fountains.[9] During the ascent the team also met several individuals who had either previously visited the springs or were on their way to bathe in the waters.

Third, the expedition provided Americans with the first scientific study and indeed the first extensive description in English of the varied landscapes and the flora and fauna of the Ouachita River region of northern Louisiana and southern Arkansas. Many of the documented species no longer exist in the Ouachita River valley. Creatures such as buffalo, mountain lions, swans, whooping cranes, and red wolves are no longer found in the bottomlands and hills of the region. In several cases, Dunbar and Hunter's sightings are some of the earliest records in English of particular species; in other cases, they are among the final records of such habitation. These and other unique features of the journals confirm that this second of President Jefferson's Louisiana Purchase explorations resulted in the accumulation of important data concerning the southern areas within the new territory.

8. Congress established the area around the springs as the Hot Springs Reservation on April 20, 1832. On March 4, 1921, Congress changed the name of the Hot Springs Reservation to Hot Springs National Park.

9. Dunbar journal entries for December 10–12.

A fourth significant outcome centers on the experiments by the two Scottish explorers conducted at the hot springs. Their tests and observations were the first attempts to understand the natural systems and components found in the boiling waters. Not only did Hunter and Dunbar record temperatures and compositions of the various springs, but they also discovered microorganisms living in that hostile aquatic environment. Their descriptions of small crustaceans called ostracods and possibly smaller microscopic life thriving in these waters provide a very early record of thermophile or extremophile organisms. For modern scientists, the thorough study of these organisms did not begin in earnest until the 1960s and 1970s. Dunbar rightly stated while examining the small creatures,

> I have been fortunate in another research; I have always thought it probable that minute animalcules might be found in this water & have always looked attentively to that object; at length I found this evening upon the green matter a minute shell animal, shaped like a muscle or kidney, it is about the size of the smallest grain of sand, the colour of the shell is purplish brown, it opens the shell & thrusts out two articulated & very sharp clawed legs before; two more behind . . . it probably feeds upon the green matter.[10]

Dunbar also stated that finding life in the extreme environment of the hot springs should be "proof of the wonderful powers of nature in the production of animal & vegetable life in temperatures which have been hitherto thought sufficient to extinguish the vital principle."[11] Among the four expeditions sent by Jefferson into the Purchase, only the Dunbar-Hunter trip described such organisms.

Finally, Dunbar and Hunter were the first to report their findings to President Jefferson. Because the trip ended well in advance of the conclusion of the Corps of Discovery journey, the detailed notes, journals, and specimens compiled and collected by the travelers became the earliest reports Jefferson received that described to Americans the unfamiliar landscapes and varied resources within the United States' new territory. Although their abbreviated journey into the new American lands in no way rivaled the scope and the detail of the Lewis and Clark exploration, these two explorers obtained useful geographic, demographic, and environmental information for the Jefferson administration.[12] As explorers com-

10. Dunbar journal entry, December 28.

11. Ibid., December 12.

12. Rowland, *William Dunbar*, 9; Flores, *Jefferson and Southwestern Exploration*, 15; Flores, *Southern Counterparts to Lewis & Clark*, 15, 18, 19.

missioned by the president, the two Scottish immigrants combined finely honed skills and a wide knowledge base to complete the expedition. They contributed their own expertise to the analysis of the new territory. Together, Dunbar and Hunter offer a fairly comprehensive record of their short journey up the Red, Black, and Ouachita Rivers.

The story of William Dunbar is unique. Born in a castle in Elgin, Morayshire, Scotland, in 1749, the youngest son of Sir Archibald and Anne (Bain) Dunbar, William studied astronomy and mathematics in Glasgow and London. His studies in Scotland and England ignited a lifelong interest in all areas of science and discovery. In April 1771, at the age of twenty-two, he traveled to Philadelphia, where he engaged in trade with the Indians of the Ohio River Valley in and around Fort Pitt. During the next two years, he formed a partnership with a fellow Scottish merchant from Philadelphia named John Ross.[13] Around 1773, he moved to a site near present-day Baton Rouge, Louisiana, called Richmond (also called Richmond Settlement); there he began farming operations by growing indigo and making barrel staves.[14] Dunbar purchased a fifteen-mile section of land near the convergence of the Iberville and Mississippi Rivers. In 1785 he married an English woman named Dinah Clark, and together they had several children. Because of William Dunbar's frequent business trips, Dinah tended to the business of their plantations for many extended periods. In the abundant correspondence between Dinah and William, it is evident that she was not only a plantation wife but also a partner in all matters pertaining to the daily maintenance of their various business ventures.[15]

Dunbar eventually traveled to Jamaica to purchase slaves, but when he returned, his fortunes began to turn. Between 1775 and 1779 some of his slaves were

13. Prior to departing England, Dunbar secured trading supplies for the Indians from the company Hunter and Bailey for £1,000. Some biographies of William Dunbar confuse John Ross with an Alexander Ross who settled with Dunbar near Natchez and died in 1806. John Ross died in 1800, and in 1801 Dunbar dissolved his partnership through Ross's son Charles. Postlethwaite to Archibald Dunbar, Forest near Natchez, November 17, 1810, in Rowland, *William Dunbar*, 383–384; William Dunbar, "Memoir" (unfinished), Dunbar Papers, Jackson (hereinafter cited as Dunbar Memoir); Dunbar Papers, Ouachita Baptist University; Riley, "Sir William Dunbar," 85–86. According to the Dunbar Memoir, the young Dunbar called Philadelphia "one of the wonders of the Globe."

14. William Dunbar to Dinah Dunbar, New Orleans, March 13, 1794, Dunbar Papers, Jackson. Staves were the wooden slats (various widths and lengths) that could be bound together by metal strips to make barrels of differing sizes.

15. Dunbar Memoir; Dunbar Papers, Jackson; Dunbar Papers, Ouachita Baptist University. In his memoir Dunbar described himself as "little above the medium size, rather stoutly built, but well proportioned."

seized for participating in a rebellion and his home and farm became the target for plunderers: first an American marauder and then Spanish raiders.[16] After several years at Richmond, he settled near Natchez, Mississippi, where he built a fairly modest home known as "the Forest" in an area nine miles south of Natchez called Second Creek. By 1803 Dunbar owned around four thousand acres on two separate pieces of land—one the Forest, the other called "the Grange," both southeast of Natchez.[17] In addition, he acquired lots within Natchez as grants given to him for his service to the Spanish government in "laying off & measuring the lots of this City, forming plans of the elevations in the vicinity of the Fort, copying charts, and several Journies preformed by order of your Excellency."[18]

Dunbar became known throughout the lower Mississippi Valley for his many interests and endeavors. For example, at his farm at Richmond and later at the Forest, he became one of the first persons in the area to experiment with indigo and cotton. He also promoted the many uses of cottonseed oil, and after he had established his expansive cotton plantations around Natchez, he invented a screw press that created some of the first square bales. Dunbar extensively surveyed lands for the Spanish government. In 1798 Spanish governor Manuel Gayoso de Lemos appointed Dunbar as surveyor general of the District of Natchez and requested his services in determining the line of demarcation between Spanish West Florida and U.S. territory south of Natchez. Dunbar, Stephen Minor, and Gayoso, along with Andrew Ellicott, Thomas Freeman, Peter Walker, Andrew Ellicott Jr., and a Mr.

16. Captain James Willing, a commissioned Continental Army officer, committed the first offense, in which Dunbar stated he lost "£200 sterling value." Spanish soldiers conducted the raid. Some of Dunbar's correspondence of the time is addressed "Richmond Settlement." It was also referred to as "New Richmond." Following an agreement with Spain in April 1785, the drawing of new borders placed American lands adjacent to Spanish territory in the southern Mississippi valley (the border ran near the present-day city of Vicksburg). This may have also prompted Dunbar to make the move closer to the new U.S. territory near Natchez. Natchez and the lands south to 31° latitude did not become part of U.S. territory until the Pinckney Treaty of 1795. "Dunbar Diary"; Claiborne, *Mississippi*, 117; Riley, "Sir William Dunbar," 87; Kukla, *Wilderness So Immense*, 57–58.

17. The home at "the Forest" was described as a modest "cottage" having the living quarters above the ground with a porch extending across the front. The structure was expanded extensively after William Dunbar's death by his wife Dinah Clark Dunbar. The expansion began in 1816 and included porches with Doric columns on four sides, ninety-one doors, and thirty-two windows. A new staircase that Dinah Dunbar installed cost $1,500. John Newman, "A Short Account . . . Plan of Settlements Near the Natchez, Particularly William Dunbar, 1803," Dunbar Papers, American Philosophical Society, Philadelphia, 1; Chancery Records, 1810; Carpenter, "Note on the History of the Forest Plantation," 130–135; Davis, *Way through the Wilderness*, 91–92.

18. Manuel Gayoso de Lemos to governor, Natchez, April 19, 1797, Dunbar Papers, Jackson.

Gillespie as American representatives, surveyed the 31° latitude, which today continues as one of the borders between Mississippi and Louisiana. Ellicott came to admire William Dunbar and later stated that his fellow surveyor was "a gentleman whose talents, extensive information and scientific acquirements would give him a distinguished rank in any place or any country."[19]

As a scientist, Dunbar owned one of the first telescopes in the lower Mississippi Valley. He built an observatory near his Natchez home at a place called Union Hill and opened the structure to the public. In 1803 Dunbar and others established the Mississippi Society for the Acquirement and Dissemination of Useful Knowledge. The organization sought to collect, through scientific inquiry, all relevant information concerning areas along or near the Mississippi River.[20]

Dunbar the planter, though an enlightened gentleman scientist, was very much a product of the late eighteenth and early nineteenth centuries, adhering to the southern slaveholding norm. He practiced strict discipline and subjected his slaves to the extremely hard daily labor of clearing fields and tending to the wide variety of crops grown at the Forest, such as tobacco, corn, cotton, indigo, fruits, and vegetables. Because of the proximity of vast Mississippi frontier forests, Dunbar continued the practice of producing barrel staves and headings.[21] Like many of his contemporaries, the Scottish planter dealt harshly with unproductive or un-

19. De Lemos served as governor-general from October 28, 1796, until July 18, 1799, when he died from yellow fever. Before his appointment as governor-general, de Lemos held the position of commandant at Natchez, and he became acquainted with Dunbar during that time. Glen S. Greene, Ernest Russ Williams, Lorraine Heartfield, Mitchell Hillman, and John R. Humble, *The Search for Fort Miró: A Colonial Spanish Fort in the Ouachita Valley*, Popular Series, no. 1 (Monroe: Northeast Louisiana Archeological Society, 1975), 57. Andrew Ellicott to Sara Ellicott, June 19, 1798, Andrew Ellicott Papers; Mathews, *Andrew Ellicott*, 129; Rowland, *William Dunbar*, 11; Hamilton, "Running Mississippi's South Line," 161–168; Riley, "Sir William Dunbar," 93; McDermott, "Philosophic Outpost on the Frontier," 3–9. Ellicott also trained Meriwether Lewis in the use of various navigational and surveying instruments. Freeman later led the Red River Expedition along with Peter Custis in 1806.

20. Rowland, *William Dunbar*, 10; Dunbar Papers, Jackson; James, *Antebellum Natchez*, 231–232.

21. While Dunbar lived at Richmond Settlement in Louisiana, his slaves began producing about five thousand staves per week and approximately seventeen thousand per month. At one point, Dunbar even made the laborious task of making the staves a contest, which he called a "trial of skill." The process would prove which slave could produce the most staves in a week. At the time of the Ouachita River expedition, George Hunter reported that William Dunbar's slaves at the Forest annually produced about one hundred bales of cotton. "Dunbar Diary"; "Dunbar Slave Journal," private collection, Natchez, Mississippi, copy in Special Collections, Ouachita Baptist University (hereinafter cited as "Dunbar Slave Journal"); "Hunter Official Report." Hunter reported during his 1804 visit to the Forest that Dunbar owned "about fifty" slaves. By 1810 Dunbar owned 128 slaves. "Hunter Official Report," 5; Chancery Records, 1810; Bailyn, *Voyagers to the West*, 488–492.

ruly slaves. In addition to employing the lashes and ankle chains administered at other plantations, while he lived at Richmond Settlement Dunbar also devised a unique detention structure he labeled "the Bastile." Apparently this was a four-walled brick structure sunken into the ground. Slaves could be imprisoned in the structure for various lengths of time according to the severity of their accused offenses.

However, William Dunbar also tended to his slaves as a man who desired to protect his property and investments. He became one of the first in the region who inoculated his slaves against smallpox; and he fed them with what seemed to be sustainable portions and regularity and clothed them adequately. Unlike Jefferson, Washington, and some of his other contemporaries, Dunbar never seemed to be disturbed by the burden of owning other human beings; in fact, he viewed his slaveholding as a privilege and an appropriate means to success, status, and station.[22]

Dunbar's home became a haven for the gentleman scholars of the day. Men like the naturalist William Bartram and the renowned ornithologist Alexander W. Wilson sought Dunbar's company and welcomed his hospitality at the Forest as a respite from the often rustic conditions of the lower Mississippi valley. Dunbar also corresponded with some of the most learned minds of the period, including Wilson, Bartram, William Herschel, Benjamin Rittenhouse, Henry Muhlenberg, and Benjamin Rush.[23] The most fruitful correspondence of Dunbar's life however, came from Thomas Jefferson.

Daniel Clark, who served as U.S. consul for New Orleans, first introduced William Dunbar to then vice president Thomas Jefferson through a letter in 1799, saying that "for Science, Probity, & general information [he] is the first Character in this part of the World."[24] In 1800 Jefferson recommended and secured Dunbar's election to the American Philosophical Society in Philadelphia. Shortly following his induction, the two men began a six-year correspondence over topics as varied

22. "Dunbar Diary"; "Dunbar Slave Journal."

23. Allen, "Jefferson and the Louisiana-Arkansas Frontier," 45; Riley, "Sir William Dunbar," 91. Herschel was the court astronomer to King George III. He is credited with the discoveries of the planet Uranus and of infrared radiation. Benjamin Rittenhouse became a maker of scientific and navigational instruments. A renowned naturalist, botanist, and Lutheran clergyman, Henry Muhlenberg later received some of Dunbar's botanical specimens from the expedition. Dr. Benjamin Rush was a renowned Philadelphia physician who was educated in Edinburgh, Scotland, and stressed the need for purging and bleeding victims as cures. An ardent patriot, Rush also signed the Declaration of Independence. Gardner, *Who's Who in British History*, 418.

24. Dearborn to Jefferson, February 12, 1799, Jefferson Papers.

as botany, astronomy, geography, meteorology, surveying, philosophy, western exploration, and the characteristics of American Indians.[25]

By 1803 these two intellectuals had become sufficiently well acquainted and aware of each other's strengths that Dunbar was the key figure in Jefferson's discussions and plans to explore the southern Louisiana Purchase. In 1801 Jefferson praised the Natchez planter as "a person of Great worth & wealth there, and one of the most distinguished citizens of the U.S. in point of science. He is a correspondent of mine in that line in whom I set great store." The president relied on the aging Scotsman's advice and his propensity for getting things done in the less-than-organized Mississippi valley frontier. After several attempts were made to secure a leader for a southern expedition, Jefferson turned to his old and respected correspondent.[26]

Dunbar's coadjutor, George Hunter, has an equally interesting story. He was also born in Scotland, but unlike the aristocratic Dunbar, Hunter was born into a lower-middle-class family in Edinburgh, where his father worked as a cooper. For financial and educational reasons, his father apprenticed the young Hunter to be a "druggist." In 1774 the nineteen-year-old Hunter decided to use his skills gained as an apprentice to acquire work in the American colonies. In that year he traveled with his family and new stepfather to Philadelphia, and he quickly secured employment as an apothecary. During the American Revolution, Hunter served as an enlisted soldier and later as an apothecary in the Hospital Department of the Continental Army. The pivotal days of late 1776 and early 1777 found him present at the Battles of Trenton and Princeton. He later resigned the army and for three years worked as a "Surgeon" aboard a ship named *Betty Bound to Teneriffe.*[27]

After the war Hunter married Phoebe Bryant of Philadelphia on December 28, 1786, and he began to develop a reputation in the northeastern United States as a

25. Rowland, *William Dunbar,* 111, 122–194, 329.

26. Jefferson to Claiborne, July 13, 1801, in Ford, *Writings of Jefferson,* 8:72; Dunbar to Jefferson, letter entitled, "Gentile," January 17, 1804, Jefferson Papers; Dunbar to Jefferson, May 13, 1804, Dunbar Papers, Jackson; Jefferson to Dearborn, Washington, February 14, 1807, Jefferson Papers; Rowland, *William Dunbar,* 11, 122, 126, 130. Andrew Ellicott had suggested to Jefferson that Peter Walker and Mr. Gillespie would be good choices to lead the expedition, and William Dunbar concurred. Neither was available to participate in the venture. Flores, *Southern Counterparts to Lewis & Clark,* 40; Hamilton, "Running Mississippi's South Line," 162.

27. Hunter began work at the druggist firm of Christopher and Charles Marshall. He also served for three years as an "assistant apothecary." During his time at sea, he was held for six weeks as a prisoner aboard an "English Prison ship." Hunter Papers. The Hunter papers include a section of notes in which Hunter outlined his life. McDermott, "Western Journals of Hunter," 5.

chemist, druggist, and successful businessman.[28] Between 1796 and 1802, Hunter, then going by the title "Dr.," began a series of explorations into the Illinois and Kentucky backcountry, presumably searching for areas with various mineral deposits, speculating for western lands, or both. He also engaged in various business dealings, including a short venture in the river transportation of trade goods. By 1803 George Hunter had created a comfortable living for his growing family, had acquired experience in traversing pioneer regions, and had established himself as a well-known scientist. In fact, his scientific reputation reached the White House. In 1804 Thomas Jefferson said of the forty-nine-year-old chemist, "in the practical branch of that science [chemistry] he has probably no equal in the U.S."[29] Although they came from very different socioeconomic backgrounds, George Hunter and William Dunbar would soon become partners in the labors of Louisiana Purchase exploration.

In his April 15, 1804, letter to Dunbar, Jefferson not only asked the prominent Natchez resident to lead an expedition into the Louisiana Purchase, but he also informed him that he had assigned Dr. Hunter as "fellow-labourer and counsellor" for what the two men began to call the "Grand Expedition." For Dunbar, Hunter, and Jefferson, the proposed Grand Expedition would be a trip along both the Red and the Arkansas Rivers. Such a trip, if conducted, would have rivaled the one being planned by Meriwether Lewis and William Clark along the Missouri River.[30]

Jefferson held the common belief that, like the Arkansas River, the Red River's source lay deep in the Rocky Mountains. The sources of these two rivers, according to maps of the period and conjectured accounts, lay close to each other, and thus an overland trek between the two would be feasible. In fact, a distance of approximately 450–500 miles lay between the Arkansas River's source in central Colorado and the two forks that eventually form the Red River in eastern New Mexico and in what is now the Texas panhandle. Jefferson knew that not only would this journey result in scientific and geographic information; the trip into the central and southern reaches of Louisiana would also skirt and in some places enter lands claimed by Spain. The president may also have wanted to use the Grand

28. Hunter Papers; *Record of Pennsylvania Marriages*, 574. In 1783 Hunter also became an original or founding member of the Society of Cincinnati. Metcalf, *Society of Cincinnati*, 28.

29. "Journal Kept By George Hunter Of A Tour From Philada. To Kentuckey & The Illinois Country, 1796"; and "Journal From Philada. Towards Lexington Kentucky By George Hunter Senior & Junior Begun Augt. 19th. 1802," both in Hunter Papers. Jefferson to Dunbar, April 15, 1804, Jefferson Papers; McDermott, "Western Journals of Hunter," 8–9.

30. Quotations in Dunbar to Jefferson, October 15, 1804, Jefferson Papers; Dunbar to Jefferson, May 13, 1804, Dunbar Papers, Jackson; Rowland, *William Dunbar*, 130–132.

Expedition as a means to test Spanish resolve and to acquire a glimpse of the vistas owned by another North American power.[31]

The correspondence between Jefferson and Dunbar before and after the trip to the hot springs reveals the interest of both men in understanding the extent of Spanish settlement and strength. The subsequent explorations of Freeman and Custis along the Red River, and later those of Zebulon Pike into the Rockies and the Southwest, also obtained a better understanding of the extent of Spanish control. Jefferson used all three of his southern explorations to test the Spanish; and as the team began to organize for their Grand Expedition, originally set for spring 1804, Dunbar and Hunter became aware of this political and military side of their journey.[32]

Following an appropriation of three thousand dollars by Congress, the preparations began in earnest. Hunter was given the charge to procure a boat and supplies in Pittsburgh while Dunbar continued exchanges with the president, Secretary of War Henry Dearborn, and men well acquainted with the area such as Peter Walker and the commandant in New Orleans, Colonel Constant Freeman.[33] However, during the initial planning stages, both Jefferson and Dunbar became fearful over reports they had received concerning the warring activities of a splinter group of Osage Indians in what would become Arkansas and Oklahoma. The group, led by a chief called the Great Track or Big Track, had broken away from the main tribe. Because of his concerns for the safety and success of the expedition, Jefferson wrote to Dunbar that he feared these Osage would hinder their travel along the Arkansas River "and perhaps do worse." Both Jefferson and Dunbar also had apprehensions regarding possible Spanish resistance above the Bayou Pierre in northwestern Louisiana and northeastern Texas. Appropriately, the president postponed the trip.[34]

31. Arrowsmith, "Map of North America"; Cohen, *Mapping the West;* Flores, *Southern Counterparts to Lewis & Clark,* 23, 33–34.

32. Jackson, *Jefferson and the Rocky Mountains,* 224–241, 234, 242–267.

33. The itemized $3,000 budget included $1,400 salary for Dunbar and Hunter, $500 for instruments, $300 for outfitting, $600 for presents for the Indians they would encounter, and $140 cash. Dearborn to Jefferson, April 13, 1804, Jefferson Papers. Peter Walker had worked with Dunbar on the 31° latitude survey, and he was about to embark on a trek along the Red River. Lieutenant Colonel Constant Freeman served as a member of a military group called the Artillerists. Freeman aided the Dunbar-Hunter expedition in securing supplies and altering Hunter's boat prior to the departure. Rowland, *William Dunbar,* 124–141.

34. Jefferson to Dunbar, July 1804, Jefferson Papers. Other names for the Great Track included Cashesegra or Cashecegra and Makes-Tracks-Going-Far-Away. Flores, *Southern Counterparts to Lewis & Clark,* 44; Jackson, *Thomas Jefferson and the Stony Mountains,* 237, 246.

In June of 1804, Dunbar wrote to Jefferson asking for permission to attempt what both men initially considered a trial run up a tributary of the Red River, a smaller stream called the "Washita" (Ouachita). Dunbar told Jefferson in an August 17, 1804, letter that there were many "curiosities" along the Ouachita—in particular something he called "the boiling springs or fountain."[35]

Jefferson agreed to the change, and after several weeks of planning and preparations by both Hunter and Dunbar, the group departed from St. Catherine's Landing on the Mississippi River on October 16, 1804. The expedition team consisted of Dunbar and Hunter, thirteen enlisted soldiers, Hunter's teenage son, two of Dunbar's slaves, and his servant.[36] The nineteen men occupied a strange-looking "Chinese-style vessel" that had been designed and commissioned by Hunter in Pittsburgh several months earlier. As to the design of the vessel, Hunter explained:

> This boat is fifty feet long on deck, 30 feet straight Keell, flat bottom somewhat resembling a long Scow in use to ferry over waggons . . . the sides are about 3½ feet wide upon deck where broadest . . . She is covered with light boards from the Stern 32 feet foret, so as to give good accomodations to the passengers & furnished with a Stout Mast 36 feet long [and] a Sail 24 feet by 27[,] in the Chinese stile, fastened to a yard 24 feet & boom 29 feet & spread by 5 sprits the whole width of the Boom, Has 2 large setting poles & 2 side oars[37]

35. Dunbar to Jefferson, Natchez, June 9, 1804, Jefferson Papers; also in Rowland, *William Dunbar,* 133–135.

36. St. Catherine's Landing was located near the confluence of St. Catherine's Creek and the Mississippi River, approximately 12 miles below Natchez. For more, see Dunbar's and Hunter's journal entries for October 16 and Hunter's entry for January 31. The roster of the expedition listed the names of the soldiers as Sergeant Bundy, Peter Bowers, John White, Robert Wilson, Matthew Boon, William Skinner, William Little, William Tutle, Manus McDonald, Jeremiah Smith, Edward Rylet, William Court, and Jeremiah Loper. Dunbar referred to his servant as "my own domestic." Hunter journal entry, October 18; Dunbar to Jefferson, November 9, 1804, Jefferson Papers; McDermott, "Western Journals of Hunter," 10–11, 12, 65; Dunbar journal entry, January 18–19; Dickinson, "Don Juan Filhiol," 133–135; Dickinson, "Early View of the Ouachita Region."

37. Hunter to Dearborn, June 14, 1804. A rough draft of the letter is included in "Hunter Journals," vol. 1. Dunbar, unhappy with the vessel, sent Hunter and several other men to New Orleans to have it made more suitable by alterations and to obtain supplies from Colonel Constant Freeman. A commissioned officer (Lieutenant Wilson) had been assigned to participate; however, Dunbar sent him back to New Orleans, stating, "I did not find my self authorised to deprive the Service of a Commissioned officer upon this little expedition." Dunbar to Jefferson, October 15, 1804, Jefferson Papers. There are also hints that the officer may have had disagreements with Hunter. McDermott, "Western Journals of Hunter," 10, 65.

On the same date that Dunbar and Hunter left Natchez, Lewis and Clark reached the upper Missouri River in the Dakotas. They were approaching the Mandan villages where they would spend the winter of 1804–1805. The Corps of Discovery had already traveled twice as far as their southern counterparts would venture during the entirety of their expedition.[38]

As Dunbar and Hunter ascended the Red, Black, and Ouachita Rivers, the journals of both men became replete with descriptions of soil types, water levels, flora, and fauna, as well as their daily astronomical and thermometer readings. To construct the most accurate map possible, William Dunbar used a chronometer, an instrument called a circle of reflection, pocket and surveying compasses, and other instruments.[39] Hunter attempted to use a sextant he had purchased in Philadelphia, but he often found it to be cumbersome and less than accurate.[40] In addition to the scientific recordings, they often kept copious notes in their journals that documented the human drama of their adventure.

The drama was occasioned in part by the size and design of Hunter's "Chinese-style" boat. The draft proved far too deep for inland rivers, and this created difficulties in navigating the sand bars and submerged logs that were plentiful in the Black and lower Ouachita Rivers. On many days the soldiers spent hours pulling the boat through shallow areas or digging channels to allow for passage. Their journey was also becoming arduous because of the constant complaining among the soldiers. No officer was present to lead them; only a man called Sergeant Bundy had been assigned to the expedition to guide the soldiers and maintain discipline. According to both Hunter and Dunbar, Sergeant Bundy actually encouraged discontent by refusing to temper the men and by his own complaints directed toward the leaders.[41]

38. Moulton, *Journals of Lewis and Clark,* 175–176.

39. The explorers' list of navigational and scientific instruments also included an octant, artificial horizons, a surveyor's chain, an acrometer (an instrument that measured acceleration), several pocket watches, and a microscope. Dunbar's circle of reflection (a reflecting circle) was an instrument supported by a tripod pedestal that could be used with an artificial horizon when a true horizon could not be seen. The circle of reflection was made by Troughton of London. Dunbar and Hunter almost daily took numerous solar and lunar observations. "Dunbar Trip Journal, Vol. II," December 16, 1804; Hunter journal entries, October 19; November 14; December 13, 27, 31; Dunbar, "Journal of a Geometrical survey." For more detail, see "Explanation of Navigational Techniques" in this volume.

40. Dunbar to Jefferson, November 9, 1804, Jefferson Papers. Dunbar wrote this letter from Fort Miró to give the president some news of their progress.

41. Dearborn to Hunter, March 30, 1804, Jefferson Papers; "Dunbar Trip Journal, Vol. I"; Rowland, *William Dunbar,* 134; McDermott, "Western Journals of Hunter," 103. Dunbar journal entry, November 4; Hunter journal entry, November 4. Sergeant Bundy's first name is not recorded.

As the team continued its rigorous ascent, the soldiers captured a runaway slave who called himself Harry. Hunter referred to him as a "stout fellow," and the team leaders decided to keep him with them until they returned to Fort Miró or until they could locate his owner. Just ten days later, a man by the name of Innes called from the banks and claimed to be the slave's owner. The explorers turned Harry over to him and immediately continued their voyage.

On November 6, after great difficulty in ascending the river in Hunter's Chinese-style vessel, the group reached the site of Fort Miró or the Poste du Ouachita. The fort, first established by the French around 1784, had been turned over to American control only seven months before, in April of 1804. The new American commander of the site, Lieutenant Joseph Bowmar, welcomed the explorers and treated them to what hospitality he could muster in the primitive surroundings.[42] The crew also received some much-deserved rest from the rigors of the first two hundred miles.[43]

Realizing that attempting to travel any farther in their cumbersome boat might lengthen their trip and possibly cause a mutiny among his men, Dunbar began to inquire at the fort about renting a new vessel. By November 9 they had secured a large flatboat with a cabin on deck, and they hired an experienced guide named Samuel Blazier. Both leaders described their new guide as a man who had traveled the Ouachita River many times and who had, along with many others, visited the hot springs region. Blazier's familiarity with the area above Fort Miró may be what enabled both Dunbar and Hunter to name and elaborate on many sites above this final outpost. After obtaining additional supplies and selling others to local inhabitants, the following day the group resumed its journey.[44]

42. For more on the Poste du Ouachita or Fort Miró, see Dunbar entry for November 6. Born in Tennessee, Bowmar had entered army service in 1798. On April 2, 1802, he became a second lieutenant, and he was promoted to captain only four days before the expedition left Natchez on October 12, 1804. Apparently, news of Bowmar's promotion had not reached Fort Miró by the time that Dunbar and Hunter arrived on November 6. Bowmar resigned his commission on June 20, 1806. Heitman, *Historical Register and Dictionary of the U.S. Army,* 255. For more on Joseph Bowmar, see Dunbar's entry for November 6.

43. "Dunbar Trip Journal, Vol. I," November 6, 7, 8, and 9, 1804; Hunter journal entries, November 6, 7, and 10; Rowland, *William Dunbar,* 227–235; Jackson, *Jefferson and the Stony Mountains,* 225–226; Arnold, *Colonial Arkansas,* 20. For more on Fort Miró, see Dunbar's and Hunter's journal entries for November 6.

44. For more on Samuel Blazier, see Dunbar's entry for November 12 and Hunter's entry for November 13. Dunbar also wrote a letter to Jefferson on November 9 to report on the progress of their ascent. See Dunbar's entry for November 9. According to Dunbar's journal, the flatboat or barge was 50 by 8½ feet and "well formed." Blazier had traveled the Ouachita and Red Rivers between Fort Miró and

As the expedition crossed into present-day Arkansas on November 15, 1804, the landscape began to change from predominately pine forests to bottomlands mixed with various hardwoods. The team also observed a greater abundance of sweet gum and birch trees along the banks. The waterfowl became more plentiful, and Dunbar and Hunter recorded the sighting of a small alligator (the last they observed) sunning itself on a sandbar. Some of the soldiers in a hunting party returned with a large swan, which they skinned and preserved. The soil of the eroded banks also began to change from the cream-colored loam of the southern Ouachita, Black, and Red Rivers to what Dunbar described as more sand and gravel areas.[45]

When they neared Ecore à Fabri (Camden, Arkansas), the former site of the Spanish outpost, two significant events occurred. First, the explorers found a tree with curious carved Indian hieroglyphics on its trunk. Both men described the glyphs as consisting of several human and animal figures. One symbol represented two people holding or shaking hands.[46]

The second event almost ended the expedition. On November 22, as George Hunter cleaned his pistol on the flatboat, the gun discharged. The bullet ripped through his thumb and lacerated two other fingers. The projectile continued through the brim of his hat, missing his head by only fractions of an inch. Hunter remained in severe pain and danger of infection for over two weeks. The explorer's eyes received burns, and he could not see to record entries in his journals. The wounds eventually healed, but Hunter could render little service to the team during that time.[47]

Above Ecore à Fabri the forests began to change once again. Large stands of "Birch, Maple, holly, ironwood, dogwood, Ash and Sweet Gum . . . white, black and red oak" timber dominated the landscape. Large and dense groves of cane lined some portions of the banks, creating a visual barrier to the vistas beyond. The eroded cliffs also were higher—an average of six to twelve feet above the water. Sand and gravel bars became more numerous, and travel for the team slowed. Even with Hunter's odd boat, the team had traveled an average of ten to twelve miles per day until they reached Fort Miró. As the expedition moved farther into what would become Arkansas, the daily travel averages decreased to two to six miles. The lower daily temperatures, combined with the earlier delays created by the Chi-

Natchez extensively. He had also apparently made a few trips to the Hot Springs area. "Hunter Official Report."

45. Dunbar journal entries, November 16–26.

46. For more on the hieroglyphs, see Hunter's entry for November 21.

47. Ibid., November 21, 22.

nese-style boat, also meant that the explorers might not complete their voyage before the harshest part of the winter.[48]

Near the current site of Arkadelphia, Arkansas, the team met a man named Paltz and his hunting party. Dunbar referred to him as an "old Dutch hunter," and the elderly gentleman told the leaders that he knew the area well. In an entry that revealed the extent of European penetration into the area, Dunbar recorded that Paltz had "resided 40 years on the Washita and before that period [had] been up the arcansa river, the white river, and the St. Francis." The hunter told the explorers about a salt spring nearby as well as other natural features. Hunter, Paltz, and a small group investigated the "salt spring," or "saline," which turned out to be of a substantial nature. The chemist conducted gravity experiments on the saline water and discovered that it contained a high concentration of what he called "marine salt."[49] In addition, both men recorded that by the time they reached this site, the wildlife and landscape had begun to change in a more radical manner. No longer was the river lined with what Dunbar called a vegetable brown soil; they were now seeing embankments that contained more large rocks and gravel. The team also came upon more frequent and larger flocks of turkeys and herds of deer, as well as raccoons and signs of buffalo. The forests became predominantly stands of enormous oak and beech trees.[50]

On December 3, 1804, Dunbar and Hunter confronted the greatest potential obstacle to their ascent. Near what is now Malvern, Arkansas, an enormous series of rocky rapids, called "the Chutes" by the two men, stretched almost one mile before them. Dunbar described the formations as looking like "ancient fortifications and Castles." The Scottish planter compared the roar made by "the Chutes" to the sound of a hurricane he had experienced in New Orleans in 1779. Through strenuous cordelling efforts,[51] rocking the vessel from side to side, and more or less dragging the flatboat between and over rocks, the team finally traversed the maze of boulders. As they emerged from "the Chutes," the crew could see that the rolling

48. For more on Ecore à Fabri, see Dunbar's journal entry for November 22; Rowland, *William Dunbar*, 252–253, 267.

49. Dunbar journal entries, November 28, 29; Hunter journal entry, November 29. The term *Dutch* was sometimes used during this era to denote people of German origin. In the "Hunter Official Report," Hunter wrote that Paltz was "of German extraction." McDermott, "Western Journals of Hunter," 98.

50. Hunter journal entry, November 29; "Dunbar Trip Journal, Vol. I," November 28, 29, 1804.

51. Dunbar journal entry, December 3. The 1779 New Orleans hurricane is referred to as "William Dunbar's Hurricane" in honor of the testing and observations he conducted during the storm. Roth, "Louisiana Hurricane History." Cordelling is a method of pulling a vessel up a river using lengths of rope.

foothills of the Ouachita Mountains began to rise on either side of the river. According to their calculations and the stories of previous travelers, they expressed hope that their final destination of the hot springs was only a few miles upstream.[52]

By December 7 the group had reached the closest point along the Ouachita River to the hot springs—a site called "Ellis camp" at the mouth of present-day Gulpha Creek in Garland County. Several men wasted no time and immediately began a nine-mile walk to examine the springs. The men returned the next afternoon with vivid descriptions of their experiences. They stated that they had discovered an empty cabin and some plank shacks thought to be used by people coming to use the purported healing waters of the springs. The men also described the water emerging from the springs as tasting like "Spice-wood tea" and being very hot to the touch.[53]

The following day, Dunbar and Hunter traveled to the springs and began a four-week study of the water's properties, as well as the area's geological and biological features. The group found an open log cabin as well as several flat-board shacks about one-half mile away from the springs. The team members occupied and repaired those dwellings during their month-long residence.[54]

As they surveyed and studied the site, the explorers determined that there were "four principal" and "two inferior" springs in the geologic complex. They also measured the water temperature. Both explorers determined the temperature to be between 148° and 150°. Hunter also cataloged the numerous limestone deposits, and Dunbar discovered a cabbagelike plant he labeled "Cabbage raddish of the Washita."[55] Dunbar described small microorganisms living in the hot water and on some of the nearby moss. This finding may be one of the first North American reports of such creatures (thermophiles or extremophiles) existing in such hostile aquatic environments. As to the taste of the water, Dunbar's description was much less fanciful than those of his men. He simply stated that it tasted like hot water.[56]

52. Dunbar and Hunter journal entries, December 3; Rowland, *William Dunbar,* 261–262; McDermott, "Western Journals of Hunter," 100. The explorers also describe a pair of swans on the water near "the Chutes."

53. Dunbar and Hunter journal entries, December 7.

54. Dunbar journal entry, December 9.

55. This may have been an Arkansas cabbage (*Streptanthus obtusifolius*), which occurs in the Hot Springs area and throughout the Ouachita Mountain Range. It may also have been a spring cress (*Cardamine bulbosa*). However, the spring cress occurs throughout the state. Dunbar and Hunter describe the species as being very localized. Dunbar and Hunter journal entries, December 13.

56. Hunter journal entry, December 10; "Dunbar Trip Journal, Vol. I," December 10; "Dunbar Trip Journal, Vol. II," December 12, 28, 1804. The microorganisms may have been ostracods, minute crus-

Dunbar and Hunter also took separate treks into the surrounding mountains, where they described the slopes, valleys, creeks, and vistas that they viewed and traversed. The men sighted more swans, deer, and raccoons, as well as additional signs of buffalo in the vicinity of the hot springs.[57]

Toward the end of their examination of the springs, Dunbar began to speculate in his journal about the nature and source of the water's constant temperature. Seeing no volcanic formation in the surrounding area, he surmised that a compound called "schystus" might be the cause. According to Dunbar's analysis, when this mineral decomposed, it generated enormous amounts of heat. Large deposits of the mineral, Dunbar speculated, must be located beneath the springs. Despite their hypotheses and experiments, however, both men revealed that the true sources of the heated waters remained a mystery to them.[58]

Following a brief snowstorm, the continual drop in daily temperatures, and a week of waiting for the river to rise, the expedition finally began the return trip on January 8, 1805. The ascent of the Ouachita had taken fifty days of grueling work, but the descent took the team less than half that time. While descending, the explorers met a group of Indians who were possibly Quapaw. Hunter called them "Indians . . . from the river Arkansa." The Indian party was led by a French trader named LeFevre,[59] who apparently accompanied the expedition as far as Fort Miró. LeFevre provided Dunbar and Hunter with a wealth of knowledge concerning the region. Probably the information given by the French trader enabled both men to include in their journals much more detail as to the place names, the river sources, and the lore about adjacent western lands.

As the exploration team continued their downstream journey, they observed and described an eclipse of the moon. After a three-day stop at Fort Miró to reluctantly retrieve Hunter's Chinese-style boat, they resumed their homeward voyage. Dunbar separated from the group at Fort Miró, taking with him one soldier and his servant. He hoped to make better time by utilizing a canoe and then traveling by horseback to Natchez. Dunbar arrived at the Forest on January 26,

---

taceans that are found throughout the world. For more on these organisms, see Dunbar's entry for December 28 and Hunter's entry for December 25. On the taste of the water, see Dunbar journal entry, December 10.

57. Dunbar journal entry, December 21. The last buffalo herd in southern Arkansas was killed in the Saline River bottoms around 1808–1809; however, some animals continued to survive near the White and Cache Rivers in eastern Arkansas until the mid-1800s. Sutton, *Arkansas Wildlife,* 27–28.

58. Dunbar journal entry, December 24. For more on the possible sources, flow patterns, and temperatures of the hot springs, see Dunbar's entry for December 24, 1804.

59. Hunter journal entry for January 10.

1805; Hunter and the remaining crew members did not arrive in Natchez until January 31.[60]

During the following weeks, William Dunbar and George Hunter settled their accounts and began to work on their reports to President Jefferson. Between February and July 1805, Dunbar wrote the president a series of letters that contained in a piece-meal fashion his journals, calculations, maps, and a few specimens from the excursion. Jefferson in turn forwarded the information and specimens to other scientists, scholars, and government officials. Dunbar explained to the president:

> Dor. Hunter and myself are just returned from the Washita; time does not permit the preparation of a short report of our researches, before the departure of this mail, the objects which have presented themselves to us are not of very high importance, it must however be acknowledged that the hot Springs are indeed a very great natural Curiosity: the temperatures of the water of this spring is from 130 to 150° of farheneit's thermometer; the heat is supposed to be greater in summer particularly in dry weather. In water of 130° which was comparatively in a state of repose, or one side of the spring run, I found by the aid of an excellent microscope, both vegetable and animal life the first a species of moss the latter a testaceous bivalve of the size of the minutest grain of Sand. I do not despair of being able to reannimate these as soon as I can procure a little leisure: The meanders of the river have been carefully taken as far as we went the lat. was ascertained everyday & the longitude at consec. stages exclusive of the courses and distances, will be forwarded as fast as it can be transcribed . . . I am concerned that the rigourous Season & other circumstances have so much retarded our return that I fear that this report will arrive just before the breaking up of congress. [T]he season was unfavorable for botanical researches, had we been better qualified in the practical part of that Science: it is believed never the less that something new has been found, particularly a species of mountain dwarf cabbage, which partakes of the nature of the cabbage & the raddish & is very agreeable to the taste, the root is White and tastes like horse raddish but much milder . . . I shall only now mention that from our analysis of the water of the hot Springs it appears to contain with a little excess of Carbonic acid, lime & minute Portion of iron: this is indeed visible

60. Hunter journal entry, January 16, 19, 20, 31; Dunbar journal entry, December 16–19, 1804, January 26, 1805. Dunbar arrived in Natchez on January 26, 1805, six days before the rest of the company arrived.

> upon the first view of the springs; an immense body of Calcareous matter is accumulated upon the side of the hill, by the perpetual depositions of the waters & the bed of the run is colored with red oxid or iron, or rather carbonated iron.[61]

Dunbar's journals arrived on the president's desk more than a year before Lewis and Clark returned from their trip to the Northwest. The Dunbar journal and later the Hunter journals provided Jefferson his first glimpse into the new territory from a commissioned exploration team.

After spending an evening at the Forest, George Hunter and his son delivered the odd-looking boat to New Orleans and then returned to Philadelphia by way of New York. On February 14, 1805, an interview with George Hunter appeared in the *Orleans Gazette,* in which he presented a grandiose view of the natural potential of the lower Louisiana Purchase and touted the medical virtues of the hot springs:

> The doctor gives a flattering account of the country, generally, through which he passed. He found a variety of soil and situation—sometimes a low flat country, whose whole surface is overflown by the river in the wet season—sometimes high and eligible, and at others broken, and tolling—but generally fertile and capable of the highest cultivation. He ascended the river about five hundred miles, and found it uniformly gentle and beautiful . . . The country abounds in salt springs, some of which are of equal strength with the water of the ocean . . . He visited the hot springs of Ouachitta, and found them amongst the greatest natural curiosities in the country . . . The doctor is of the opinion that they possess extraordinary medical virtues.
>
> From the information we have obtained from Doctor Hunter . . . we are induced to believe that there are few parts of Louisiana, that hold out greater temptations to emigrants.[62]

Both men fully expected their time at home would be brief and that the Grand Expedition would be reorganized in 1805. However, the War Department informed

61. Dunbar to Jefferson, undated, in Rowland, *William Dunbar,* 142–143; see also Jefferson to Dunbar, May 25, 1805, ibid., 175. Dunbar apparently neatly copied from his journal for this letter to Jefferson. The copy sent to the president is housed at the American Philosophical Society in Philadelphia.

62. Hunter journal entries, February 16–April 1, 1805; *Orleans Gazette,* February 14, 1805; McDermott, "Western Journals of Hunter," 15.

Dunbar on May 24 that Hunter would not be part of the next expedition. The chemist discovered that his business affairs had suffered greatly in his absence, and he probably did not want to neglect them for what would be another lengthy trip. By that time, the fifty-six-year-old Dunbar had also decided that another arduous journey was not physically feasible.[63]

Nevertheless, during 1805 and the early months of 1806, Dunbar, Jefferson, Secretary of War Henry Dearborn, and a Natchitoches, Louisiana, citizen named Dr. John Sibley began exchanging a series of letters concerning the further exploration of the Arkansas and Red Rivers. In these letters Dunbar emphasized that the Arkansas River might be the better avenue for ascent and the Red the better one for descent. The explorer reasoned that the Spaniards would be less likely to detect a boat traveling quickly downstream through territories they claimed. However, Jefferson decided on and pushed for an ascent of the Red River despite Dunbar's calm warnings that any party trying such a journey might be halted by the Spaniards above Bayou Pierre. The president had come to believe that the Red River, "next to the Missouri, the most interesting water of the Mississippi" was the key waterway in the southern Purchase. He also believed that the United States had some claim to part of the Texas territory held by Spain. A trip up the Red, Jefferson believed, might test Spanish resolve and place pressure on Spain during the newly opened negotiations with the United States concerning areas such as West Florida and the true extent of the Louisiana Purchase. Because of failing health, Dunbar informed the president that he would not be able to lead such a venture.[64]

Jefferson nonetheless gave his old acquaintance supervisory authority to organize the 1806 Red River expedition of Thomas Freeman and Peter Custis. Dunbar met with Freeman and gave advice on traversing the southern rivers, dealing with possible Spanish and Indian threats, and the completion of a supply list. In the end,

63. Dearborn to Dunbar, May 24, 1805, in Rowland, *William Dunbar,* 152–153; McDermott, "Western Journals of Hunter," 16. Hunter had stated in the "Hunter Official Report," April 1, 1805, that his business had indeed suffered during his previous trip up the Ouachita.

64. Jefferson to Dunbar, May 25, 1805; Dunbar to Dearborn, February 25, 1806; Jefferson to Dunbar, March 18, 1806; May 25, 1805; all in Rowland, *William Dunbar,* 174–175, 329–330, 192–193, 177; Dunbar to Dearborn, March 28, 1806, Jefferson Papers. Quotation in Jefferson to Dunbar, April 15, 1804, Jefferson Papers. John Sibley (1757–1837) served as an Indian agent for the Jefferson administration and lived and worked in the early settlement of Natchitoches, Louisiana, as a physician. He had a good understanding of the lower Louisiana Purchase, having made several excursions throughout what is now northern Louisiana and along the Red River. Dunbar had written Sibley a letter, which is reproduced in Whittington, "Dr. John Sibley of Natchitoches," 468–475; Garrett, "Dr. John Sibley and the Louisiana-Texas Frontier"; Flores, *Southern Counterparts to Lewis & Clark,* 24–28.

much as he had warned the president, the Freeman-Custis expedition was halted by Spaniards just north of present-day New Boston, Texas, on July 29, 1806.[65]

Upon his return to the Forest, William Dunbar resumed the daily maintenance of his lands and began to diligently prepare his report to the president. By the time of his death five years later, he had published twelve papers in the American Philosophical Society's journal on subjects as varied as natural history, astronomical observations, and Indian sign language. During his final, ailing years, he devoted almost his entire daily schedule to scientific inquiry. He amassed a significant collection of data concerning Indian vocabulary, seasonal river levels, fossils, and astronomical phenomenon. He also invented boats expressly suited to the inland rivers in the trans-Mississippi region, including one used on the Freeman-Custis Red River expedition of 1806, and he developed a method of finding longitudinal coordinates without the use of a timepiece.[66] Despite his desire to continue his scientific inquiries, William Dunbar, the scientist, planter, and explorer, died on October 16, 1810, and was buried in the family cemetery near his Natchez home.[67]

In 1815, Hunter moved his entire family to New Orleans, where he ran a steam distillery called Hunters Mills until his death on February 23, 1823. Hunter became a minor celebrity in New Orleans, and in 1821 John James Audubon sought the advice of the elderly explorer before embarking on his own journey to the Red River

65. Dearborn to Dunbar, March 25, 1805; Dunbar to Jefferson, March 17, 1806; Dunbar to Dearborn, March 18, 1806, all in Rowland, *William Dunbar,* 150–152, 330–333; Flores, *Jefferson and Southwestern Exploration,* 199.

66. Dunbar to Dearborn, May 4, July 13, 1805; Dunbar to Jefferson, December 17, 1805; Dunbar to Freeman Esq. and his associates, April 28, 1806; Jefferson to Dunbar, March 14, 1805; Dearborn to Dunbar, May 24, 1805; Dunbar to Jefferson, July 6, 1805, all in Rowland, *William Dunbar,* 148–150, 151, 185–188, 339–340, 152–153; Riley, "Sir William Dunbar," 98–100. At his death, Dunbar's estate was valued at ninety-one thousand dollars. McDermott, "Philosophic Outpost on the Frontier," 3.

67. Rowland, *William Dunbar,* 12. One of the inscriptions on his obelisk tombstone reads,

> His Wife
> Laments a tender Husband
> His Children,
> An affectionate Parent;
> His Friends,
> A valuable Acquaintance;
> His Country,
> A most useful Citizen;
> Science,
> A distinguished Votary. (William Dunbar Tomb Inscription)

Postlethwaite to Archibald Dunbar, Forest near Natchez, November 17, 1810, in Rowland, *William Dunbar,* 383–384.

country. In *Journal of John James Audubon Made during His Trip to New Orleans in 1820–1821*, Audubon called Hunter "the renowned Man of Jefferson," though he also noted that by this time George Hunter could communicate only in a very broken fashion.[68]

Thomas Jefferson included Dunbar's and Hunter's accounts of the Ouachita River expedition in his message to Congress, and in 1806 the details of the journey were published in a work entitled *Message from the President of the United States, Communicating Discoveries Made in Exploring the Missouri, Red River, and Washita, by Captains Lewis and Clark, Doctor Sibley and Mr. Dunbar with A Statistical Account of the Countries Adjacent*. This narrative summary also included a list of the flora the Scottish planter had cataloged, in a section entitled "Of the medical properties of the Hot Springs" and in an appendix from Dunbar that presented the "List of Stages and Distances, on the Red and Washita rivers, in French computed Leagues."[69] Jefferson's message on February 19, 1806, stated:

> Having been disappointed, after considerable preparation, in the purpose of sending an exploring party up that river [Red], in the summer of one thousand eight hundred and four, it was thought best to employ the autumn of that year in procuring a knowledge of an interesting branch of the river called Washita. This was undertaken under the direction of Mr. Dunbar, of Natchez, a citizen of distinguished science, who had aided, and continues to aid us, with his disinterrested and valuable services in the prosecution of these enterprizes. He ascended the river to the remarkable hot springs near it, in latitude 34° 31′.4″, 16, longitude 92°. 50′. 45″. West of Greenwich, taking its courses and distances, and correcting them by frequent celestial observations. Extracts from his observations, and copies of his map of the river, from its mouth to the hot springs, make part of the present communications. The examination of the Red river, is but now commencing.[70]

68. McDermott, "Western Journals of Hunter," 18. Audubon commented on Hunter's extremely feeble state, explaining, "The Phisician May have been a Great *Doctor formerly* but Now deprived of all that I Call Mind I found it Necessary to leave to his Mill's Drudgery." Audubon, *Journal Made during His Trip to New Orleans*, 149. McDermott, "Western Journals of Hunter," 18, 5.

69. Jefferson, *Message from the President of the United States*, 3–165; Also see *Discoveries Made in Exploring*, 160–169; Jefferson's report basically summarizes the Ouachita River expedition using information from Dunbar's letters and his journals. Flores, *Jefferson and Southwestern Exploration*, 45; *Documents Relating to the Purchase and Exploration of Louisiana*, 7–9.

70. *Documents Relating to the Purchase and Exploration of Louisiana*, 3–4.

What did William Dunbar and George Hunter accomplish in their abbreviated expedition up a Red River tributary? To be sure, they were not the first to travel the meandering waters of the Ouachita River, drink from the boiling fountains of the hot springs, or describe the region in journals or publications. They did, however, complete the first scientific mapping and description of the Ouachita River and its confluences.

They became the first U.S. citizens to officially inform others of a vast new territory. And they compiled information that, when assembled today, presents an important view of the early environs of the southern Louisiana Purchase. Their journals reveal an active European and white American presence in the region and show that numerous small settlements, individual homesteaders, hunters, and traders had been utilizing the natural resources of the area for decades. The place names that are identified in the two men's daily entries offer further indication of a region well-known and used by these same people.

They identified animal and plant life that in some cases no longer exists in the Ouachita River Valley. They also conducted and recorded the first scientific experiments on a natural hot springs west of the Mississippi River. The discovery of the microscopic life in these springs may also be the first to be documented in the Louisiana Purchase and indeed in the western United States.[71]

In the two centuries since Dunbar and Hunter's exploration, towns and communities have emerged and grown along the river's adjacent areas, and bridges have spanned the eroded banks that Dunbar and Hunter once described. Paddlewheel riverboats plied the unpredictable river channels until the early twentieth century, and the river channel has been dredged to eliminate the numerous obstacles that the expedition once experienced. Locks and dams have been built, forever halting the water's natural flow and lessening the frequency and the effects of the inundations so often referred to by Dunbar. The site where the ancient waters of the hot springs emerged and where the team tested the properties found in the heated fountains became a National Reserve in 1832 and later, in 1921, Hot Springs National Park.[72]

In one of his final letters to William Dunbar, Jefferson poignantly expressed his thoughts on the short journey up the Ouachita River and conveyed his appreciation to the elderly explorer. The president wrote:

71. Dunbar's "Appendix," which is found in the Dunbar Papers, contains a list of 187 different plants that he had cataloged during the trip. The "Appendix" was also printed in *Discoveries Made in Exploring*, 113–164.

72. *Act authorizing the governor of the territory of Arkansas to lease the salt springs.*

> The work we are now doing is, I trust, done for posterity in such a way that they need not repeat it. For this we are much indebted to you, not only for the labor and time you have devoted to it, but for the excellent method of which you have set the example and which I hope will be the model followed by others. We shall delegate with correctness the great arteries of this great county. Those who come after will extend the ramifications as they become acquainted with them, and fill up the canvas we begin. With my acknowledgments for your zealous aid in this business, accept my friendly salutations and assurances of great esteem and respect.[73]

The Dunbar and Hunter voyage did not rival in scope that of the Corps of Discovery, but their journey up the Red, Black, and Ouachita Rivers, along with those of Freeman and Custis, and Zebulon Pike, resulted in important accounts that complete the story of Louisiana Purchase exploration. Some of the scenes they recorded are forever lost. Some remain. By reading and studying their journals, we can imagine the natural world they experienced and the rich exchange of cultures present in the lower Mississippi Valley during the early nineteenth century.

73. Jefferson to Dunbar, May 25, 1805, Jefferson Papers.

Journal of a voyage commencing at St. Catherines Landing on the East bank of the Mississippi, proceeding downwards to the mouth of the Red River, & from thence ascending that river, the black river and the Washita river, as high as the Hot Springs in the proximity of the last mentioned river.

by

WILLIAM DUNBAR

Adams Co, Mississippi

Journal of an Excursion from Natchez on the Mississippi up the River Ouachita, 1804–1805

by

DR. GEORGE HUNTER

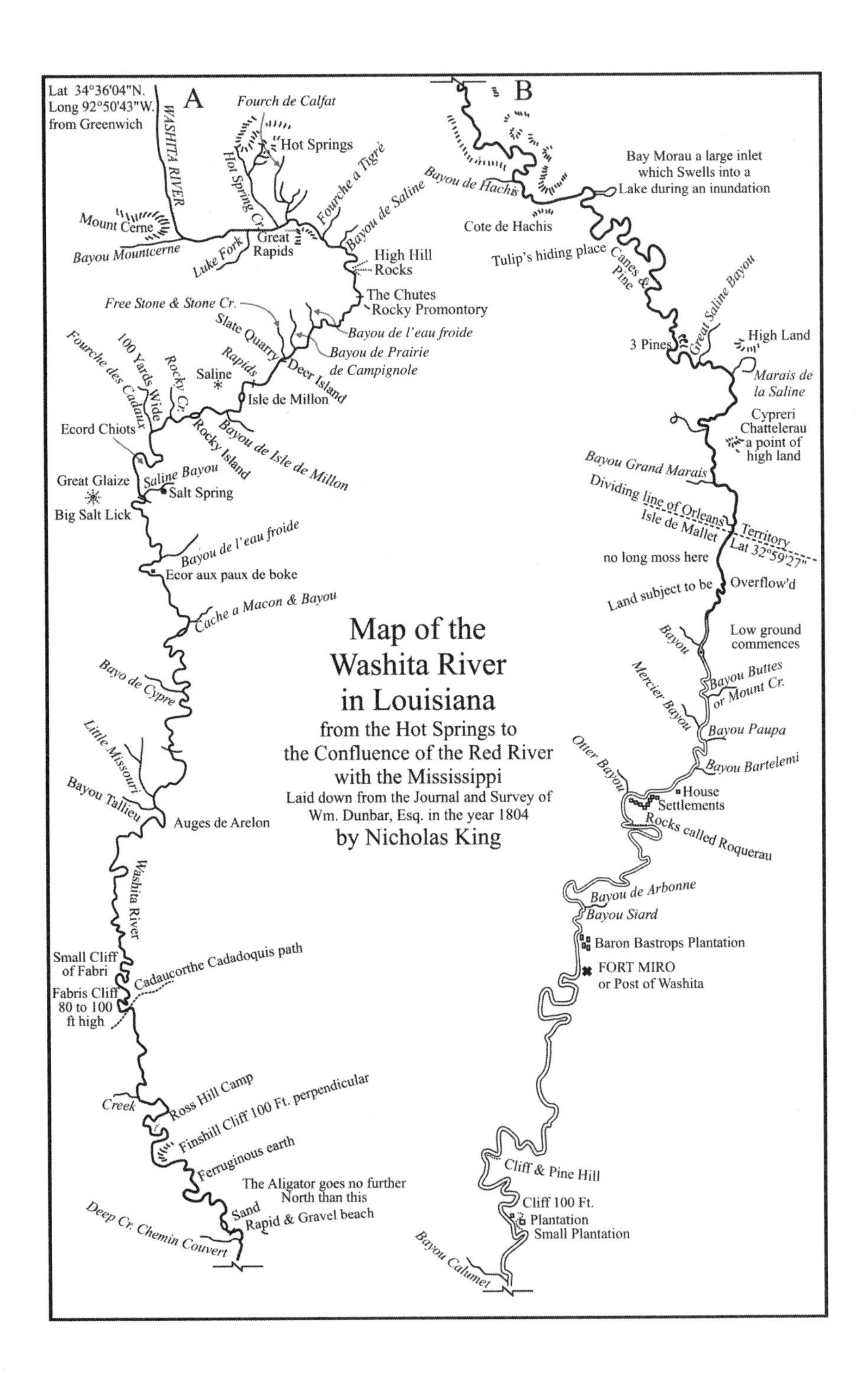

Lat 34°36'04"N.
Long 92°50'43"W.
from Greenwich
A
B
WASHITA RIVER
Fourch de Calfat
Hot Springs
Hot Spring Cr.
Fourche a Tigre
Bayou de Saline
Bayou de Hachis
Bay Morau a large inlet which Swells into a Lake during an inundation
Mount Cerne
Bayou Mountcerne
Luke Fork
Great Rapids
High Hill Rocks
Cote de Hachis
Tulip's hiding place
Canes & Pine
The Chutes
Rocky Promontory
Free Stone & Stone Cr.
Slate Quarry
Bayou de l'eau froide
Bayou de Prairie de Campignole
Great Saline Bayou
3 Pines
High Land
Marais de la Saline
Fourche des Cadaux
100 Yards Wide
Rocky Cr.
Saline
Rapids
Deer Island
Isle de Millon
Cypreri Chattelerau a point of high land
Ecord Chiots
Rocky Island
Bayou de Isle de Millon
Bayou Grand Marais
Dividing line of Orleans Territory
Isle de Mallet
Lat 32°59'27"
Great Glaize
Saline Bayou
Salt Spring
Big Salt Lick
no long moss here
Bayou de l'eau froide
Ecor aux paux de boke
Land subject to be Overflow'd
Cache a Macon & Bayou
Bayou
Low ground commences
Map of the Washita River in Louisiana
from the Hot Springs to the Confluence of the Red River with the Mississippi
Laid down from the Journal and Survey of Wm. Dunbar, Esq. in the year 1804
by Nicholas King
Bayo de Cypre
Mercier Bayou
Bayou Buttes or Mount Cr.
Bayou Paupa
Little Missouri
Otter Bayou
Bayou Bartelemi
Bayou Tallieu
House
Settlements
Auges de Arelon
Rocks called Roquerau
Washita River
Bayou de Arbonne
Bayou Siard
Baron Bastrops Plantation
Small Cliff of Fabri
Cadaucorthe Cadadoquis path
FORT MIRO or Post of Washita
Fabris Cliff 80 to 100 ft high
Creek
Ross Hill Camp
Finshill Cliff 100 Ft. perpendicular
Ferruginous earth
Cliff & Pine Hill
The Aligator goes no further North than this
Cliff 100 Ft.
Sand
Rapid & Gravel beach
Plantation
Small Plantation
Deep Cr. Chemin Couvert
Bayou Calumet

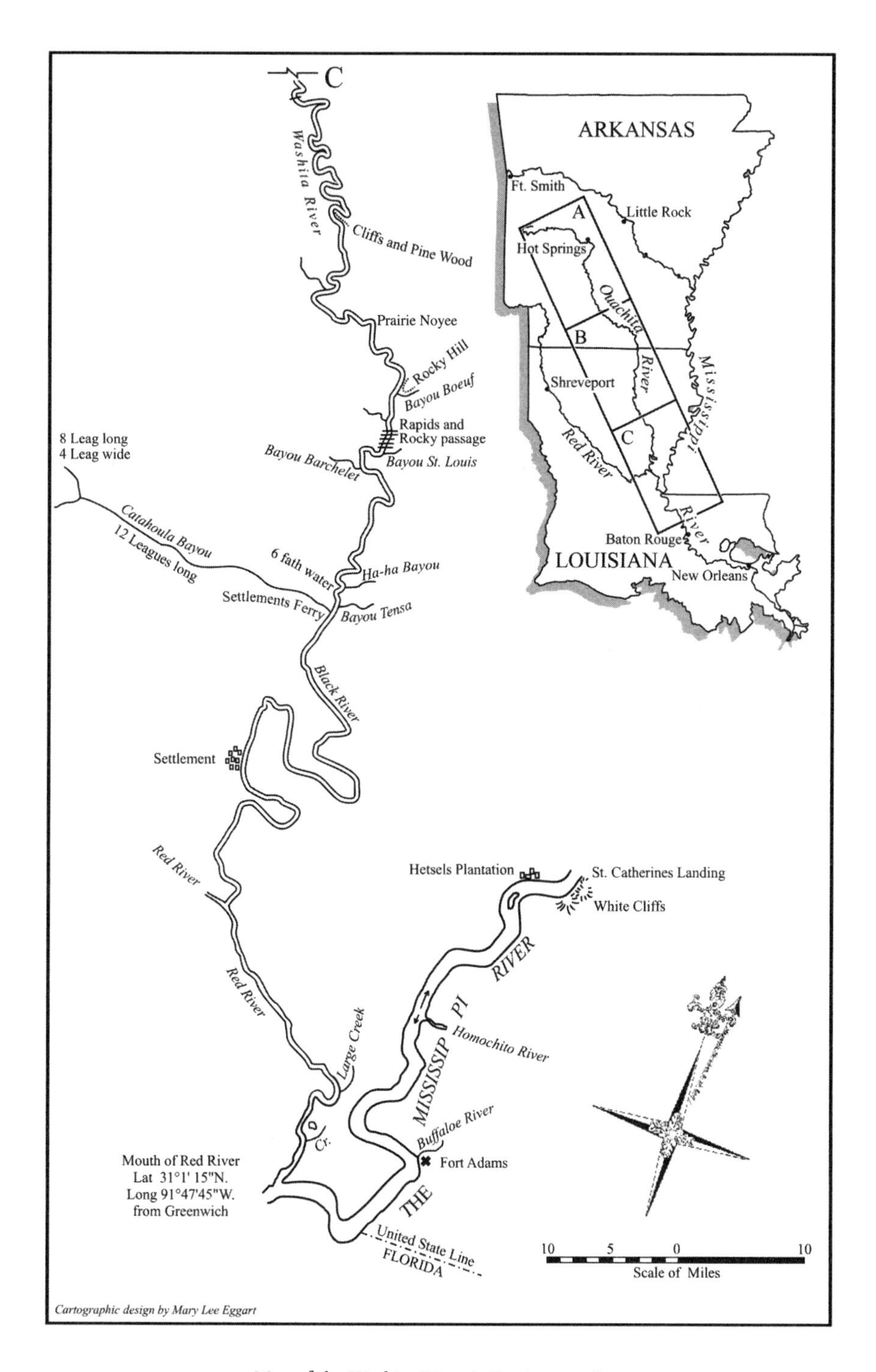

Map of the Washita River in Louisiana, 1804

William Dunbar, undated painting. No picture of George Hunter is known to exist.
Courtesy of Riley-Hickingbotham Library, Ouachita Baptist University.

siderable length & at least 150 yards wide, and
flowing with a full Current from bank to bank.
We found a considerable number of unknown to us plants
some of them very handsome, but our very limited knowledge in ~~botany~~ practical botany, did not enable
us to discover what they were, particularly as they were
not in flower. Made this day 13 miles 39 perches.
Thermom.r at 8h p.m. 66° Extremes 54°-71°.

Wednesday 28th. } Therm.r 68° – river water 60°. fallen 4 inches in the night – Cloudy – calm. Set off at 7h 5' and continued our voyage, meeting the same species of obstacles as yesterday – the river appears to increase in width being some times 170 yards broad, flowing at this time with a full tide from shore to shore, the Current is in some places extremely rapid, that is where the depth of the Channel is diminished & the bed contracted, in such situations we are under the necessity of catching hold of the willows &c & hauling up along shore, oars & poles being insufficient to stem the violence of the torrent; in other situations for miles together

Dunbar journal entry for November 28, 1804. *Courtesy of Riley-Hickingbotham Library, Ouachita Baptist University.*

Dunbar's journal, compass, glasses, and pen. *Courtesy of Riley-Hickingbotham Library, Ouachita Baptist University.*

October 1804

Tuesday 16th

DUNBAR

Set out from St. Catherine's landing in the afternoon. The Latitude of this place is 31° 26′ 30″ North; and Longitude 6h 5′ 56″—west of Greenwich.

A little below are the white cliffs 5 leagues below the Natchez the face of the cliffs is chiefly white sand surmounted by pine; the cliffs are from 100 to 200 feet high; when the waters are low the basis of the cliffs are uncovered consisting of clay of different colours and some beds of ochre covered here and there by a thin lamina of iron ore; small springs possessing a petrifying property flow over the clay and ochre; numberless logs and pieces of timber converted into stone are strewed about the beach. Fine pure argil of various colours chiefly white and red is found here. Encamped at night upon an Island 7 miles below the place of departure.[1]

HUNTER

On the 16th Octr. 1804, when we had dropped down the Mississipi as far as St. Catherines Creek which is 15 miles below Natchez about 1 p.m. Wm. Dunbar Esqr. came on board, & in about two hours we set sail & proceeded a short distance when the wind came ahead, we then took to our Oars & a little before sunset encamped

The opening inscription appears on the first page of the "Dunbar Trip Journal, Vol. I" and also at the head of the "Dunbar Report Journal." Hunter's journal title is as it appears on the document we refer to as "Hunter Journals."

1. Natchez is located on a high bluff on the Mississippi River. It is the county seat of Adams County. The expedition used the French league (*lieue commune*) as their means of measuring nautical distance. One French league equals 2.4 nautical miles. Nautical miles may also be defined as the average distance on the Earth's surface that represents one minute of latitude. Since the explorers were entering territory with no mile markers, the use of latitude and nautical miles was the most practical system of measuring distance. Chardon, "Linear League in North America"; Dunbar appendix, in *Discoveries Made in Exploring;* Dunbar, "Journal of a Geometrical survey."

Throughout his journal, Dunbar's knowledge of the lands, soils, flora, fauna, astronomy, surveying, and other branches of science is apparent. The journal is a testament to his passion for scientific investigation and understanding.

on an island on the west side of the Mississippi about twenty four miles below Natchez, where we staid all night, it rained and blew with the wind from the Northeast the first part of the night, & grew very cold, the other part was clear.[2]

Wednesday 17th

DUNBAR

Set off; passed Fort Adams,[3] and six miles farther the line of demarcation, and arrived at the mouth of the red river about nine miles below the line of demarcation; encamped just within its mouth; the waters of this river have a red appearance from a rich fat earth or marl of that colour born down by the floods from which it derives its name;[4] the mouth of the river is about five hundred and fifty yards wide:[5] here we commenced taking the meanders of the river by course and time depending upon the log to inform us of our rate of going as well as the velocity of the Current; there is however no sensible Current at the mouth: the banks on both sides are here clothed with willows, the land is low and subject to inundation to the

2. The confluence of St. Catherine's Creek and the Mississippi is now approximately 12 miles below the city limits of Natchez. On current maps it is labeled Old St. Catherine Creek. An early traveler on the Mississippi named Zodak Cramer estimated in 1804 that the confluence of St. Catherine's Creek and the Mississippi was then 18 miles below Natchez. The landing was used regularly by Dunbar and other planters. Cramer, *Ohio and Mississippi Navigator;* McDermott, "Western Journals of Hunter," 71; Louisiana Quadrangle Maps, from U.S. Geological Survey, Denver, Colorado, and Department of Public Works, Baton Rouge, Louisiana, 1983.

3. Constructed in 1799, Fort Adams became the U.S outpost that guarded the territory below the 31° line following the Spanish withdrawal from the Natchez district. A site was chosen by a Captain Isaac Guion 37 miles below Natchez on a ridge and along a river or creek called "Roche a Davion" (Loftus Heights) after a French missionary who worked with the Tunica Indians in the area. Thomas Freeman, who led the Red River expedition in 1806, served as the construction engineer for the fortifications. Fort Adams was named for John Adams, U.S. president at the time. Mathews, *Andrew Ellicott,* 128.

4. The "line of demarcation" refers to the 31° latitude line, or the line between Spanish West Florida and U.S. territory (today the border between Louisiana and Mississippi). The line was surveyed by a joint American and Spanish team including Andrew Ellicott, Thomas Freeman, Stephen Minor, William Dunbar, and Spanish governor Manuel Gayoso de Lemos from 1791 to 1799. Minor, Dunbar, and de Lemos were commissioned by the Spanish government. Interestingly, Ellicott helped train Meriwether Lewis in surveying and navigational techniques prior to the departure of the Corps of Discovery. Ellicott to Sarah Ellicott, Darling's Creek, June 19, 1798, in ibid., 129–130; Hamilton, "Running Mississippi's South Line," 158–168.

Marl consists of clay and calcium carbonate. Uses of marl include fertilizing lime-depleted soils.

5. The mouth of the Red River is today located at the corner of Pointe Coupee and Concordia Parishes, about 45 miles below Natchez and 150 miles above New Orleans near the community of Lettsworth. Louisiana Quadrangle Maps.

height of 30 or more feet above the present level of the waters, the mouth of the red river is accounted to be 75 leagues from New-orleans and 3 miles above the exit of the Chafalaya or Opelousa river which was probably the continuation of the red river, when perhaps its waters did not unite with those of the Missisippi excepting during the inundation. M de Ferrer has settled the Latitude and Longitude of this place;[6] the first at 31° 1′ 15″ N. and the last at 6h 7′ 11″ west of Greenwich.

HUNTER

Set out before sunrise[,] Sailed & rowed alternately according to circumstances, Our Boat was made somewhat in the form of a ferry flat, with a mast fixed to strike occasionally, & were provided with a large sail, manned with 12 men & a Sergeant, rowed twelve oars, was 50 feet long & about 8 feet beam on deck at the mast which was her extreme breadth, tapering to the stern. had a cabbin abaft & a pavilion amidships for the accommodation of the Officers & crew, with tarpaulins & curtains to keep off the weather, & every thing fitted for the expedition with about 3 months provisions on board. About 9 a.m. came to Loftus Heights, Fort Adams where a Corporals guard is kept. This is on an high Bluff commanding the passage of the River situate about 6 miles above the former Line between the U States & the Spaniards, we waited here about an hour & then set out for the Red River, where we arrived on the same day in the afternoon about 5 P M Latitude 31°.1′.10″ This river is of a reddish muddy colour owing to a clay or Marle of that colour suspended in its waters, which is said to give an amazing degree of fertility to the ground overflowed by it. Here at our first entrance the appearance of the face of the country seems changed, every vegetable puts on a fresher green. The banks as yet are all overflooded in time of freshes; the timber is small & that in sight cheifly Willows & Cottonwood which resembles the Lombardy poplar, & on each side the ground appears to be but of recent formation.[7] We have already seen

6. Dunbar's description of the Red River confluence is basically accurate. Today, the U.S. Army Corps of Engineers keeps the waters of the Red, the Atchafalaya, and the Mississippi from spilling together by a series of locks and dams. There are even some theories that the Atchafalaya may have been the ancient channel of the Mississippi River.

"M de Ferrer" is Jose Joaquin de Ferrer, who served as a Spanish surveyor and geographer. He determined the mouth of the Red River to be at "31 deg. 1 min. 15sec. N.L. and 91 deg. 47 min. 45 sec. West of Greenwich." Ferrer's work was published as "Astronomical Observations."

7. For the recorded roster of the expedition members, see the introduction, note 36. The term *abaft* means the aft or rear of the vessel. The black willow (*Salix nigra*) is also called swamp willow or gooding willow; the cottonwood (*Populus deltoides*) is also called eastern cottonwood or southern cottonwood. The black willow may grow 50 feet high, and its leaves, 3–6 inches long, are tapered and finely toothed. Little, *Field Guide to Trees*, 106, 335.

a great many flocks of wild Geese & Brandt, altho still shy, a few ducks, many large Alligators. Here we picked up a few shells of mother of Pearl Muscels, very light, thin & transparent. Encamped at 6 p.m. at a bank covered with pea Vine, the ground very rich composed of fat earth very deep soil.[8]

Thursday 18th

DUNBAR

Set off up the river, remarked vegetation to be surprisingly luxuriant along the banks owing no doubt to the rich red marle yearly deposited by the floods of the river—willows grow to a good size, but other forest trees are much smaller than those seen upon the banks of the Mississippi, which may be owing to the newly formed soil or its excessive richness. The river narrows gradually as we advance: at noon it was about 200 yards wide. Got out the instruments, which requiring a good deal of adjustment we were unable to make perfect observations. The Latitude 31°.8′.54″.6, perhaps accurate enough to correct the traverse of the river. Note. The place of observation was at the extremity of the course N32° E 17″ to a pt. on the left. The banks of the river are luxuriantly clothed with peavine and several kinds of grasses yielding seed, of which geese and ducks are very greedy:[9] got our log line prepared and divided into perches—hove the log and found we went at the rate of 4 perches in half a minute. i.e. 1½ mile per hour—very slow—Soldiers do not exert themselves at the oar; came to, for the night having made nearly 13 miles—hove the lead in the middle of the river and found 11 fathoms. There are generally willows growing on one side of the river, and on the other the same small growth of forest trees continues. consisting chiefly of black oak, packawn, hickory,

8. What Hunter refers to as a "Brandt" is a brant (*Branta bernicula*), a goose species that is almost exclusive to the northeastern United States. The brant has a black head, neck, and chest, a gray back and wings, and a whitish-gray underside. These birds only use the Atlantic Flyway and do not venture into the Mississippi Flyway. It is more likely that Hunter was seeing Canada geese (*Branta canadensis*), which are 25–43 inches long with a gray and white body, a black head and neck, and a white chin strap. Peterson, *Field Guide to the Birds,* 44–45. The "Hunter Official Report," 11, stated, "took observations for the Latitude, Longitude &c. This was our constant rule during the whole excursion." McDermott, "Western Journals of Hunter," 71.

9. Dunbar and Hunter carried several surveying and navigational instruments, including a sextant, several pocket compasses, a surveyor's compass, an acrometer (to measure acceleration), artificial mercurial horizons, a circle of reflection, an octant, and a surveyor's chain. See the introduction, note 39 and the text it refers to. See also "Dunbar Trip Journal, Vol. II," December 16 and 26, 1804; Hunter journal entries for December 13 and 27 and January 2. The "peavine" could have been any of the many varieties within the *Leguminosae* or *Fabaocae* genus (the pea family). Thieret, Niering, and Olmstead, *Field Guide to Wildflowers,* 525–550.

elm &c.[10] The Trees are so exceedingly grand & lofty upon the banks of the Missisippi, that by comparison those bordering on this river seem dwarfish, and appear to bear a kind of proportion to the magnitude of their own river. The extremes of temperature were from 46° to 48° of Fahrenheits thermometer. made this day 12 55/60 miles.

HUNTER

Got under way at 6 a.m. after a pleasant cooll night no mosquitoes. & scarce any current. The wind sprung up fair & we set our sail, but the breeze being but light went only at three miles pr hour for ¼ hour when we took to our oars. Having fixed a Logline & reell & marked our line in perches viz a knot at every perch[,] we hove the Log using an Acrometer that told seconds very distinctly. by which it appears that we row only 1 ½ miles pr hour. The river is very crooked, we set our course with a surveyors Compass, & instead of points as they do at sea, take it down in the degrees; so that it sometimes happens that we go four or five courses in half an hour. As we ascend this river the banks rise by degrees a little higher, a great Variety of trees appear but still the banks are lined with willows. This River which was about 500 yards wide at the Mouth is now reduced to about 200—& appears to be very deep.—We now sounded & found it to be 11 fathoms in the Middle about 15 miles from the Mouth. encamped on the left bank going up at a pleasant place, where there are plenty of Poccon trees;[11] The men made a large fire on the top of the Bank & slept by it under the shade of the trees, covred by their Mosquito Curtains, we that is Mr Dunbar Myself & Son with Mr Dunbar's Servant & his two Slaves slept on board the boat.[12] This night was also cooll, The Thermometer stood

10. One perch equals 1 linear rod or 5½ yards, or 16½ feet (5.03 meters); 1 fathom equals 6 feet (11 fathoms = 66 feet). A sounding line was usually employed to make these measurements. The black oak (*Quercus velutina*) ranges from 70 to 100 feet in height and has moderately lobed leaves 4–10 inches long that are shiny above and hairy below. Dunbar's "packawn" is a pecan tree (*Carya illinoensis*), with a height of 50–150 feet and leaves of 9–17 inches; it has 5 to 7 leaflets per leaf, which is lance-shaped. The "hickory" could be one of more than seven species within the *Carya* genus, and the "elm" could be one of four species in the *Ulmus* genus. Petrides, *Field Guide to Trees and Shrubs*, 88, 206–208, 252–253, 298.

11. "Logline and reell" is a reference to measuring devices used in river travel. Hunter's "Acrometer" may be a crude type of accelerometer that measures speed. "Poccon" trees are pecan trees.

12. Dunbar refers to the presence of these people only in two short references. On December 28 he states in his journal, "after breakfast set out upon a geographical tour round the Hill of the hot spring; young Mr. Hunter, with one of the people and my negro servant attended." The second is in his January 18–19 entry, where he writes, "I determined after being disappointed in procuring horses, to take the canoe with one soldier and my own domestic and push down to Catahoola."

in the morning at sunrise at 42 & yesterday afternoon at 84. Got no observation this day.—No mosquitoes, made 12 miles 293 ½ perches.

Friday 19th

DUNBAR

Continued our rout up the river; having given the Soldiers this morning a few words of advice and encouragement, they improved considerably in activity and cheerfulness, hove the log and found we went 7 perches per half minute, the Current yet continues so moderate as to offer no impediment to our rowing along shore therefore not worth estimating: landed before 12 to observe and for dinner. Latitude 31° 14′ 50″.1 after dinner caught a runaway negro; proceeded on to the confluence of red and black river in Latitude 31° 15′ 48″ which by our reckoning appears to be 26 1/3 miles from the Mississippi, the Contrast of the two rivers is great, the red river being charged with red marly earth and the other a clear river gives it by comparison a dark appearance, hence the name of the black river—Each river is about 150 yards and when united about 200 yards wide. sounded in the black river and found 20 feet black sand, little or no current. Took specimens of the red marl of red river bank. The water of the black river is rather clearer than that of the Ohio and of a warm temperature, probably owing to the waters which flow into it from the valley of the Missisippi particularly from the Catahoola. Made 15 miles 102 perches.[13]

HUNTER

Set out at 6 a.m. which is now here only just day light. Hove the log by which we rowed at the rate of 1½ miles pr hour. stopped one hour to breakfast to rest the Men. I shot 2 ducks but got but one of them, afterward in our course up the river met with & at several times killed the whole flock which consisted of 5. They proved to be good. As we stopped to take an observa[tion] observed a Canoe hauled ashore & a black man who left it at our approach, the canoe was empty & we suspected that the person we had just seen was a run away slave, accordingly when we had dined, we left two men hid in the bushes, & pushed off our boat & went on. We had not proceeded far when we were hailed by those we had left behind, & perceived they had got a black man a stout fellow who called himself Harry, with nothing but his shirt & trowsers on, who gave no satisfactory account of himself, said

13. The Black River is the lower portion of the Ouachita River just below Jonesville, Louisiana, and the confluence with the Tensas River between Catahoula and Concordia Parishes. The Catahoula River flows into the Black River near present-day Harrisonburg in Catahoula Parish. Louisiana Quadrangle Maps.

he was free, but had nothing to show for it. We took him into the Boat, he was half famished, we gave him plenty of ham & biscuit to eat, which he devoured with a voracious appetite, said he was pleased to go with us, but still never gave any satisfaction as to whom he belonged to. &cc—We came to the Black River in about one ½ mile further Latitude 31.15.48. having the water differently coloured from the red River, being quite clear; Still scarce any current, Sounded in 3½ fathoms water. made 15 miles. 102 perches this day.

The banks of the black river are of fine black garden mold producing a Variety of Forrest trees & a great burthen of herbage & grasses, here & there patches of Reeds or Canes, shewing that the banks are not so long nor so frequently overflowed as below. The game wild fowls appear not so often yet as in the Red River neither are the Alligators so plenty nor so large. Went this day by our Log about 15 miles. Thermometer 46° in air. were delayed about one hour waiting for one of the men Skinner who had go leave to hunt along the left bank of the river going up.[14]

Saturday 20th

DUNBAR

Continue ascending the river; Thermometer 47° Temperature of the water 73°—a spring issuing from the river bank 66°—Forest trees on the banks chiefly red and black oak interspersed with ash, paccawn, hickory, some elms, pirssimon &c; several kinds of grass and many humble plants in flower, so that even at this season our country affords employment for the Botanist. great luxuriance of vegetation along the shore, grass very rank, and a thick curtain of shrubbery of a deep green;[15] the soil black marl mixed with a moderate proportion of sand, resembling much the soil on the Missisippi banks, yet the forest trees are not lofty like to those on the margin of the great river, but resembling the growth on the red river. I omit-

14. "Cane" (*Arundinaria gigantia*) may grow up to 40 feet high. "Skinner" is William Skinner. Dunbar and Hunter seem to have compared notes on the daily distance traveled, astronomical observation, and the temperature readings. The two usually were in agreement in their journal entries.

On this date, Hunter expounded in his official report on certain matters including the medicinal purposes of a "China Briar," or Chinabrier (*Smilax bona-nox*), a climbing evergreen vine that produces a small bluish-black fruit and has small prickles on the vine. Hunter extolled the virtues of the plant's root ball, which he said could be made into edible cakes after washing, drying, and mashing. Indians used the plant as a bread cake to treat urinary tract ailments, as a tonic, and to wrap cigarettes. "Hunter Official Report," 13–14. Grimm and Kartesz, *Illustrated Book of Wildflowers and Shrubs*, 390; Moerman, *Native American Ethnobotany*, 533. Dunbar also discussed the Chinabrier in his entry of December 1.

15. Common persimmon (*Diospyros virginiana*): height from 20 to 70 feet; leaves 2–6 inches, egg-shaped, and not toothed. Petrides, *Field Guide to Trees and Shrubs*, 372. "Rank" here means vigorous or tall.

ted mentioning in its proper place, that the last single inundation of the red river appears to have deposited on the high bank a stratum of red marl above ½ inch thick now dry; some specimens were taken. Took a meridian altitude of the Sun, from which the Latitude deduced was 31° 22′ 46″.6—observed Canes growing on several parts of the right bank, a proof that the land is not deeply overflowed, perhaps from 1 to 3 feet: the banks have the appearance of stability, very little willow or other productions of a newly formed soil being seen on either side: the solid high bank being deeply shaded by vegetation from the humble creeping plant to the spreading oak.

Encamped at sun-set. Sounded; 5 fathoms—black sand—Extremes of the Thermometer 47°–80°. made this day 13 miles 40 perches.

HUNTER

Set out ¼ past 6 A. M. This day saturday, The banks grow up very gradually in height as we ascend producing very luxuriantly on the margin of the river a large Variety of barren kinds of vines hanging in festoons amongst the shrubs & trees, & also pea vines of different kinds.[16] In the interior I found many briars which rendered it difficult to walk thro them. The banks here as well as in all the rivers in this country subject to be annually overflowed are generally higher next the rivers & descend as they leave the river, owing to the mud brought down suspended in the flouds being the heaviest part of it deposited there & as the water leaves the river it gradually clears itself & thus at a distance the grounds rise by much slower degrees than near.—For the whole of this country appears to be newly formed & forming & growing gradually more & more elevated, dry & healthy. Sounded in 5 fathoms water. The river continues falling about 4 Inches perpendicular in 24 hours, scarce any current, the men row very indolently. Made this day, 13 miles, 46 perches.

Sunday 21st

DUNBAR

Thermometer before sun-rise 60°. Continue ascending; no current to imped us, for altho' there be a feeble current along the principal thread of the stream, yet as this is deflected from bend to bend, we easily avoid its influence by directing our course from point to point or rather passing a little under the points, and in fact where there is any current, a compensation is found by the counter current or eddy under the points. The river is now only 80 yards wide; the timber becomes

16. A "festoon" refers to a chain or ribbon of foliage, sometimes suspended between two points.

larger, the banks in some places 40 feet high, yet liable to inundation, not from the floods of this small river, but from the intrusion of its more powerful neighbour the Missisippi: The lands decline rapidly (as in all alluvial countries) from the margin to the Cypress swamps, where more or less water stagnates all the year round. The current of the river is still so insensible even in the thread of stream, that we take no account of it; at 8ha.m. we arrived at an Island, small but elevated, said to be the only one in this river for more than 100 leagues ascending. on the left bank near the Island is a small settlement commenced by a man and his wife: a Covered frame of rough poles without walls serves for a house, and a Couple of acres of indian corn had been cultivated, which suffices to stock their little magazine with bread for the year; the forest supplies Venison, Bear, turkey &c, the river fowl and fish; the skins of the wild animals and an abundance of the finest honey being carried to market enables the new settler to supply himself largely with all other necessary articles; in a year or two he arrives at a state of independence, he purchases horses, cows & other domestic animals, perhaps a slave also who shares with him the labours and the productions of his fields & of the adjoining forests. How happy the contrast, when we compare the fortune of the new settler in the U.S. with the misery of the half starving, oppressed and degraded Peasant of Europe!!—The banks here are not less than 40 feet above the present level of the river water and but rarely *overflowed;* the nearest road to the high lands at the Rapid-settlement on the red river, nearly west is said to be 40 miles thro' an inundated alluvial country;[17] it is probable the direct distance does not much exceed one half, the numerous lakes in the overflowed lands rendering the road very circuitous: both banks are clothed with rich Cane-brake, pierced by many creeks fit to carry boats during the inundation: saw many Cormorants and the stately Hooping Crane: Geese and Ducks not yet abundant; they arrive in myriads with the rains & winter cold: Landed before noon to observe: we had been disappointed at the hour of breakfast by clouds in making observations for the magnetic variation and for regulating the time & rate of going of the watch, preparatory to the lunar observation, & now apprehended the same disappointment, the heavens being loaded with flying clouds: just before the Sun was expected on the meridian, a dense cloud concealed him from view, when he reappeared he was already dipped a little; the lat-

17. The reference is to a settlement that eventually became Alexandria, Louisiana, in Rapides Parish. As early as 1723–1724, the French had established a fort at the site, called Poste du Rapides. The fortifications were built to protect the numerous French migrants to the area, who had come as early as 1718. At the time of the Dunbar-Hunter expedition, the site may have had a population in excess of 700 people. Fortier, *Louisiana,* 1:29, 2:346; Stoddard, *Sketches of Louisiana,* 186; Whittington, "Rapides Parish, Louisiana."

itude deduced is undoubtedly too far North 31° 37′ 52″.5 the sun had therefore not attained his meridian altitude.[18]

This afternoon found the shore favorable for tracking, (i.e.) running along shore & towing the boat; rate of going by log a little improved 5 perches pr .½ minute. At 3h p.m. thermr. 83°—The banks have a regular shelving slope from the top to the water's edge & are totally covered with the most luxuriant herbage consisting chiefly of 5 or 6 kinds of strong grass yielding vast crops of seed nearly mature, upon which Geese and Ducks get surprisingly fat: we shot some water fowl of the Duck kind, whose web-foot was partially divided, the body covered with a bluish or lead coloured plumage; they were extremely fat and excellent, resembling in taste the Canvass-back.[19] The teal of these rivers is also very fat and fine. Wind S.S.E. and cloudy. Encamped. Extremes of the thermometer 60°–83°. Made this day 14 miles 59 perches.

HUNTER

Thermometer before sunrise 60°. Cloudy Observed for the first time in this river a small Island containing a few acres of land in it. The wind being fair, we set our sail & went on for a while at a brisk rate. A small settlement. Landed to Observe the suns Altitude, but were prevented by clouds from doing it to satisfaction.—Extremes of the Thermometer 60°. to 83° cloudy Wind S.S.E. Made this day 14 miles 59 perches.—The Banks resemble those passed yesterday except rising in height as we ascend. The water good & pleasant to drink.

18. In this area and along the Black and Ouachita Rivers, crescent-shaped lakes (oxbow lakes) border the waterways on both sides. These continue to near Friendship, Arkansas, in Hot Spring County. It is most likely that Dunbar was seeing a sandhill crane (*Grus canadenis*), which is 40–48 inches tall and has a gray body with a red crown, or a whooping crane (*Grus americana*), which is 50 inches tall and has a white body with a red face. Both were abundant species at this time. The "Cormorants" could have been double-crested cormorants (*Phalacrocorax auritus*) or American anhingas (*Anhinga anhinga*). Peterson, *Field Guide to the Birds,* 40–41, 106–107.

19. The partially web-footed duck was probably an American coot (*Fulica americana*), which is 15½ inches long with a slate-colored body. The canvasback (*Aythya valisineria*) is 20–24 inches long and has a white body, a black chest, and a red head. The green-winged teal (*Anas crecca*) is 14 inches long with a chestnut-colored head and a green ear patch; its body is brown, grey, and pale yellow with a green wing patch. The blue-winged teal (*Anas discors*) is 14–16 inches long with a slate-colored head and a white crescent shape in front of each eye; the body is brown to tan with a blue wing patch. Peterson, *Field Guide to the Birds,* 52–53, 58–59, 64–65.

Monday 22nd

DUNBAR

Thermometer before sun-rise 65°. Wind S.S.E. cloudy. A few drops of rain before day: set off as soon as we could get the men ready & on board.—Soldiers slow in their movements—continues cloudy & threatens rain. Green matter floating on the river, supposed to come from the Catahoola and other lakes and bayoos of stagnant water, which when raised a little by rain flow into the black river. Saw also many patches of an aquatic plant resembling little Islands, some floating on the surface of the river, and others adhering to or resting on the shore and logs; examined the plant & found it to be a hollow jointed stem with roots of the same form; extremely light with very narrow willow shaped leaves projecting from the joint, embracing however the whole of the tube extending to the next inferior joint or knot; the extremity of each branch is terminated by a spike of very slender and narrow seminal leaves from one to two inches in length and *1/10* or less in breadth, producing its seed on the under side of the leaf in a double row, almost in contact, the grains alternately placed in perfect regularity: I have not been able to detect the flower, so as to be able to determine the class and order to which the plant belongs, it is not probably new; I at first supposed it might be the same which is described by Mr. Bartram as occupying large portions of the surfaces of rivers in East Florida, but upon examination I found it to be entirely different.[20]

The day continued cloudy; at noon it rained, we had consequently no observation for the Latitude. At 3h p.m. thermr. at 79°—the afternoon continued cloudy. The current is yet insensible as to any opposition made to our progress. Sounded in the evening, found 3½ fathoms, the river being now considered very low. Extremes of the thermr. 65°–79°. Wind S.S.E. Cloudy—made 13 miles 76 perches.

20. William Bartram (1739–1823) was a renowned Philadelphia naturalist during this period and an acquaintance of Dunbar. During one of his cataloging excursions into the South, Bartram stayed with Dunbar at the Forest. Bartram was widely published but mostly known for his work *Travels through North and South Carolina, Georgia, East and West Florida*, published in 1791. Riley, "Sir William Dunbar," 91. The plant Dunbar describes may have been from the *Elodea* family. An example of such a species is the waterweed (*Elodea canadensis, Elodea densa*, or *Elodea nuttallii*). It occurs throughout the Southeast and the lower Mississippi valley. Grouped, it does tend to form what would appear as small islands of floating growth. Waterweed can grow to lengths of 1–10 feet. Smith, *Keys to the Flora of Arkansas*, 254–256.

HUNTER

Thermometer before sunrise 65° cloudy Wind S.S.E. It rained in the night & a little in this day. could make no observation. extremes of the Thermometer 65° to 79° Made this day 13 miles, 76 perches.[21]

Tuesday 23rd

DUNBAR

Thermometer 68°—the river for several nights past has fallen about 3 inches perpendicular each night: observed a great number of muscles and periwincles along shore: the muscle is of the kind commonly called pearl-muscle, & by means of its long tongue makes considerable progress along the bottom & upon the beaches of the river when under water: our people had a quantity of them dressed and found them to be agreeable food: to me they were tough and unpalatable.[22] The wind altho' a head but not strong, we got along pretty well; but towards 11h a.m. it became much stronger, and we made little way. Notwithstanding the cloudy state of the atmosphere we were fortunate in getting a good meridian observation, by which it appears we were in Lat: 30° 36′ 29″ nearly 3 miles higher than the town of Natchez: after dinner proceeded to the mouth of the Catahoola on the left and landed to get information from a french man settled here: he has a grant of land from the Spanish government, has made a small settlement and keeps a ferry-boat for crossing men & horses traveling to or from Natchez and the settlements on red river and on the Washita river: the Country here is all alluvial; in process of time the rivers shutting up ancient passages & elevating the banks over which their waters pass, no longer communicated with the same facility as formerly; the consequence of which naturally is that many large tracts formerly subject to annual inundation are now entirely exempt from that inconvenience: such is the situation of a most valuable tract upon which this french man is settled: his house is placed

21. The similarities of the statistical information in the two men's entries suggest that Hunter may have been periodically copying Dunbar's entries.

22. These "periwincles" may have been a species of periwinkle called running myrtle (*Vinca minor*) or a plant called bluestar (*Amsonia tabernaemontana* or *Amsonia illustris*). These latter two species grow in moist bottomlands and along stream banks. Although periwinkles bloom during the warmer seasons, sporadic flowers can appear on plants as late as December in the lower Mississippi valley. Dr. Henry Robison, Biology Department, University of Southern Arkansas, interview by Berry, January 7, 2004; Dr. Tim Knight, Biology Department, Ouachita Baptist University, interview by Berry, January 9, 2004. Numerous varieties of mussels (*Lampsilis, Actinonaias,* and *Villosa* families) are native to the Ouachita and Black Rivers. There is no way to identify the exact species from Dunbar's description. One of the most common shoal mussels in the Ouachita River is called a mucket (*Actinonaias ligamentina*). Robison interview.

upon an Indian mount with several others in view: there is also a species of rampart surrounding this place & one very elevated mount; all of which I propose to view and describe on my return, our situation not now admitting delay:[23] the soil here is equal to the best Missisippi bottoms; the proprietor says the high mount is not less than 80 feet perpendicular, of this we shall form some estimate at our return. We obtained from him the following list of distances from the mouth of the red river to the Post on the Washita called Fort Miro.

From the mouth of Red river to the mouth of Black river	10 Leagues [24 miles]
To the mouths of Catahoola, Washita & Tenza	22 [52.8 miles]
To the River Ha-Ha on the right1	[2.4 miles]
To the Prairie de Villemont on the same	5 [12 miles]
To Bayoo Louis on the same—rapids here	1 [2.4 miles]
To Bayoo Boeufs on the same	4 [9.6 miles]
To the Prairie Noyee (drowned Savannah)	3 [7.2 miles]
To Pine point on the left	4½ [10.8 miles]
To the Bayoo Calumet	3½ [8.4 miles]
To the Coal mine on the right & Gypsum on the opposite shore	3 [7.2 miles]
To the 1st Settlement	12 [28.8 miles]
To Fort Miro	22 [52.8 miles]
	Leagues[24] 91 [219.31 miles]

The accounts of the low state of the river we receive here are rather discouraging, as it appears, that on the first rapids, seven leagues distant there are only 22 inches of water, and we now draw at the stern 30 inches or more.—Went on and encamped within the mouth of the river Washita. This river derives its appellation

23. Dunbar later refers to the "french man" as "Hebrard." This might have been Don Juan Hebrard De Baillion, who obtained a land grant on March 22, 1786, as well as the right to build a ferry at the site of present-day Jonesville, Louisiana, in Catahoula Parish. Hunter calls the same man "Cadi" or "Cadet" in his journal entry of this date; Robin, *Voyages dans l'intérieur de la Louisiane,* vol. 2, *Paris,* 309–314; McDermott, "Western Journals of Hunter," 82. The site Dunbar describes is known to archaeologists as the Troyville site. It once stood near the town of Jonesville, Louisiana. For more on this site, see Hunter's entry for January 24, where he gives a fairly good description of the complex of mounds. Also see Dunbar's entry for January 21–22.

24. This is the French league, equal to 2.4 nautical miles. Hunter journal entry for this date; Chardon, "The Linear League in North America," 143–147. For more see Dunbar's entry for October 16 and note 1 to that entry.

from the name of an indian tribe formerly resident on its banks, but now no more to be found; it is said that the remnant of the nation went into the great planes to the westward & either compose a small tribe themselves, or are incorporated into another nation.[25] The Junction of the Washita[,] with the Tenza and the Catahoola a little below, all together form the black river, which last here, loses its name, altho' our maps represent it as taking place of the Washita, the Tenza and Catahoola are also names of ancient tribes now extinct: the latter is now the name of a Creek or bayoo 12 leagues long, which is the issue of a lake of the same name 8 leagues in length & 2 leagues generally in breadth, it lies west of the this place & communicates with the Red river during the time of the great annual inundation; it receives at the West or N.W. angle a Creek called little river, which preserves a channel with running water at all seasons, meandering along the bed of the lake; but all other parts of it superficies during the dry season from July to november & often later, are completely drained & become clothed in the most luxuriant herbage: the bed of the Lake then becomes the residence of immense herds of Deer, of Turkeys, Geese, Ducks, Cranes &c &c feeding upon the grass and grain; the Duck species being generally found on or near the little river. The Bayoo Tenza serves only to drain off a part of the waters of the inundation from the Missisippi low lands which here communicate with the black river during the season of high waters.[26] By reference to our Latitude at Noon we find the mouth of the Washita to be in Lat: 31° 37′ 57″—Extremes of the thermometer 68°–73°. Sounded—found 6 fathoms—muddy bottom. Made this day 9 miles 77½ perches.

HUNTER

Extremes of the Thermometer 68° to 75° Wind WNW The river fell 3 Inches during the night. Set out at sunrise & came this day nine miles 77½ perches to the Mouth of the Catahoula a long lake on the left hand & nearly opposite on the right

25. The Ouachita Indians were a dwindling tribe by 1690–1700. Some scholars believe that they merged with the Natchitoches Indians. The Spanish explorer and founder of Fort Miró, Don Juan Filhiol (Jean Baptiste Filhiol), said he believed that the tribe had moved to a northern region. Swanton, *History and Ethnology of the Caddo Indians*, 12–13; Dickinson, "Early View of the Ouachita Region," 12; also see Dickinson, "Historic Tribes"; Hiram F. Gregory, "The Caddo Indians of Louisiana," Anthropological Study no. 2, 1978, Louisiana Archeological Survey and Antiquities Commission, Baton Rouge, LA, 29.

26. The confluences of the Little and Tensas Rivers are basically across from one another near Jonesville, Louisiana. The Tensas River enters on the east bank of the Ouachita and the Little River on the west bank. Catahoula Lake lies just west of Jonesville, Louisiana, in Catahoula Parish. The Tensas River also enters the Black River across from Jonesville. The Tensas and Black Rivers form the border between Catahoula and Concordia Parishes. Louisiana Quadrangle Maps; Dickinson, "Early View of the Ouachita Region," 13.

is the entrance of the Bayu Tenza, The river Ouachita laying in the middle or rather the Quachita is the main branch of the black river which here loses its name in the three above mentioned waters. Sounded 6 fathoms water in the mouth of the Ouachita. We landed at Monsr. Cades here, where he keeps a ferry boat to carry over travellers to & from Natchez to fort Miro.[27] Here we staid till evening to get the necessary information to enable us to pursue our rout, which we took down in writing & is as follows

From the mouth of Red to mouth of Black river ten leagues	10 [24 miles]
from thence to Catahoula or Ouachita	22 [52.8 miles]
(by our reckoning from the mouth of Red River to Ouachita)	77 1/8 miles
from thence to Bayu or river ha-ha	1 [2.4 miles]
to prairie Villemont opposite Pine Point	5 [12 miles]
to Bayu Louis & the rapids on the right	1 [2.4 miles]
to Bayu Beauf on the right	4 [9.6 miles]
to the drownded prairie	3 [7.2 miles]
to pine point on the left	4½ [10.8 miles]
to Bayu Calumet three & an half	3½ [8.4 miles]
to the Coal Mine on the right & the Plaster of Paris on the left	3½ [8.4 miles]
Olivots first settlement, (Shoal & rapid)	12 [28.8 miles]
To Fort Miro	22 [52.8 miles]
	french leagues 90½ [217.2 miles][28]

Latitude of the mouth of the River Ouachita 31, 37, 57.

27. For more on "Monsr Cades," see Dunbar's entry dated January 21–22; and "Dunbar Trip Journal, Vol. II," January 21 and 22. In his official report, Hunter says of the area of Tensas confluence and "Cades" settlement, "This has been & shortly will be a place of importance, it is a short pass from Natchez, to the settlements at the rapids on Red River, & Fort Miro on the Ouachita." "Hunter Official Report," 11.

28. Hunter's calculation of total leagues is one-half league less than Dunbar's estimates. The difference lay in the distance between "Bayu Calamet" (Bayou Calamus in southern Caldwell Parish) and the coal and gypsum mines. Hunter marked the distance at 3½ leagues, while Dunbar measured it at 3 leagues. There is no way to know which is correct, since the exact location of the mines has not been determined. Louisiana Quadrangle Maps. McDermott stated that the name origin of the "Olivots first settlement" may be Peter and Boston Olivos, who owned property along the Ouachita. McDermott also discovered that in 1803 the population of the settlement "above Catahoula" was approximately "fifty and sixty families." D'Anemours, "Mémoire sur le district de Ouachita," 20; McDermott, "Western Journals of Hunter," 85.

This Monsr Cadi lives on an Indian mount about an acre in extent which is the only place near him that is not overflowed in the great freshes, & he seems to express a satisfaction that he has no bad neighbors. The ground here is very rich & if it were to be defended by a dike or Bank would be inexhaustibly fertile. River gentle scarce any perceptable current.—Memorandum[29]

Wednesday 24th

DUNBAR

Thermometer before sun-rise 54°—Wind North—Cloudy—Temperature of the river water 71°. No current to impede our progress worth estimating. made slow advancement as usual with our oars; found the shore favorable for tracking or towing, which mode we continued nearly all day making at the rate of five perches pr. ½ minute, which is about half a perch more than by rowing: a boat properly constructed for an expedition of this nature ought to advance with more than double our velocity.[30] The wind was contrary all day otherwise we might have gone at the rate of 6 perches which is equal to 2¼ miles per hour, more might be performed, but our Soldiers seem at certain times to be without vigour & now and then throw out hints that they can work only as they are paid.

The high lands on both sides have now the appearance of being above the inundation; the timber is such as is generally produced upon high lands[:] chiefly Oaks, red, white & black; interspersed with a variety of others; the magnolio grandiflora is absent ; its presence is an infallible sign of lands not subject to inundation. We observed today along the banks the strata of solid clay or marl (not recent but apparently ancient) to lie in very oblique positions, some making an angle of nearly 30° with the horizon & generally inclined with the descent of the river, altho' in a few cases the position was contrary; timber was also seen projecting from under the solid bank, which last seems to be in some measure indurated;[31] it

29. Either no memorandum was attached here by Hunter, or it did not survive.

30. "Tracking" is similar to cordelling, the simple but arduous task of pulling a vessel up a stream or river.

31. The "red" oak could have been the southern red oak (*Quercus falcata*), the northern red oak (*Quercus rubra*), or the nuttall oak (*Quercus nuttalli*). The original range of the southern magnolia (*Magnolia grandiflora*) extended across the gulf coastal regions and into Georgia and South Carolina, but it did not extend much farther north than Fort Miró (Monroe, Louisiana). Today these trees may be found as far north as the Ouachita Mountains in west-central Arkansas. Little, *Field Guide to Trees*, 388, 402, 407, 440. Dunbar's assertion that the presence of Magnolia trees would have proved the lands were not subject to flooding is not correct. Magnolias can survive in areas of frequent inundation along waterways in the southern regions of North America. Jim Taylor, Biology Department, Ouachita Baptist University, interview by Berry, January 16, 2004. Induration is the hardening of rock or soil by heat or pres-

is unquestionably very ancient presenting a very different appearance from the recently formed soil: the river is here about 80 yards wide. The Bayoo Ha-ha comes in unexpectedly from the right about a league above the mouth of the Washita,[32] and is one of the many passages or issues thro' which the waters of the great inundation penetrate & pervade all the low countries, annihilating for a time the currents of the lesser rivers in the neighbourhood of the missisippi. Vegetation is extremely vigourous along the alluvial banks; the twining vines entangle the branches of the trees & expand themselves along the margin of the river, in the richest and most luxuriant festoons, and often present for a great extent a species of impenetrable Curtain varigated and spangled with all possible gradations of Color from the splendid orange to the enlivening green down to the purple & blue and interwoven with bright red and russet brown. A carpet of the finest shrubbery overspreads the elevated margin, composed of a variety of elegant vegetables, to many of which probably no names have yet been assigned by the Botanist; and in positions where the shade is not too deep, the surface is enameled with thousands of humbler plants in full blossom at this late season.

The day has continued cloudy but begins to clear away about 11h a.m. we therefore landed before noon to observe & found our latitude to be 31° 42′ 30″.5—The timber of the higher grounds is still remarked to be inferior in size and height to that on the Missisippi; but here it may be accounted for by a less fertile soil, not apparently (at most rarely) subject to inundation. The wind still continues in the N. or N.N.W. but the clouds are disipating and tomorrow we expect fair weather, for making observations. Extremes of the thermometer 54°–68°. Encamped after completing a poor days voyage of 14 miles 48 perches. Thermr. at 8h p.m. 54°.—

HUNTER

Therm. before sunrise 54° Temperature of the river 71.

Set out about day light & in about an hour passed a large Bayu called Ha-ha on the right & some highland. Observed on the left the strata of clay obliquely down the river inclining about 30°. The river is still gentle with little current Landed to observe on the right side of the river & found the O.d. Apt. mer. Alt 92.4.50″ Ind error +0.13′,45″ Lat found. 31°. 42′. 31.5″ After dinner passed some highland to

---

sure. It can also refer to the creation of hardpan—a hard soil horizon caused by a chemical action. The term is sometimes used for the process that produces shale. Induration is also called lithification or lithifaction. Parker, *Dictionary of Geology and Mineralogy*, 129, 150; Bates and Jackson, *Dictionary of Geological Terms*, 258–259, 298.

32. Ha Ha Bayou enters on the east bank of the Ouachita River about 2 miles north of Jonesville, Louisiana, in Catahoula Parish. Louisiana Quadrangle Maps.

the right, & towards evening a large Bayu going to the left in a S.W. direction.[33] We tracted the greatest part of this day. The river sometimes about 80 yards wide. Made this day 14 miles, 48 perches—

Thursday 25th.

DUNBAR

Thermr. in air 49°—in river water 68°. Wind north. Cloudy. Continued & passed Villemont's prairie on the right & pine point opposite:[34] the prairie obtained its name in consequence of its being included within a grant under the french Government to a gentleman of that name; some of the family & name yet remain at New Orleans but I have not heard of any claim for this land; many other parts of the Washita are named after their early proprietors: the french people projected & began extensive settlements upon this river, but the general massacre planned & in part executed by the Indians against the french, and the consequent massacre of the Natchez tribe by the french, broke up all those undertakings & they were not re-commenced under the french government.[35] Those prairies are planes or savannahs without timber, generally very fertile, producing an exuberance of strong thick and coarse herbage. When a piece of ground is once got into this state

33. This "Bayu" may have been Bushley Bayou, just south of the present-day town of Harrisonburg in Catahoula Parish, or it may have been Bayou Bacheloi, or, as Dunbar calls it in his "Journal of a Geometrical survey," "Bayou Barchelet"; Louisiana Quadrangle Maps.

34. "Villemont's prairie" may possibly be named for Henri Martin Mirbiaze, Sieur de Villemont, who brought his family to Louisiana in 1719. According to Charles Gayarre in *History of Louisiana,* 1:281–282, a Villemont served as a French army officer in John Law's failed colony, and he later received a land grant for his service. There was also a Captain Melchior de Villemont who served as the commandant of the Arkansas Post from 1794 to 1802. Whayne et al., *Arkansas,* 61–62, 63, 65, 69, 73.

35. Dunbar's phrase "the general massacre planned & in part executed by the Indians" refers to the November 28, 1729, attack by the Natchez Indians on French settlers at Fort Rosalie (the current site of Natchez). At least two hundred settlers were killed and many (mostly women, children, and slaves) became prisoners. The French sought retribution in 1731 with the help of Choctaw, Tunica, Colapissa, and Ouma Indians, and several hundred Natchez were either killed or sold into slavery in Haiti. Governor Perier to Count de Maurepas, no date, in Rowland, *Mississippi Provincial Archives,* 1:54–56, 72–73, 77–81, 87, 117–126, 129, 176. An account of the French defeat of the Natchez Indians may also be found in "Relations of the Defeat of the Natchez by Mr. Perier, Commandant General of Louisiana. Attached to the Letter of Mr. Perier, governor of Louisiana, of March 25, 1731," "Pre-1801 Collection," Library of Congress, Washington, DC. Some of the Natchez later fled across the Mississippi River to lands along the lower Ouachita and Black Rivers. Some also later joined the Cherokee, the Creek, and the Chickasaw. In 1911 anthropologist John R. Swanton declared that the Natchez were "practically extinct." *Indian Tribes of the Lower Mississippi Valley,* 231–236, 251–257; Wilds, Dufour, and Cowan, *Louisiana Yesterday and Today,* 4–5. Green, "Governor Perier's Expedition."

in an indian country, it can have no opportunity of re-producing timber; it being an invariable rule to fire the dry grass in the Fall or winter, to obtain the advantage of attracting game when the young tender grass begins to spring; & thus the young timber is destroyed, & annually the prairie gains upon the wood land; it is probable that the immense planes known to exist in America may owe their origin to this practize. The planes of the Washita lie chiefly on the East side, & being generally formed like the Missisippi lands sloping from the bank of the river towards the great river, they are more or less liable to the influence of inundation in the rear, which has been known to advance so far in certain great floods, as to be ready to pour over the margin into the Washita river; this however has now become a very rare case & it may generally be estimated that from ¼ mile to a whole mile in depth will remain exempt from inundation during the high floods: and this is pretty much the Case with those lands nearly as high as the Post of the Washita, with the exception of certain ridges of primitive high land; the rest being evidently alluvial, altho' not now subject to be inundated by the Washita river, (which has originally cause their formation), in consequence of the great depth, which the bed of the river has acquired by abrasion.

We saw a good deal of high land today on either bank producing pine and other timber not the growth of inundated lands. About a league beyond Pine point we arrived at Bayoo Louis on the right, being the commencement of the rapids or rather shallows:[36] Sent people into the water to search the best channel, and after being frequently aground and dragging the boat we got up into a situation about a mile higher, where we were in a manner embayed, being shut in by a gravel-bar upon which there was scarsely in the deepest part a foot of water: finding the men fatigued by being so much in the water at hard labor, we thought it best to rest for the remainder of the day and consult upon what was best to be done.—The bar being of inconsiderable breadth & no rock in the bottom as we had been taught to expect, it was thought best to cut a channel sufficient for the passage of the boat, which we supposed would take less time than unloading, transporting & reloading at a considerable distance from our present station.—The weather continued damp and disagreeably cold all day: we had no observation at noon. Extremes of the Thermr. 49°–60°. Wind at North. Clearing up—many stars to be seen in the evening: made 3 miles 120 perches.

36. Bayou Louis meets the Ouachita about 4 miles northeast of Harrisonburg, Louisiana, in Catahoula Parish. Louisiana Quadrangle Maps.

HUNTER

Set out half past six A.M. from a few miles below the Rapids,[37] where we arrived at dinner time, being stopped by the shallows, after various efforts we passed all of which but one, where there was only about 1 foot water & as our Boat drew two & an half by the stern & less by the bow we brought her upon an even keell by moving part of the loading forward & as the men were much fatigued by wading in the water & dragged the boat thro the strong current, it was thought best to let them rest & dry & warm themself for the rest of this day. The Lower part of these rapids are formed by several small bars or Islands formed of gravel & mud & the upper part by a ledge of soft rocks which seem to be formed of indurated sand & clay, a kind of bad free stone, which acquire an iron brown colour externally by exposure to the air tho white within. On these Bars we found plenty of clams which we eat & found not unpleasant; & their shells were of the Mother of Pearl some white & others of a beautiful purple in the inside, very thin & semitransparent. These rapids are but a triffling obstruction to the navigation of the river & only at times & seasons like the present when the waters are very low; They might with very small expense be made at all times passable for boats drawing six feet or perhaps 8 feet water: For when the river is high, there is plenty of water for any Vessel. Opposite to the ledge of rocks the ground rises about 100 feet high, forming broken ground, producing pines Oaks &cc, & in the gulleys between cypress[38]— & various other trees; under which even to the tops of the ridges abundance of grass grows fit for a range for Cattle.—I here waded into the water & with an assistant staked out the narrowest part of the bar thro which it was determined to cut a passage for the boat with her loading. It was about 36 yards in length & 44 in breadth & marked in such a way as to receive the aid of the current to assist in sweeping out what we dug, the shallowest part was about 6 Inches deep—

Friday 26th

DUNBAR

Thermomr in air 40°. in river water 65°—Wind N.W. light clouds. The morning being very cool, it was thought best for the people to take an early breakfast before going into the water to work. After breakfast commenced digging the cannal which was required to be about an hundred feet long: this business went on heav-

37. At the site of Bayou Louis.

38. By "clams" Hunter must mean mussels. Also see the Dunbar entry for October 23. The bald cypress (*Taxodium distichum*) is also called swamp cypress; it is 100–120 feet tall and has needle-leaves. Little, *Field Guide to Trees*, 302.

ily & slowly as usual, and it was not untill noon that it was made barely of the depth which it was supposed might pass the boat.

The day being fine made some observations for the regulation of the watch & for the magnetic variation, and at noon had a fine observation, from which the Latitude of this remarkable place was ascertained to be 31° 48′ 57″,5—a little way up the river ¼ of a mile there is a high ridge of primitive earth studded with an abundance of fragments of rock or stone, which appears to have been thrown up to the surface in a very irregular manner, the stone is of a friable nature, & some of it has the appearance of indurated clay; without it is blackish from being exposed to the air, and within of a greyish white: it is said that within the hill, the strata are regular, & that good grind-stones may be obtained.[39] After dinner the boat was moved into the channel, where she stuck fast. Cables, ropes, and pulies were got across and fixed to trees; handspokes were used to raise & push her along and we made some way thro' the bar, but evening coming on we were obliged to desist in hopes of being able to get over in the morning. Extremes of the thermomr. 40°–70°. Wind N.W. Clear star light. Discovered a barge coming up behind us; she also grounded & sent her people out to search for the channel.

HUNTER

This morning being raw & cold, it was thought best that the men should have their breakfast before they should go into the water to dig out the channel to let the boat pass, as we immagined they would soon do it & then we should go on, without stopping till midday, but we reckoned without our host, for when midday come the channel was but half done; The men seemed jaded or unwilling to work at it any more & it was concluded to try to force the boat thro it with hand pikes. This we attempted & got thro only a few feet when we were obliged to stop for want of force. I then got a runner & tackle fixed to a tree on the opposite bank, to obtain which were obliged to shipp the Mast & use our Haulliards & all the spare rope on board. We then divided our force, set six of our strongest men to use the hand pikes upon the boat in the water & the rest to the tackle ashore & and by working all together we got about half over when night came on.—In the forepart of the day whilst the men were employed digging the channel I went with an assistant & sounded the river for the best passage from where we were to the end of the rapids & found to our great satisfaction that tho the current was very strong yet there was plenty of water for our boat. saw a large flock of wild Turkeys, some

39. Dunbar is probably describing a shale formation. Shale is friable (easily crumbled) and ranges in color from black and grey to brown and red. Bates and Jackson, *Dictionary of Geological Terms*, 460.

plover, & many wild geese this day, which seemed to be not quite so shy as farther down the river.[40] Here at the ledge of rocks to the right is a bayou which is now dry at the entrance.

Saturday 27th

DUNBAR

Thermometer in air 32°. in river water 64°. Wind N. Clear above. a fog upon the river, occasioned by the condensation of vapor arising from the surface of the river: the morning being very cold with a hoar-frost, the people were directed to get their breakfasts and prepare to use their exertions in getting the boat over the shoal; the day proved very fine with an agreeable warm sunshine, but it was 1h p.m. before we got entirely over into floating water on the opposite shore, the men having upon this occasion exerted themselves to my entire satisfaction.[41] The occupation of this day prevented us from making any astronomical observations.—After dinner we pushed on and arrived at the last of the rapids at this place; here we found a ledge of rocks across the entire bed of the river, but having previously sounded and discovered the best channel, we got over into deep water after grounding and rubbing two or three times: The river became again like a mill-pond without current, excepting a motion barely perceptible along the concave shore, the velocity was nevertheless very considerable upon the shoals where the depth of water was small. The whole of those first shoals or rapids embraced an extent of 1½ miles; that is, the obstruction was not continual, but felt at short intervals along this space: Encamped about 1½ mile above the last rapid. Extremes of the thermr. 32°–73°. The evening proves fine & mild. Thermr. at 8h p.m. 62°. Wind North. High pine land on the right—breadth of the river 100 yards.

HUNTER

Began to work again after breakfast & by the addition of another block to the tackle & assistance of the crew of another boat which came up & was also stopped

40. The similarity of the first sentence in each explorer's entry on this day again points to the possibility that one of the men may have been copying from the other. A "runner & tackle" is a block-and-tackle apparatus. "Haulliards" (or "halbiards," halyards, or "habiers") are ropes used to haul items or lines used in hoisting objects. The average draft of a keel or flatboat was 12–18 inches. The draft of the Chinese boat was 2½ feet. The canoes used were probably small dugout canoes. The wild turkeys would have been *Meleagris gallopavo.* Hunter's "plover" was possibly a killdeer (*Charadrius vociferus*), a year-long resident species and member of the plover or *Charadriidae* family. It has a brown back and a white underbelly with two large black breast bands. Peterson, *Field Guide to the Birds,* 120–121.

41. A "hoar-frost" is a white, covering frost.

by the shallows we forced our way over about 2 p.m. We then dined & set forward & soon got thro the rest of the rapids; where we found the river as before a smooth, & peaceful stream with scarce any currents; The banks still rising gradually in height on each side as we advanced, The land on both sides is now more composed of sand intermixed with the black vegetable mold than formerly, & on the right bank coming up observed frequent prairies of seeming small extent, with trees scattered thro them. The Timber assumes now a larger size & growth than lower down. From the bar where we stopped to the ledge of rocks which terminates the shoals here may be about ¼ mile & we found the best water on the right side very near the shore, even where the rippling seems to indicate otherwise. Here is a Bayu on the left.[42] Extremes of the Thermr. 32° to 73 hoarfrost. This day made 2 miles 77 perches.

Sunday 28th

DUNBAR

Thermometer in air 40°—in river water 63°.—Wind N.W. Clear—fog on the river. Continued our voyage & made some observations for the Longitude & magnetic variation at the hour of breakfast. High lands and a large Savannah seen on the right in the morning passed a rocky hill soon after and 'Bayou aux boeufs' on the right about 4 leagues from the rapids.[43] At noon got a good observation, Latitude deduced 31° 53′ 35″, 5—at 3h. p.m. the thermomr. was at 78° in the shade; the day was warm and the sun powerful: observed some more planes to the left: the river made several returning courses today, to the southward of west. Thermomr. at 8h. p.m. 56°.—Extremes 40°–73°. Sounded—3 fathoms—mud & sand. Made this day 12 miles 116 perches.

HUNTER

Set out at half past six a.m. The morning very foggy on the river & not so cold as yesterday. The banks still rising in height by slow degrees & the land more & more intermixed with sand, at least it appears so by the banks. This day towed all

42. Dunbar mentions the arrival of another "barge" behind them in his October 26 entry. He also says that the new barge has been "grounded." He does not, however, mention the assistance of the other crew in freeing the Chinese boat. See also "Dunbar Trip Journal, Vol. I," October 26, 27, 1804. The "Bayu" could possibly be Rawson Creek entering on the west bank in Catahoula Parish.

43. The "Bayu Beouf" or Bayou Boeuf, which forms the border today between Catahoula and Franklin Parishes, may be translated from the French as "Bayou of the Ox" or "Bayou of the Bison." Louisiana Quadrangle Maps. Hunter calls it "Bayu Beauf" in his entry of this date. Dickinson, "Early View of the Ouachita Region," 14.

the way; found on the bank a young Fawn just killed by a Panther, the throat being tore very much.[44] we took it on board & made a hearty meal of it, or two for all hands, Trees increasing in size. Thermr. 40°. in air & 63 in the river water in the morning. A prairie on the left, Bayu Beauf on the right on the right & an hill composed of the white sandy stone crowned with Tall Pine trees[.] prefer tracting when the nature of the banks will permit, as it is both easier to the men & we go faster. Latitude observed 31°, 53′, 35″ depth of the midchannel of the river sand & muddy bottom 3 fathoms[.] Made this day 12 miles 116 perches.

Monday 29th

DUNBAR

Thermomr. in air 41°. in river water 62°. Wind N.W. Fog on the river. Continued our voyage—The banks of the river seem to retain very little alluvial soil; on the opposite shores we see frequently to the water's edge the high land earth, which is a sandy loam of a greyish light color with streaks of red sand & clay; the soil is not rich, bearing great numbers of pines, interspersed with red oak, hickory and dog-wood.[45] The river is now from 60 to 100 yards wide. At the hour of breakfast made three lunar observations, and one sun's altitude to regulate the watch, which with the observations of yesterday will give the rate of going of the watch proportioning for change of Latitude and departure as we advance in the progress of our voyage; I do not however think it of much importance to regard those observations untill we arrive at the post of Washita, which I suppose to be nearly the most easterly point of the river; there and at the hot-springs (the most westerly point we shall visit) we shall take time to make correct observations; all other points of the river will be ascertained with sufficient precision from our geometrical survey so frequently corrected by the Latitude. At noon we found our Latitude to be 31° 58′ 2″. Having make some advantageous alterations in the arrangement of our benches and oars, we advanced with a little better speed; about 6

44. The cougar or mountain lion (*Felis concolor*) is 72–90 inches in length and tawny to grayish in color. It was once abundant in Arkansas, but by the late nineteenth century its numbers had declined drastically due to hunting and loss of habitat. Sporadic reports of sightings continued in Arkansas and Louisiana into the midtwentieth century. In Arkansas by 1988, most reports had been labeled as escaped exotic pets. Burt and Grossenheider, *Field Guide to the Mammals*, 77–78; Sutton, *Arkansas Wildlife*, 20, 28, 88, 246–247; Sealander and Heidt, *Arkansas Mammals*, 235–236.

45. Loam is a soil comprised of a mixture of clay, silt, sand, and organic matter. The dogwood or flowering dogwood (*Cornus florida*) grows to a height of 15–40 feet and has 2–5-inch egg-shaped, clustered white flowers with four distinct petals. Bates and Jackson, *Dictionary of Geological Terms*, 301; Petrides, *Field Guide to Trees and Shrubs*, 49.

perches pr. ½ minute which however does not exceed 2¼ miles pr. hour in water without any sensible opposition from the Current. The wind came about to S.W. in the evening; Thermr. at 8h p.m. 62°. Extremes 41°–85°. Soundings—3 fathoms mud & sand—made this day 14 miles 65 perches.

HUNTER

Thermometer in Air 41° & in the river 62° in the Morning about ¼ past six when we set out. Land generally rising in height above the river which at this season of the year is at its lowest A Creek on the left.[46] Lat. ob[s]ervd 31°.58′.2″ at three P.M. Thermometer 85°. The land is generally poor, thin & sandy, timber Pines &c depth of midchannel 3 fathoms This day made 14 miles 65 perches

Tuesday 30th

DUNBAR

Thermomr. in air 47°. in river water 60°. Wind W.N.W. fog on the river. Clear above.—Continued our voyage: the land on either bank seems to be from 30 to 40 feet high and does not improve in quality: pine-trees seen in most situations—nothing remarkable occurred except a rapid we passed in the afternoon, formed by a ledge of rocks which traversed the river, narrowing the water channel to about 30 yards, but the extent between the high banks was not less than a hundred. At noon found the Latitude to be 30° 5′ 24″. It would appear from the distances run by our Log and time, when compared with the estimated distances by the french inhabitants and hunters, that their league scarcely exceeds two miles. Encamped near a sand beach favorable for hauling the sene [i.e., seine] & catched a sufficiency of fish to serve all the people for supper and breakfast. Thermr. at 8h p.m. 60°. Extremes 47°–83°. Made this day 15 miles, 150 perches.[47]

HUNTER

Thermometer in the morning in the air 47° do in the river water 60° Fog on the water, Wind WNW clear w[e]ather Set off at day light about 6 A.M. Came in

46. The "Creek on the left" may be Bayou Dan in Caldwell Parish. McDermott identified this creek as possibly being called "Bayou aux Dindes," and he noted that in Lafon's *Carte Generale du Territoire d' Orleans,* an area called the "Prairie Noyee" was across on the east bank. "Western Journals of Hunter," 84.

47. Most of the Ouachita River below Camden, Arkansas, has been dredged for more suitable river travel. Many, if not most, of the rocks and gravel bars that the expedition encountered have been cleared by this dredging. A recent visit to the river near Enterprise, Louisiana, makes it apparent that the river has been altered from what the explorers experienced. "Sene" refers to a seine or fishing net that hangs vertically in the water with floats on the top edge and weights on the bottom. It is usually pulled through the water by two to four persons.

the afternoon to a rapid where the river was only about 30 yards wide, A creek on the left. Thermometer in the afternoon 83°. Land as before said to be rather light & thin. Now & then observe high sandy hills on each side of the river, but seldom opposite to each other, The Banks still increase slowly in perpendicular height, & appear not subject except in particular place to inundations. distance made this day 15 miles & 150 perches The Master of the runaway Negro hailed us from the bank coming up & we were relieved from the charge of him to his Master's great satisfaction. His name is Innes, he is a planter at the rapids of Red River originally from New York, he has been in this country 11 years.—[48]

Wednesday 31st

DUNBAR

Thermomr. in air 44°. in river water 62°. Wind N.N.W. Clear—fog on the river—Continued our voyage. This morning met with shallow water & strong currents, our rate of going, deducting the velocity of the stream was reduced to 2 perches: got upon shoals about 8h. a.m. which detained us greatly, and impeded us more or less untill the afternoon; at noon we had a good observation; Lat: found 32° 10′ 13″—at 2 h p.m. got over the last shoal for this day & went on in good water untill the evening, the channel was very narrow, the sand bars at every point extending so far into the bend as to leave little more than the breadth of the boat of water sufficiently deep for her passage, altho' the water often covered a breadth of 70 to 80 yards upon the shoal: in the afternoon passed a little plantation or settlement on the right and at night came up with three others joining each other: here is a plane or prairie upon which those settlements are placed;[49] from the regular slope of the land from the river bank towards the eastward, we may be assured the soil is alluvial, yet the bed of the river is now so deep that it is no longer subject to that inconvenience, but in the rear the Missisippi advances & sometimes leaves dry but a narrow stripe along the banks, it is however now more common

48. McDermott identified Innes as possibly Alexander Innes. This man owned property along Bayou Rapides (in Rapides Parish today and maybe near the "rapids Red River settlement"), along Bayou Castor in DeSoto Parish, and along other portions of the Red River. It is interesting that Dunbar makes no reference to returning "Harry" to his purported owner. Instead he writes, "nothing remarkable occurred." McDermott, "Western Journals of Hunter," 85; Dunbar journal entry of this date; "Dunbar Trip Journal, Vol. I," October 30, 1804.

49. The "plane or prairie" could have been Boeuf Prairie in Caldwell Parish. The "extensive cypress swamps" are probably those between the Boeuf River and Turkey Creek in present-day Franklin Parish, and the "high lands" to the west could be a reference to the Bayou Dan Hills in Caldwell and Catahoula Parishes. Louisiana Quadrangle Maps.

that the extent of the fields cultivated (from ¼ to ½ mile) remain dry during the season of the inundation: the soil here is very good but not equal to missisippi bottoms; it may be esteemed second rate. At a small distance to the East are extensive Cypress swamps, over which the waters of the inundation always stand to the depth of 15, 20 & 25 feet. On the west side after passing over the Valley of the river, whose breadth is various from ¼ to 2 miles or more, the Land assumes a considerable elevation from 100 to 300 feet and extends all along to the settlements on the Red river, those high lands from report are poor & badly watered, being chiefly what is termed a pine-barren: there is here a ferry & a road of Communication between the Post of the Washita and the Natchez & a fork of this road passes on to the Settlement called the rapids on Red river, it is distant from this place by computation 150 miles.[50]

From the experience we have had of this river and the information obtained, it appears that the present is the least favorable season for ascending this river with a boat of so considerable a draught of water as ours; the spring of the year is the most advantageous, the Missisippi then flows up into the beds of the inferior rivers, raising their waters sometimes within a few feet of the top of the banks; the small current is then often in favor of the ascending boat: this objection would vanish if light boats were used drawing 6 or 8 inches of water and if well constructed might make with ease 12 leagues or even 40 miles pr. day; such ought to be the kind of boats for an expedition fitted out to explore; as little time as possible ought to be lost in moving, that more may be left for observation and research: in our actual situation our dayly progress seldom equals 14 or 15 miles, which is a sad drawback upon the accomplishments of the objects of an exploring expedition. On this part of the river lies a considerable grant of Land conceded by the Spanish Government to the Marquis of Maison rouge a french emigrant, who bequeathed it with all his property to M. Bouligny son of the late Colonel of the Louisiana regiment & by him sold to Daniel Clark; it is said to extend from the post of the Washita with a breadth of two leagues including the river down to the bayou Calumet,[51]

50. Hunter earlier (in his October 23 entry) identifies this settlement with the ferry and the road as "Olivots first settlement." The settlement Dunbar names as the "rapids on the Red River" is today Alexandria, Louisiana, in Rapides Parish. Flores, *Southern Counterparts to Lewis & Clark*, 40, 111.

51. McDermott said of the Maison Rouge grant that it "extended on the west bank of the Ouachita from a point five arpents below the mouth of Bayou Cheniere au Tondre to the mouth of the Bayou Calumet and on the east of the Ouachita from Point L'aine, two leagues below Fort Miro, down to the Prairie de Lé [Lait]. Two other parcels were located on Bayou la Loutre, Bayou Siard, and Bayou Barthelemi and on the right bank of the Ouachita from the Mouth of Bayou Barthelemi down to Bayou la Loutre." Maison Rouge died in 1799 in New Orleans and "made Louis Bouligny his universal heir and

the computed distance of which along the river is called 30 leagues, but said to be not more than 12 in a direct line. Extremes of the thermomr. 44°–84°. Made this day 6 miles 165 perches.

HUNTER

Set out at ½ past 6 a.m. The current was here pretty strong for the first part of this day, Thermometer in the morning in air 44° & in the river 62° Our Course generally N a little Westerly, got on a shoal & were for some time embarrassed by shallows, the Land is no longer here subject to innundations. found our latitude by observation to N. 32.10′.13″ At 3 p.m. came to a Settlemt consisting of one house. Thermometer in the afternoon 84°. Here were informed that Dan. Clark of Orleans had purchased the Land on both side of the river from the Post, or Fort Miro to Bayu Calumet, a distance of 30 leagues. from the Heir of one Marquis De Maison Rouge. The price is said to be ten cents pr acre & the width supposed to be 40 acres deep on each side of the river which is about 1 ½ mile on each side.—The size of the french Acre 180 french feet square, by our measure 15 french feet are equal to 16. english feet. This large body of land is said not to be very good, generally, tho capable to produce wheat, corn, cotton &cc Were obliged to stop just below this shoal, where we slept all night & purchased a few Vegetables & a Canoe to lighten our boat when crossing shoals, by giving a small canoe which we had picked up in the Mississippi & six dollars for it. By estimation of settlers here, it is 20 leagues from this Shoal to the Fort Miro or Post of Ouachita & there are said to be several rapids & shoals between them, particularly one on which there are only said to be six Inches of Water at this season. Distance made this day six miles 165 perches, & that with a great deal of exertion and labor.

---

the latter, by acts of sale . . . conveyed all his rights to Daniel Clark." "Western Journals of Hunter," 85; also see Maison Rouge Land Claims.

A Spanish officer, Francis Bouligny led settlers in 1779 to an area on Bayou Teche called New Iberia. He had also proposed establishing a Spanish settlement along the Ouachita River as early as 1778, but other political figures, including Governor Bernardo de Gálvez, overruled Bouligny's plans. Bouligny to Bernardo de Gálvez, New Orleans, June 23, 1778; Maison Rouge Land Claims; Din, *Francisco Bouligny*, 92–98, 111.

In 1804 Daniel Clark (1760–1813) served as U.S. Counsel in New Orleans. He had been the U.S. representative in New Orleans during the Louisiana negotiations. In 1806 he was elected as a Louisiana territorial delegate to Congress and was later identified as a participant in the Burr Conspiracy. Clark died suddenly in 1813 and left the remaining portions of the Maison Rouge lands to his mother, Mary Clark. Gayarre, *History of Louisiana*, 3:120, 211, 267, 311, 341–342, 397, 405, 471–472, 584, 607–613; 4:19, 80, 144, 161.

Dunbar's "bayou Calumet" is probably Bayou Calamus in southern Caldwell Parish.

November 1804

Thursday 1st

DUNBAR

Thermomr. in air 48°. in river water 62°—Calm—clear above, a little fog on the river. Having sounded last evening a shoal upon which there is 18 inches water in the deepest place, we prepared, by unloading part of our Cargo, to cross it: we obtained the use of two Canoes, which with a great deal of trouble enabled us to get over about noon: finding a Canoe so useful & being informed of other rapids and shoals before us, we bartered away a smaller canoe with a little cash for the larger of the two we had borrowed, proposing to put two of our best hunters into the empty Canoe by which she might keep a head & procure some game, & be ready on all emergencies to assist the Barge. Dined & continued our voyage; met with several retardments from shoals. made only 4 miles 155 perches. Extremes of the thermomr. 48°–85°. at 8h. p.m. 64°. Weather extremely fine & agreeable, the slow progress of our boat being the only circumstance of regret, as tending to disappoint our prospects.

HUNTER

Thermometer in the morning in air 40°, water 62° Calm Clear weather except a little fog on the river which was dissipated by 9 A.M.

Having examined & sounded the Channel of the rapid over the shoals & marked out our best passage across, where we found it to be 18 Inches deep, our boat drawing 2½ feet by the stern, we found that by bringing her on an even keell draft of water would be reduced to two feet. Therefor on deliberation it was determined to leave out such part of the provisions & baggage as would reduce her draft of water to 18 Inches & carry it over afterwards in our Canoe purchased for that purpose at several times. This we effected which occupied our time all the forenoon. & immediately after dinner set out again on our voyage; we passed several shoals & shallow sand barrs which embarrassed us not a little, sometimes wading in the water up to the middle & dragging the boat thro & over the bad & difficult places, sometimes rowing & then tracting as it seemed to answer best. so that all this day we made only after much fatigue 4 miles & 115 perches—About 3 p.m. passed a sandy cliff about 100 feet perpendicular above the water near which [we] went ashore to examine a stratum of blackish substance looking like stone coal, but which proved to be only an indurated clay colored with iron, easily pulverant between the fingers.[1]

Some of the land here appears very sandy producing pine, Oak, Hickory &cc & other parts are Prairies, level clear of all sorts of trees & shrubs tho of no great extent, at least those on the banks which we saw in our way, These prairies do not seem to be here in much esteem altho they are generally surrounded with timber land. Altho it is said to be twenty leagues by water to Fort Miro yet they call it but 12 or 14 miles by land. The river where there are sand barrs appears to be about 50 yds wide, in other places twice as much, altho it gradually grows narrower & the banks higher as we assend it, tho with many exceptions—Above all innundations

No observations

Friday 2nd

DUNBAR

Thermomr. in air 48°. in river water 62°. light clouds—Wind S.S.E. a little fog on the river.—Continued our voyage with immense sand bars in view at every point: the utmost care in steering was necessary to keep clear of shoals and sunken logs, which latter were frequently very embarrassing: we suffered much detention this day from those causes, being twice fast upon a sunken log under water, and our boat being so unwieldy & heavy, there was no getting her off by any exertion of poles &c which could be made on board, a rope was carried ashore from the stern, & by that means she was hove backwards & cleared of the log: we lost 1½ hour each time by two such accidents, & several times upon shoals which delayed us greatly: light flat boats proper for the navigation of shallow waters would pass over all such obstacles without touching, & when they do touch, being light, they are easily pushed back; external keels are very improper for any boat upon the missisippi or any river where logs are to be encountered: our boat to her other inconveniences was provided with a keel, which added to her draught of water, made her much more difficult to get over a log or shoal, it being impossible to clear her by pushing latterally. Thermr. at 8h. p.m. 78°. Extremes 48°–84°. Made this day 8 miles 104 perches.[2]

1. Hunter uses the word "pulverant" or pulverulent to mean a substance that is easily crumbled to powder or dust.

2. "External keel" refers to the central connecting board that forms the spine in the hull of a vessel. The external keel protruded along the bottom of the vessel from bow to stern. The advantages of

HUNTER

Thermometer 48°. in Air & in the river 62° [.] at 3 p.m. 84°& at 7 p.m. 64. Cloudy Wind S.S.E.

Set off at near seven a.m. & rowed cheifly northerly all this day inclining to West. The banks continue sandy; The river more narrow & here & there very shallow, so that we are often obliged to go from one side of the river to the other to pick a passage fro the boat; The timber grows here large many tall pines on the highlands & here & there Cypress & swamp white oak in moist places by the water sides. The inland has many fine oaks hickory &cc Saw [for] the first time a flock of Pelicans; on the left appearances of half formed stone in thin strata by the waters edge; caught a few fish in the evening with the net & a fine soft shelld turtle by hook & line.[3] The fish in these fresh water rivers are not so good to eat as those near the sea, they are soft & comparatively insipid. In our way this day got on several sunken logs which cost us several hours labor to extricate ourselves from The land on each side here, as well as on the Mississippi now & then slips down in considerable portions into the river carrying the trees with it, & sometimes the earth washed away & carried down by the current leaving the trees standing in the water, these in time lose their tops & many of their branches but their trunks remain for a long time as chevaux de frise in the water & thereby stop other drift wood in its passage down & often injure the navigation of the river. The aligators are not so frequent as usual, & it is said they do not go beyond Fort Miro.—Made this day only 8 miles 104 perches.—In general the current is very gentle, tho here & there in narrow places it is more rapid—scarcely averaging ½ mile pr hour from the Mississippi to this place.

Saturday 3rd

DUNBAR

Thermr. in air 52°. in river water 64°. some light clouds. continued our voyage with very little variety, a great sameness appears as to the river and its banks. Al-

---

the external keel design included better directional stability as well as the enhancement of the vessel's overall strength.

3. The bald cypress (*Taxodium distichum*) is 80–140 feet tall and has needles one-fourth—seven-eighths inch long. The white oak (*Quercus alba*) grows to 80–100 feet, and its leaves are rounded lobes 4–9 inches long. The swamp white oak (*Quercus bicolor*) is not native to Louisiana. Little, *Field Guide to Trees*, 382–384. The pelicans were probably white pelicans (*Pelecanus erythrorhynchos*). The brown pelican (*Pelecanus occidentalis*) rarely strays from the Gulf of Mexico and the eastern coastal regions. Peterson, *Field Guide to the Birds*, 78. The smooth softshell turtle (*Trionyx muticus*) inhabits the rivers of Louisiana and Arkansas. Wernert, *North American Wildlife*, 165.

tho' we got several times aground we were not so unfortunate as yesterday; immense sand bars or beaches with steep banks on the opposite shore continued to be the objects of our view, very little alluvial land except at some points opposed to Cliffs, was to be seen: along the margin of the river, many humble plants are to be seen in flower at this late season, altho' the leaf falls from the trees of the forest: the great variety of tints which the foliage assumes before it separates finally from the parent stock, presents to the Eye an infinitude of beautiful landscapes, and if critically examined is perhaps not without its use: it will be found that the leves of the same tree are all changed to the same Color, which is probably occasioned by the oxigen of the atmosphere acting upon vegetable matter deprived of the protecting power of its vital principle, & thereby calls forth its latent colorific properties: I have always remarked that the leaves of such trees whose barks and woods are known to produce a dye, are changed in autumn to the same Colour, which is extracted in the Dyer's vat from the woods more especially by the use of alumn or other mordant; whose predominant principle yields oxigen: thus the foliage of the hickory & the oak yielding the quercitron bark is changed before its fall to a beautiful yellow; other oaks assume a fawn colour, a liver or blood colour, and are also known to yield dyes of the same complexion: I am persuaded from the few observations I have made that this rule will be found general, and may therefore serve as an excellent guide to the Naturalist who directs his researches to the discovery of new objects for the use of the Dyer.[4]

At noon we found ourselves in Latitude 32° 17′ 17″—nothing remarkable occurred in the afternoon, except a discovery made by Dr Hunter (walking along the riverside) of a substance resembling mineral Coal: I suppose, from its appearance, that it is the Carbonated wood described by Kirwan and other Chemists: some specimens were preserved; it does not easily burn, but on being applied to the flame of a Candle, it seemed to encrease it & yielded a faint smell resembling, in a slight degree that of the gum-lack of common sealing wax.[5] In the evening passed over some rapids and shoals; bottom stone & gravel. Thermomr at 8h p.m. 72°. Extremes 52°–86°. Made this day 11 miles 140 perches.

4. Dunbar's explanation of leaf color change is loosely correct. The loss of chlorophyll causes the change. This may be what he refers to as being "deprived of the protecting power of its vital principle." Alum is the mineral aluminum sulfate $KAl(SO_4)_2 \cdot 12\ H_2O$. It has a sweet-sour taste. Alum is used today mostly in pickling. Bates and Jackson, *Dictionary of Geological Terms,* 16–17. A mordant is a substance used to fix the coloring in the process of dying fabrics.

5. Hunter's discovery was possibly lignite, or a lower grade of coal. Richard Kirwan (1733–1812), an Irish chemist, mineralogist, and geologist, was quite a prominent scientist during the early nineteenth century. His best-known work, *Essay on Phlogiston* (1787), supported the "fixed air" principle. Charles Coulson Gillispie, ed., *Dictionary of Scientific Biography* (New York: Scribner's, 1981), 387–389.

HUNTER

Therm. in Air in the morning 52° & in the river 64° [.] at 3 p.m. 86 & in the evening at 6 p.m. 72° Set of[f] before sunrise. Our course this day generally Northwestward, tho sometimes by the turns of the river even south easterly. The land on the banks much the same as yesterday, & still as we asscend the river, it assumes a more ancient appearance & rises in height a little more than the stream does, yet by slow & scarce perceptable degrees. In the afternoon the left bank assumed more of clay than formerly & less sand, consequently the growth of timber larger & no pines. The greatest part of this days journey the river was in many places narrower & more rapid, with frequent falling in or rather slipping down of very large portions (say from 1/4 to 1½ acre at a time) of ground which sometimes chocks up the channel as to make it difficult for boats to find depth of water to pass over. Were frequently aground on rough coarse gravel barrs this day which delayed much of our time & cost us some exertions to get off, so that we made but about 11miles 140 perches this day. Encamped on the left bank on a bed of gravel, under which were several small strata of fine bluish clay & one of a black substance of about 2½ Inches thick resembling mineral coal. It is light, friable, soft has no grit in it. & when held in the flame of a candle seemed to increase the flame yet did not kindle, it sent forth at the same time a smoke resembling in smell that of sealing wax.[6]

Latitude by observation 32°.17′.17″.—

In the forenoon went ashore on the right bank & walked with my riffle along the bank ahead of the boat, found the land of a thin sandy soil, yet the timber large, came to an opening sending out a small rivulet, followed it up for a mile. it terminated in a prairie of about a mile in length in the center of which was a small Lake, now almost dry in which were a number of wild Geese, ducks & hoopping Cranes, but all so shy that I could not get a shot at them. Saw many tracts of Deer. In the afternoon went ashore as before, but on the left bank, here the land seemed somewhat more fertile & sent forth such a quantity of underwood, small briars & vines of various kinds as made it difficult to pass thro them. saw no game tho many deer tracts. tasted many of the springs that run out of the bank into the river which are chiefly ferriginous & deposit an ochry yellow mud, oxid of iron.[7] These appearances are observed all the way from the mouth of the Red river, even where the

6. Dunbar credited Hunter with the discovery of the "mineral coal." Dunbar called it "carbonated wood" in the "Dunbar Trip Journal, Vol. I," November 3, 1804.

7. The whooping crane (*Grus americana*) did exist in Louisiana and Arkansas during this time. Hunter may have seen a whooping crane, or he may have been viewing a sandhill crane (*Grus canadensis*), also common to the region. Baerg, *Birds of Arkansas*, 56; Howell, *Birds of Arkansas*, 27; Peterson, *Field Guide to the Birds*, 106–107; James and Neal, *Arkansas Birds*, 166. A "ferriginous" (ferruginous) soil or clay is iron-bearing and the color of iron rust.

banks seem but of a very recent date. It would seem that the water as it drains out of the banks after the inundations, dissolves a small portion of Iron which it deposites whenever it is exposed to the atmospheric Air.—In my tour I found this coally substance & it being late, we determined to stay there till morning when we might give it more attention.

Sunday 4th

DUNBAR

Thermomr in air 54° in river water 64°—Clear. This has been an unfortunate day; the morning and afternoon were spent upon shoals and rapids with stoney & gravelly bottoms, the Men having been a great part of the time in the water. Got a good observation at noon; Latitude found 32° 21′ 10″. Made only 4 miles 233 perches. Thermomr at 8h p.m. 63°. Extremes 54°–83.[8]

HUNTER

Therm. at sunrise 54° in air & in the river 64°[.] at 3 pm 83° & at 7 pm. 63.

Set out before sunrise as usual but had proceeded but a little way when we found ourselves surrounded by shoals & fast on a gravel bar, we sounded the water in all directions to find a channel deep enough for the boat, which we at last effected by moving some of the loading forward to bring the boat on an even keell, all hands then wading in the water & forcing thro it; This took till breakfast time; set out again & pushed forward, some times rowing, some times wading & dragging the boat over the shallows & then again tracting according to circumstances. The greater part of this day were embarrassed by rapids & shoals very often getting aground, & then delayed till a person would wade forward & across the river, a head of the boat in all probable directions in order to find the deepest water, before we could venture to proceed again. The men, or rather some of them often grumbling & uttering execrations against me in particular for urging them on, in which they had the example of the sergeant who on many occasions of triffling difficulties frequently gave me very rude answers, & in several instances both now & formerly seemed to forget that it was his duty in such cases to urge on the men under his command to surmount them rather than to show a spirit of contradiction & backwardness.—In the same spirit this day when at the helm he steered inshore too much[,] altho I cautioned him to keep out[,] & run under a projecting Tree & carried away our Mast which cost me so much pains to procure at Pittsburg & fix to strike at Orleans.—made 4 miles 233 p

8. Hunter was equally frustrated with the day, and his descriptions of the ordeal of navigating the river are more detailed.

Monday 5th

DUNBAR

Thermr. in air 52° in river water 62° heavy fog & damp air. We were obliged this morning to take out part of our loading to enable us to pass over a shoal carrying only 18 inches water, which detained us untill near 10h a.m.—In the course of the day got upon several shoals of inferior note, but upon the whole we were more fortunate than usual, the water being generally deeper and with little current. we remarked a greater appearance of fertility as we approached the Settlement [Fort Miro]; the trees are of larger dimensions, & there is a due proportion of shrub or underwood, which was absent in the poorer lands; some fields of Cane began to appear, which is a sure indication of a fertile soil: we had also leisure to admire the beautiful tints assumed by the foliage of the vegitable world: it was apparent that the external leaves most exposed to the light & to a freer circulation of air, exhibited the first changes of Color, while those of the same plant under a thick shade still retained their deep verdure.[9] The Willow tree pendent over that water, presents a fine deep yellow along the outline of the plant, from whence may be traced a regular gradation, thro' the admired lemon color down to the soft and delicate summer's green, which last in the shade, retains its full verdure: on other trees may be seen a deep blood color inclining to black, descending by regular shades to the palest pink mingled with green & from these by similar gradation to the usual summer verdure of the plant: Leaves plucked from the tree at this season & preserved in the shade will retain their beautiful colors for a great length of time.

The river continues of the same general breadth. i.e. from 80 to 100 yards, but the water channel is often confined to 30 yards. The Atmosphere had this day a smokey or misty appearance; the Sun broke forth a little in the afternoon, but shone with diminished lustre. This smokey or misty appearance which in our Country is common in the months of november and december is attributed to a common practize of the Indians and Hunters, of firing the woods, planes or savannahs; the flames often extending themselves some hundred miles, before the fire is extinguished; it is observed that rain always follows these conflagrations; sometimes the condensation of the smoke occasions a fine rain resembling a fog or thick dew, but at other times the rain is impetuous accompanied by thunder & lightening & immediately after it clears up fine, but not always without a continuation of the blue misty appearance to the Atmosphere.[10]

9. "Verdure" means a green or a fresh flourishing condition.

10. Indians used fire as a means to clear the land and as a method of hunting. This was a common

Soft friable stone is frequently seen and great loads of gravel and sand upon the beaches; reddish Clay appears in strata much indurated and blackened by exposure to light and air.—The water of this river is extremely agreeable to drink and much clearer than that of the Ohio; in this respect it is very unlike its two neighbours the arcansa and red rivers; whose waters are extremely charged with earthly matter of a reddish brown color, giving to the water a chocolate-like appearance; & when those rivers are low their waters are not potable, being extremely brakish, from the great number of salt springs flowing into them & very probably from the beds of rock-salt over which, (it has been reported) they flow: the inconvenience from this cause, to voyagers, is not so great as might be apprehended, as it appears that brooks & springs of fine water falling into those rivers, particularly the arcansa, are very frequent, and may be met often in the course of a days progress.—Altho' the water of the Washita river does not exhibit any saline impregnation, yet from report there are many situations in its neighbourhood where salt may be procured by digging pits in the places called salt-licks, where water is found equally strong with sea-water; we expect to examine some of those on our way upwards. Thermomr. at 8h. p.m. 58°. Extremes 52°–68°. Wind at N.W. Made this day 11 miles 276 perches.

HUNTER

Thermometer at 6 A.M. in air 52°, river 62°[.] do at 3 pm. 68 & at 7 pm. 58

Thick fog on the river which continued more or less all this day. At 6 A.M. finding the water not deep enough to swim the boat, unloaded part which our canoe brought to us after we had passed the shoal, at two turns. This delayed us till breakfast time, which when finished we set out again about 10 A.M. & pushed on winding thro the various shoals we met with, rowing, Towing or setting with our poles occasionally.

During this days course we observed the land gradually assuming a more fertile appearance, still rising in height as we assend the river; half formed stone appears at the waters edge, seemimingly composed of clay, & sand penetrated by water possessing something of a petrifying quality. The trees are hickory Oaks, pines &cc—The woods more open not so much underbrush & briars as lower down; now & then for some distance Canebrakes show themselves on the banks. Ferruginous water still ouzes out of the banks in many places, particularly near the edge of the

practice among tribes in the eastern half of North America. William Cronon, *Changes in the Land: Indians, Colonists, and the Ecology of New England* (New York: Hill and Wang, 1983), 13, 24–30, 47–51, 57–58, 90–91, 118–119.

river. On the left observed a small rivulet sending out water manifestly much darker in appearance than the Washita. The overcast weather prevented taking an observation for the latitude this day. Encamped on the evening on a sandy Beach where we hawled the seine 2ce [twice] but caught only a few fish, viz Catfish, Buffalo fish Garrs, & a few small ones of little value.—Shot a couple of ducks, one of which proved delicious, the other rather indifferent.—The men shot a wild Turkey & caught also a few cat fish with hook & line.—

Tuesday 6th

DUNBAR

Thermomr. 45° in air—in river water 64°—heavy fog Wind W. Continued our voyage with better fortune; that is, we escaped any considerable obstructions from rapids and sand bars. no variety was to be seen in the appearance of the Country on either side of the river. at noon got a fine observation about a league below the Post of Washita; Latitude deduced 32° 28′ 58″; by the sinuosities of the river it appears we are not more than a mile to the south of it: arrived there about 3½ h p.m. and were very politely received by Lieut Bowmar, who immediately offered us the hospitality of his Dwelling with all the services in his power.[11] The Position called Fort Miro being the property of a private person, who was formerly civil commandant here, the Lieutenant has taken post about 400 yards lower and has built himself some log-houses and enclosed them with a slight stockade: this young officer exclusive of the manners of the polite Gentleman, appears to possess talents; he has formed a tollerably good chart of the river from its mouth to the Post, being the result of his own labors on the way up to take possession of the Post, this he has continued upwards from the best information he had been able to obtain;

11. Lieutenant Joseph Bowmar took possession of the settlement for the United States on April 15, 1804, from Don Vincente Fernandez Tejeiro. Bowmar did not occupy the Spanish stockade possibly because it had been built on Don Juan (Jean Baptiste) Filhiol's private property. Instead, Bowmar had a series of cabins and a small stockade built on unclaimed lands. In March of 1804 a French writer-traveler named C. C. Robin arrived at Fort Miró. He was present when Lieutenant Bowmar arrived with his small detachment of soldiers for the transfer to U.S. hands. Robin wrote that Bowmar was in such a hurry to take possession that he could hardly wait until the next morning." Robin further described Bowmar with the following passage: "He was a young man of twenty-seven years, a lieutenant or second-lieutenant, a native of some small village in the United States. His education and abilities were limited, and yet he was dispatched to this distant post, and provided with the extensive powers of the Spanish Commandants to govern the hands of families of whom some were venerable gray-beards, including veterans of military service. I was much surprised at this choice . . . What astonished me even more was the fact that he knew not one word of French and neither did any of his troops." Robin, *Voyages dans l'intérieur de la Louisiane*, 384. For more on Bowmar's military career, see the introduction.

the whole gives a satisfactory idea of the river & part of the Country; we have also obtained some further information from the former Commandant a french man, and other persons here, of all which we have made notes & shall avail ourselves in the prosecution of our voyage.[12]

Thermomr. at 8h p.m. Extremes 45°–79°. Made this day 9 miles 257 perches; amounting in the whole to 196 miles 256 perches from the mouth of the red river to the Post of the Washita; and by the old computation 90 leagues.

HUNTER

Therm. in the morning at 6. A.m. in Air 45° & in the river water 64 Foggy weather.—

The land as before described on both sides; Ferruginous springs as usual running out of the bank near the water's edge; in some places it only ouzes out & in others it bubbles & boils up from below like a fountain thro a fine light quicksand mixed with clay in a state of suspention in the water, & it hence appears that the slipping in of the bank in many places is owing to this unstable foundation & also those various inclinations of the strata so remarkable in many places of the river banks.—

I put a pole 15 feet long down thro the quicksand of some of these springs with very little difficulty & if it had been longer no doubt it would have gone down much further. In our course up the river for the last 30 miles altho there are few or no habitations we frequently observed cattle browsing on the Banks & very shy. About half past 3 p m arrived at the Military post originally called fort Miro named after a Governor of that name at Orleans formerly. The Spanish old stockade fort has been torn down, & a new small one without cannon or port holes erected by the Americans under Leuit Bowman. it is only a defense against the Indians being unfinished & scarcely that, for the spaces between the stakes that compose the fort shew the men in the inside & leave openings for expert marksmen to pick off those within. There is but an Infant settlement here; The land hitherto, or as far as we have assended, being only habitable here & there immediately on the river banks

12. Filhiol, Dunbar's "french man," had explored the Ouachita beginning as early as 1782 and later founded Fort Miró on land called Prairie des Canots in 1784. Filhiol named the site for the Spanish governor at the time, Esteban Miró (governor 1782–1792), and he was granted large tracts of land on both sides of the Ouachita River. As Dunbar states, Filhiol still lived at Fort Miró in 1804. Dunbar estimated that Ouachita Post and its surrounding area "contains only 500 persons of all ages and sexes." In the "Hunter Official Report," 26, Hunter described the inhabitants as mostly living in cabins, with the exception of two or three "wooden homes." "Dunbar Trip Journal, Vol. I," November 6, 1804; Dickinson, "Don Juan Filhiol," 133–135; Heitman, *Historical Register and Dictionary of the U.S. Army,* 1:255.

& on the Vicinity of Bayu's. The rest being the greatest part swamp overflowed every year. The old settlers cheifly Canadian French appear to have little ambition, few wants & as little industry, They live from hand to mouth & let tomorrow provide for itself. Some of them have from thirty to 100 Cows, but no milk, butter, or Cheese; Their houses are cabbins, afford but little protection against the Winter. The weather being mild there generally, they have not so much occasion for tight houses as we have further northward, & as the woods afford pasturage for their cattle in the winter, they give themselves but little trouble to feed them[;] consequently, they stray about with their calves & shift for themselves, coming to the habitations only now & then, by which means their milk is not obtained. They are supplied from the woods during the hunting season, with animal food, such as Venison, Bear meat, Buffaloe &cc, wild Ducks, Geese, Swans, Turkies, Brant in great abundance; But at other times they are often very badly off for provisions, both Animal & Vegetable; for altho the earth would produce very well, yet their want of forethought & industry leaves them in want of almost every comfort, except what is absolutely necessary for subsistance.[13]

Made this day 9 miles 257 perches. The Latitude of this place by a variety of observations appears to be 32°, 29′,57″ And on the whole, it appears by our measurement that the distance from here to the mouth of the red river by the cources of the Ouachita & black river, is 196 miles & 256 perches.—

Wednesday 7th

DUNBAR

Thermometer in air 52° in river water 64°. Clear. Finding from past experience that the boat in which we have come up, would be improper for the continuation of our voyage, we made enquiry this morning for other craft, but it appears there is no great choice of boats at this place; prepared also for astronomical observation: being greatly interrupted by visitants who came to offer services &c we were prevented from making any useful observation untill noon & even then we were incommoded: the Sun's meridian altitude gave the Lat: 32° 29′ 52″.5 but I was not perfectly satisfied with this observation; from the Causes mentioned I suspect the

13. The explorers viewed this hunter-herder lifeway only from the banks of the streams they traveled. They seemed to be unaware of the widespread nature of this subsistence lifestyle in the area. Author Frank Owsley points out that many of the antebellum travelers—recorders of life in the south wrongly portrayed these people as unproductive. He also says that these travelers failed to see the extent of the livestock held by many of the herdsmen because most may have been foraging in the nearby forests. Owsley. *Plain Folk of the Old South,* 24–25, 30–36, 48–51; also see Bolton, *Territorial Ambition,* 47–49; McDonald and McWhiney, "Antebellum Southern Herdsman"; Hilliard, *Hog Meat and Hoecake,* 158.

altitude was taken a little too late, & shall hope to correct if necessary by future observations. Thermr. at 8h p.m. 67°. Extremes 52° – 80°.

HUNTER

On enquiry were informed that there were in many places between this & the warm springs where there was but little water, many falls & rocks; Mr Dunbar concluded to hire another boat & leave our old one till our return, with such parts of our baggage &cc as could be dispensed with, & we began immediately to look out for a suitable boat. In the meantime were visited by Leiut Bowmar Commandant here, & were by him invited to drink coffee at the Garrison. He seems to be a plain, intelligent, active officer, is well liked here, has no affectation, treated us with civility & attention[,] did us all the services in his power, which we stood in need of 8th & 9th were both cloudy days & no observations made. 9th Therm. 42° to 72° in air & in the river 61°.[14]

Thursday 8th

DUNBAR

Thermomr. in air 53° in river water 58°. Cloudy. This was a disagreeable, damp and cold day: made further enquiry for small boats with little success; found only one, which with another of the same burthen might answer our purpose: no observation made this day. Upon viewing the Country on the East of the river, it is evidently alluvial; the surface is equal with a gentle slope from the river towards the rear of the plantations; the land here is of excellent quality, being a rich black mold to the depth of a foot, under which there is a friable loam of a brownish liver color, which very probably will itself become good soil when broken up & exposed to the influences of the elements. Thermr. at 8h p.m. 56°. Extremes 53° – 61°.

Friday 9th

DUNBAR

Thermomr. in air 42°. in river water 61°. Cloudy, damp & cold. Continued our search for proper vessels and heard of a flat-bottomed barge, which we expect will be very suitable, with the reduced loading we intend to carry with us, the boat will probably draw only 12 inches water: no observation, it being dark, cloudy & disagreeable all day. Extremes of the thermometer 42°–72°.[15]

14. Hunter made no entries in his journal for November 8 and 9.

15. On this day Dunbar penned and mailed a fairly lengthy letter to Jefferson to inform him about the progress of the team's ascent. Dunbar reported,

Saturday 10th

DUNBAR

Thermomr. in air 40°. in river water 58°. Clear—calm—this day having the appearance of being fine & serene, prepared for observation; and in the course of the day took altitudes of the Sun for the regulation of the watch and the magnetic variation: at noon found the Latitude by a fine observation to be 32° 29′ 35″, this differs from that of the 7th by 17″; I give the preference to the result of this day, for reasons already mentioned; In the afternoon took distances of the moon from the Sun to the west of her and in the evening took distances of the moon from ? Arietis to the east of her, which may be considered as a complete series for the determination of the Longitude.[16]

Having hired the barge and agreed to give 1¼ dollar pr day for the use of her, we had her brought along side: She is upwards of 50 feet long & 8½ feet in breadth built tollerably flat, her bottom being still a little convex & being pretty well formed for running. This boat with some improvements is probably the best form for penetrating up shallow rivers, she is undoubtedly too long, as we shall certainly meet with short turns among logs & perhaps rocks, the passage of which might be facilitated by a shorter boat: got her loaded before the evening with a view to set off early next morning. She made some water—found about bed time, that she had made a great depth of water; kept her baled all night. Thermomr. at 8h p.m. 34°. Extremes 40°–72°.

> after a voyage of trouble and retardment we are at last arrived at this place: Doctor Hunter's boat constructed after his chinese model had proved to be an unprofitable vessel . . . we have made great sacrifice of time in getting to this post & in order that we may retrieve as much as possible our past loss, we are now changing our boat for one which is handsomely formed for advancing against the Current . . . Hitherto we havenot seen any thing interesting which is worthy of being particularly communicated to you at this moment, altho' I have got to the 36th quarto page of my Journal exclusive of the Courses & distances of the river with astronomical observations. (Dunbar to Jefferson, November 9, 1804, Jefferson Papers)

16. Here Dunbar's determination for accuracy in his observations is evident. "Arietis" refers to the constellation Aries (the ram). This constellation was used frequently in fall observations. Travelers searching for longitude and not having the benefit of John Harrison's famous clocks used celestial navigation by taking observations of objects such as the moons of Jupiter (Jovial satellites) or by measuring (when possible) the distance between the Sun and the Moon during the daylight hours in conjunction with the use of a chronometer and/or a watch. After the Ouachita expedition, Dunbar devised a method for calculating longitude without a timepiece. Dava Sobel, *Longitude: The Story of a Lone Genius Who Solved the Greatest Scientific Problems of His Time* (New York: Penguin Books, 1995), 24–26, 51–53; Dunbar to Jefferson, December 17, 1805, Jefferson Papers.

HUNTER

Having procured a boat suitable for the purpose, at the rate of one Dollar & a quarter pr Day, we took every thing out of the old one, & having left it together with all our baggage & heavy articles that could be dispensed with under the care of Leuit. Bowmar, & put on board of the other, our provisions[,] Tents, Instruments, medicine chest, Arms[,] amunition & cloathing, found that we drew but one foot water. This Boat is 55ft. long 9 feet broad, has a small mast to tract by, no sail, rows 12 Oars, has no keell, a rudder & Tiller; we cut our sprits out of the sail of our Old boat & made of each of them two setting poles.[17] This boat is built like a barge with light timbers & 3/4 Inch plank on an handsome model for poling or rowing. In the course of this day took a number of observations from the sun & moon. At night before going to rest found our new boat half full of water, immediately called all hands to save the provisions from being spoiled by the water & bailed out the boat. Kept a watch all night for that purpose. I sold Leuit. Bowmar 30 Gallons Whiskey at 1¼ dollar pr Gallon, payable at my return; left with him also my two sadles & bridles, with directions to dispose of them if he could at 24$ for one & $8 for the other payable in Bear skins. vide Novr. 13th.

Sunday 11th

DUNBAR

Thermometer in air 24°. in river water 53°. Clear—calm.—Got the Barge hauled ashore and caulked, which detained us untill the afternoon; got another good observation at noon, which gives the latitude 32° 29′ 30″.5 that is 4½″ less than yesterday, and as those two observations were both very good, the mean of the two results may be taken for the truth, the latitude of the place of observation will therefore be 32° 29′ 32″.75 and as the post or Garrison lies 4 ½″ north of the place of observation, we may consider its latitude as fixed at 32° 29′ 37″.25. Set out after dinner and made 3 miles, Encamped at the plantation of Baron Bastrop.[18] It

17. The new boat had a draft of 12 inches, compared to the Chinese boat's 2½ feet. Hunter to Dearborn, June 14, 1804, Henry Dearborn Papers, Library of Congress. Sprits are small poles that cross a sail diagonally to keep the fabric taut.

18. Presumably of Dutch origin, Baron de Bastrop, or Felipe Enrique Neri, or Philip Henrik Nering Bogel (1759–1827), arrived in Spanish Louisiana in 1795 after fleeing embezzlement charges in Holland. He received permission to establish a colony along the Ouachita River just north of present-day Monroe, Louisiana. Bastrop was to settle five hundred families on a grant of land 12 leagues square. In 1799 Bastrop sold his rights to Colonel Abraham Morehouse (Morehouse Parish), and following the Louisiana Purchase and the transfer of lands to the United States, he moved to San Antonio de Bexar in Spanish Texas, where he became a businessman and politician. Gayarre, *History of Louisiana*, 3:353, 456; Charles A. Bacarisse, "Baron de Bastrop," *Southwestern Historical Quarterly* 58 (1955): 319–320; McDermott, "Western Journals of Hunter," 19, 43, 63–64, 88n, 90–91, 117–118.

appears that this small settlement on the Washita & some of the Creeks falling into it contains only 500 persons of all ages & sexes; it is reported that here is a great deal of excellent land upon several considerable Creeks falling into the Washita & that consequently the Settlement is capable of great extension, & may be expected, with an accession of population to become very flourishing: there are three merchants settled at the post, who supply the inhabitants at very exorbitant prices with their necessaries; those with the garrison & two small planters and a tradesman or two constitute the present village: a great part of the inhabitants still continue the old practize of hunting during the winter season; their peltries go to the Merchant at a low rate in exchange for necessaries; in the summer these people content themselves with making corn barely sufficient for bread during the year; in this manner they always remain extremely poor; some few who have conquered their habits of indolence (which are always a consequence of the indian mode of life) and addicted themselves to agriculture, live more comfortably & taste a little the sweets of civilized life.

HUNTER

Therm. in Air at day break 24° hoarfrost. Unloaded our boat, dried what was wet, hauled the Barge ashore, caulked the necessary seams & pushed her into the water again, to accomplish which it was necessary to have the aid of all the men of the Garrison. We then took in our loading again, Whilst this was doing some more observations were taken to determine the Latitude & longitude of this place making the lat. 32.29.37.8 but the mean Lat of the place of observation is 32°.29′.33. N. This being muster day for the Militia, about 100 men are now assembled to go thro their exercise, they are not all yet met, they say they will have one half more, but as we are all ready & eager to go on, we cannot wait to see them. Therefor we set out about half past 3 P.M. & came, after passing & rubbing on several shoals about 3 miles to a plantation of Baron Bastrop where we pitched our tents for the first time, & slept ashore comfortably with a good fire at our feet.—The land here is a sandy pine country very well timbered The Baron began some years ago a saw mill here, but owing to his various occupations, & the disorder of his affairs, it niver was finished, the works are now useless A Mr Richards now lives here of whom we purchased half a Beeve for fresh meat & issued half to the men, paid 4$ for lbs. 79. being the usual price here. Bought also 2 bbls sweet Potatoes from Mr Phiol the old Spanish Commandant for 2$ & 33 pompions for 1$.—[19]

19. According to McDermott, this Mr. Richards may be Mordecai Richards. Bastrop transferred 1,669 arpents (an old French unit: 1 arpent = about 1 acre) of land to a man by that name in October of 1804. McDermott, "Western Journals of Hunter," 90. Mr. "Phiol" is Don Juan (Jean Baptiste) Fil-

Monday the 12th

DUNBAR

Thermomr. in air 36°—in river water 54°—Clear—Calm—Got on board some fresh beef and other provisions this morning, which detained us a little. Continued our voyage with a pilot on board hired at the rate of 30 dollars pr month. Met with several shoals, but passed over them with ease, our Barge not drawing half the water of our own boat, & being also very light both in her timbers & planks; the appearance of the lands along the river is not very inviting, much pine woods upon a thin poor soil: to the right the settlements on the Bayou Barthelmi and Siard are said to be rich lands.[20] At noon got an observation; Latitude 32° 34′ 47″. Made this day 16 miles 32 perches. Thermr. at 8h p.m. 54°—This Evening a little Cloudy.

HUNTER

Therm. in air at daybreak 36° in the river 54° & at 7 p m. 54°.

Set off at half past 8 A.M. & pursued our course up the river, the Banks land & trees &cc much the same as before described, we now go faster than formerly owing to having a much lighter cargo & a more slender boat. This day passed several rapids & shoals, some formed by gravel barrs & others by ledges of those soft sandy half formed rocks. Between 9 & 10 a m. passed Bayu Siard on the right, on which are the principal settlements, & a considerable quantity of good land. This Bayu extends a considerable distance in to the country in a northeastwardly direction, till it meets with the waters of Bayu Bartholemew which also empty into the Ouachita farther up forming an Island of considerable magnitude in which are contained a part of the lands forming the grant of Marquis de Maison Rouge. The Ouachita is here about 100 yards wide but it is varying in width & depth every few miles according to the local situation. About 10 a.m. passed the Bayu Chenier. This is a very inconsiderable Bayu. & about ¼ of an hour more appears Bayu Darbone to

---

hiol. For more, see Dunbar's entry for November 6. The "pompions" mentioned by Hunter are probably pumpkins.

20. Hunter identified the pilot as Samuel Blazier. For more see Hunter's entry for November 13. Bayou Siard is 26 miles long and approximately 100 yards wide. It runs from Bayou Bartholomew through present-day Monroe, Louisiana, and meets the Ouachita at Monroe in Ouachita Parish. Bayou Bartholomew begins in southeast Arkansas near Pine Bluff in Jefferson County. It meets the Ouachita just north of Sterlington, Louisiana, at the conjuncture of Morehouse, Ouachita, and Union Parishes. Louisiana Quadrangle Maps. The bayou received its name from St. Bartholomew, and it is said to be the world's longest bayou. Clifton Birch, *Bartholomew: A Regional Stream* (McGehee, AR: n.p., 1999). In 1687 LaSalle and Joutel visited Bayou Bartholomew and called it a very deep body of water. Foster, *LaSalle Expedition to Texas*, 262n.

the left.[21] Near this place the Ouachita seems to have made a cut off & left the old channel which is now shut up forming a blind large pond, the river has made a new course for itself about 40 yards wide. The banks this day appear in many places above the inundation. see a few straggling huts; Good Timber. Latitude by observation 32°, 34′, 47″ Came this day 16 miles & 32 perches.

Pitched out tent & encamped for the night on the right bank Plenty of Timber such as Oaks, Maple, hickory Dogwood &cc Soil mold & sand, not very deep of the former.—[22]

Tuesday 13th

DUNBAR

Thermomr. in air 33° in river water 55° Fog on the river. Calm. Continued our voyage without change in the appearance of the Country: passed an Island and strong rapid at 8h a.m. & arrived at a little settlement where we halted to breakfast a little below a chain of rocks crossing the channel between an Island & the main-land called Roquerau[23]—great misery depicted in the Countenances of the Spaniard & his family inhabiting this little settlement, arising as it appears from extreme indolence: the wind at south indicates rain, with a dark cloudy sky: we find our situation greatly improved in our new barge, being able to go about 3 miles pr hour when the Men use a little exertion: we pass without difficulty over shoals of 11 or 12 inches water. The river acquires a more spacious appearance, being in most places about 150 yards wide. Lost some time on the shoals and at half an hour past noon arrived at the last settlements. Began to rain—put ashore to dine—cleared up—set out and passed the mouth of Bayou Barthelmi on the right at 4h p.m. being 12 computed leagues from the post. Here commences Baron Bastrop's great grant of land from the Spanish Government, being a square of twelve leagues to each side; a little exceeding one million french acres, which I presume is more than double of what that Government granted to all persons within the Missisippi territory.—At 11h a.m. passed Otter Bayou on the left.[24] The Banks of the river

21. For information on Maison Rouge, see Dunbar's entry for October 31. "Bayu Chenier" may be Chauvin Bayou, which today enters on the east bank of the Ouachita in the northern sections of the city of Monroe. Bayou D'Arbonne is an extensive waterway. It begins in northern Union Parish and forms Bayou D'Arbonne Lake near Farmerville, then flows southwesterly until it meets the Ouachita on the west bank, just north of Monroe in Ouachita Parish. Louisiana Quadrangle Maps.

22. "Oaks" are from the genus *Quercus*, maples from the genus *Acer*, and hickories from the genus *Carya;* flowering dogwood is *Cornus florida*. Little, *Field Guide to Trees*, 383–410, 570–579, 345–354, 615.

23. This site would have been approximately 8 miles below Sterlington, Louisiana, in Union Parish. Louisiana Quadrangle Maps.

24. The Spaniard in an apparently desperate state of existence represents many such people who

continue to be about 30 feet high, of which 18 feet from the water are a clayey loam of a pale ash colour, upon which the river has deposited an alluvion of 12 feet of light sandy soil, which appears in most places to be fertile, being of a brownish dark color. It seems that this species of land is here of small breadth, not exceeding half a mile on each side, & may be called the valley of the river Washita, beyond which there is high land clothed chiefly with pines.—The Evening is cloudy & dark. Made this day 16 miles 312 perches—Thermomr. at 8h p.m. 62°—Extremes 33°–66°.

HUNTER

Therm at daybreak in air 33 & in the river Water 55 & in the evening at 3 P.M. 66°.

Set out near 7 A.M. This day in our course, passed all the remaining settlements on this river, The Banks are generally above the inundations, if not the first, at least the secondary bank—About half past 7 A M. came to the Bayu of Black water, on the left. & about 8 were stopped a few minutes by a Shoal. Here is a small Island & a rapid. At breakfast time stopped at a Bark cabin inhabited by a Spaniard; it seemed to need no windows neither had it any, but what light passed thro the joints was fully sufficient for every purpose. It was one story high, about 15 feet square, an earthen floor, the chimney composed of mud & grass mixed; The furniture were, one bed for the whole family which consisted of the man & his wife[,] four Children, the eldest girl of about 16. The youngest at the breast, three short blocks of wood by way of stoolls, one of which was a trough to pound Indian corn in, a riffle & shot pouch; In short altho they said they had been settled these five years, there was no appearance of any crop or any store of any kind of vegetable produce, altho he had the winter before him already commenced, & a wife & 4 children to provide for.—Thus are indolence & poverty allied. The river here spreads out & coves a quantity of low ground which as it falls, it leaves bare in places forming ponds which atract multitudes of Wild Geese, Brant, Teal & ducks. These must have drawn our Spaniard hither, for we observed two small hog troughs tied together with vines, by way of a canoe to follow these wild fowls amongst the shallows.—This place is called about 9 leagues from the post.

About 11 a.m. came to another rapid & shoal formed by a bed of gravel, near which Otter Bayu puts in on the left; here is also an house. Next comes Hiccory

---

endured a day-to-day struggle for survival in the early-nineteenth-century Ouachita River valley. Hunter presents a much more detailed description of the Spaniard and the living conditions of his family in his entry for this date. Also see Bolton, *Territorial Ambition*, 44–50, 91–93, 122–124. Bayou de Loutre enters the Ouachita in Union Parish about 5 miles below Sterlington, Louisiana.

ridge Ecor aux Noyeaux on the left where the bank is about 4 feet above the inundation About 4 P.M. on the right passed Bayu Ba[r]tholemew, which is of considerable length & passes in almost a semicircle S. eastwardly till it meets with the waters of the Bayu Siard & at the same place nearly receiving a portion of the waters of the Mississippi in common high water, by which a communication is made with a branch of the Arkansa leading to Osark settlement.[25] This Bayu had much good land on it forming in a manner the center of Baron Bastrops large grant & its entranse into the Ouachita is called 12 leagues from the Post.

About a league further is the Bayu Assmine, Paw Paw Bayu on the right, an inconsiderable stream which commences at a prairie not far off. Came this day 16 miles 312 perches

I omitted to mention that before we left the Post, we were Informed that it was necessary to have a man acquainted with the river & the adjacent country, as a Pilot as well as a Hunter to explain & point out the proper manner of passing the shoals, where to get game in plenty, where we might look for salt springs, minerals &cc, And in short every remarkable object in our voyage which without his assistance might be overlooked. Accordingly, we, after sufficient enquiry hired one of the name of Blazier who resided about ten years in this country and had been several times to the hot Springs & thro that part of the Country on hunting expeditions; at the rate of 30$ pr month, we to find him provisions & liquor out of our own rations. The weather being cloudy, no observation was taken this day.[26]

Wednesday 14th

DUNBAR

Thermometer in air 44°. in river water 55°—Clear—calm. Continued our voyage, the soil seems to be thin; the growth of the timber is small. We made small progress, being opposed by a head wind. Passed the 'Bayou des buttes' in the forenoon; this Creek derives its name from a vast number of Indian mounts dis-

25. This may be a reference to Arkansas Post. For more on Arkansas Post, see Dunbar's entry for January 10, 1805.

26. Hunter recorded in the "Hunter Official Report," 26–27, that they inquired about a guide who might be familiar with the territory and river and "This being the hunting season, had no choice Therefore hired Samuel Blazier the only capable person left at home for 30 dollars pr month." McDermott revealed that Samuel Blazier held a land grant on the western bank of the Ouachita River only a short distance above the Post. McDermott, "Western Journals of Hunter," 88. Dunbar does not mention the hiring of Blazier in his entries at Fort Miró, nor does he ever refer to him by name (only as "our guide") in his journals. "Dunbar Trip Journal, Vol. I," November 8, 1804; see also Dunbar's journal entry for November 8.

covered by the hunters along its course: we were detained an hour extraordinary at breakfast, from the necessity of repairing the rudder irons damaged going over a rocky flat. The margin of the river is clothed with such timber as generally grows on inundated lands, particularly a species of the white oak called vulgarly the overcup-oak; its timber is remarkably hard, solid, ponderous and durable, and it produces a large acorn in very great abundance upon which the Bear feeds; it is also very fattening for Hogs.[27]

At noon got a good observation & found the latitude to be 32° 50′ 8″.5—after dinner passed a long narrow Island. the face of the Country begins to change; the banks are low and steep, and the river generally deeper and much contracted, being from 30 to 50 yards wide; this low Country is 2 or 3 leagues wide on each side of the river, liable to overflow 12 or 15 feet above the level of the land, the soil is a very sandy loam in the neighbourhood of the river, & covered by such vegetables as are found on the inundated lands of the Missisippi; in short this tract presents every appearance of a newly created soil, very different from what we passed below: it may be supposed that there existed a great Lake within the space now occupied by this alluvial tract, which may have been drained off by a natural Canal worn out by the abrasion of the waters, and that since that period, the annual inundations have been replenishing this space with the alluvion of its water; 18 or 20 feet of soil perpendicular is yet wanting to render it a fit habitation for Man; it appears never the less to be well peopled by the beasts of the forest, several of which presented themselves to view, but they must all retire to the high lands during the season of the inundation.[28] [W]e now begin to see quantities of water fowl which are not generally very numerous untill the cold rains and frost drive them to us from the northward. Fish is not so abundant in this river as might be expected; at the post we were informed that the river had been extremely full of fish untill the year 1799, when the waters of the inundation of the Missisippi dammed up the Washita river some distance above the Post and produced a stagnation and consequent corruption of the waters, which destroyed all the fish within the influence of this cause. The river continues to be contracted, seldom exceeding 60 yards and generally deep; no current is felt excepting in places a little shallower than the rest.[29]—Thermometer at 8h p.m. 44°. Extremes 44°–58° Clear.

27. Bayou de Butte enters the east bank of the Ouachita in Morehouse Parish. Louisiana Quadrangle Maps. The "forenoon" refers to the latter portion of the morning or the period of daylight before noon. The overcup oak (*Quercus lyrata*), in the white oak group, is 60–100 feet tall, and its leaves are 6–10 inches long with rounded lobes; the acorns have scaly cups. Moore, *Trees of Arkansas*, 46.

28. This is apparently the area near Mud and Ditched-off Lakes on the east bank of the Ouachita in Morehouse Parish. Louisiana Quadrangle Maps.

29. The Mississippi also flooded in 1796, 1809, and 1811.

HUNTER

Therm. in air 44° at daybreak & in the river 55°. & in the evening at 8h. 44°.

Set out about 7 in the morning, The weather which had been stormy & rainy accompanied with thunder, during the night, was now clear & calm. After rowing about half a mile came to Bayu Mercier on the left.[30] & whin we were coming in towards the shore to breakfast, struck a sunken log which unshipped the rudder & almost twisted off the Rudder hinges, to repair which took an hour & a quarter extraordinary: for it is to be observed that, according to the custom of the voyaguers on the western waters; the men are allowed to stop one hour to breakfast & two hours to dinner & to encamp in the evening about sunset in order to have a little time to pitch their tents, cut & collect firewood before dark & now as the days are so short, altho the men row rather with more exertions than formerly when the weather was warmer, yet all we can make is but a short distance in a day.

Still we meet with very little current to oppose our progress, except now & then when we come to a gravel barr & rapid, which are generally passed in less than one quarter of an hour, when we have still & deep water again. Near two miles past Bayu Mercier we come to an other small Bayu on the left.[31] Here we come to low lands said to extend for several days journey up on both sides of the river, so much subject to inundation in freshes as to uninhabitable, our Pilot says we will find no more inhabitants or settlers till we return. The timber is still good, the land appears rich[,] little underwood where we landed. At 3 p.m. came in sight of a small Island, we found the best channel on the left, for our canoe could not pass between it on the right. at about half past three p.m. came to another small Bayu on the left. The river here about 60 yards wide & a little above half that width contracted by a barr of gravel. About ¼ past 5. the river is only 40 yards wide. Here we encamped for the night having seen many flocks of Wild Geese, ducks, & some cormorants, but as yet all too shy to suffer one to come within gun shot of them. Yesterday one of our men shot a Deer, young but not fat, of which we recd. one quarter. The woods here besides many sorts of trees of unknown names, consist of Hickory Oak Cypress, Dogwood, Persimon[32] many sorts of grape vines, but no pines in these drownded lands. Last evening as the weather appeared so bad I thought it best to sleep on board the boat with my son rather than lay in a wet tent on wet ground by the fire.

This day made a good observation with my Sextant __ double Appt. mer. Alt. 77°,9′,20 Ind. Error 0.0.59. Latitude found 32°,50′,6″

30. This stream may be the present Boggy Bayou or Possum Bayou in Union Parish. Ibid.

31. Possum Bayou still enters the Ouachita at the location described by Hunter. Ibid.

32. The common persimmon (*Diospyros virginiana*): height 20–70 feet, leaves 2–6 inches and egg-shaped. Little, *Field Guide to Trees*, 634–635.

This day made 12 miles 303 perches.

encamped on the edge of the left bank in veiw of the boat where we slept under a double tent.

Thursday 15th

DUNBAR

Thermometer in air 38°. in river water 54°.—Clouds—Calm. Continued our voyage thro' a Country of the same appearance as yesterday. passed some rapids without difficulty—the banks still continue low; from ten to 15 feet above the present level of the river; the water marks on the trees from 15 to 20 feet. Landed to observe about 90 yards higher than the upper point of the Island of Mallet, judging that we were not far from Lat. 33° the division line between the territories of Orleans and Louisiana; we found the Latitude by a very good observation to be 32° 59′ 27″.5. The Island of Mallet is on the right of the main channel, and the place of observation being 90 yards N 45° E from the upper point of the Island. Making allowance for the breadth of the river (50 yards), Latitude 33° may be found from the above data when the Jurisdiction of the territories may require it, this Island of Mallet being very well known to the Hunters. Should time and circumstances permit on our return, a 2d meridian altitude of the Sun may be taken and a proper mark set up in Lat: 33°.[33]—In general the bed of the river along this alluvial country is fully covered by water from bank to bank & the navigation good, but to day at 3h p.m. we passed 3 contiguous sand-bars or beaches called 'le trois battures'; & at three & a half hours p.m. the 'bayou des grand Marais' (great Marsh Creek) on the right: passed also in the evening on the same side 'la Cypriere Chattelrau':[34] a

33. McDermott cautiously credits the "Island of Mallet" to the Mallet brothers. Between 1739 and 1740 the brothers attempted to open a trade route and traveled along the Platte and Arkansas Rivers. Hunter refers to the site as "Isle de Mallet." The island would have been on the present-day Louisiana-Arkansas state border. McDermott, "Western Journals of Hunter," 92; see also Hunter's entry for this date. Congress created the Orleans Territory and the Louisiana Territory on March 26, 1804, and the act took effect on October 1, 1804, only fifteen days before the expedition left St. Catherine's Landing near Natchez. Carter, *Territorial Papers of the United States* (1940), 9:202–213.

The term "meridian altitude of the Sun" refers to a method of obtaining latitude by taking a daily reading of a heavenly body's altitude as it passes due north and/or south. The specific meridian altitude is an arc of a meridian that is intercepted between a due-south point on the horizon and any point on the meridian.

34. The Grand Marais meets the Ouachita near the town of Huttig in southern Union County and within the Felsenthal National Wildlife Refuge. Today the Ouachita River forms the borders of Union and Ashley Counties as it crosses the 33 latitude. Dunbar's "la Cypriere Chattelrau" is today probably under Lake Jack Lee within Felsenthal National Wildlife Refuge in Union and Ashley Counties. Arkansas Quadrangle Maps, U.S. Geological Survey, Denver, Colorado, and Arkansas Geological Commission, Little Rock, 1981.

point of high land approaches within half a mile of the river on the right. Thermomr. at 8h p.m. 50°.

Extremes 33°–60°. Made this day 16 miles 42 perches. This days voyage was shortened by an indisposition which confined me to the tent untill the hour of breakfast.[35]

HUNTER

Therm. at day light 33. & in the river 54 & at 3 p.m. 60 & at 6 p.m. 50.

Set out at ¼ past nine (Mr Dunbar being indisposed last night) & about 1 furlong afterwards passed the Bayu de longue vue. or the Bayu of long reach on the right, an inconsiderable opening.[36] Lands still low, bank about 16 feet high; The trees shew the marks of the freshes six or 8 feet up them. The soil a black mold about one foot deep furnished by decayed vegetation on a bed of white or greyish sand mixed with a small proportion of clay in a comminuted state which lays above several strata of light coloured clay. It perhaps might have been the work of ages to have brought this part of the country into the present state, from an original Lake or perhaps sea covering all these low lands. The river appears to be about 60 yards wide, deep & scarce any perceptable current, except where the slipping in of one or both of the banks into the channel of the river which carries the lighter parts away leaving the gravel behind forming barrs or shoals, thereby damming up the water above in a degree & making the upper water almost level, of course without a perceptible current to the next bar above. The same cause produces the same effect below, leaving an apparent ripple, a rapid over the shoal or bar when the waters are low as they are at present. How these will appear when the waters are high we will know better on our return.

Made an observation about 90 yeards above the N.E. point of the Isle de Mallet on the left shore going up, & found the latitude to be 32°.59′.27″ ½ N. by which it appears that the Line of division between the Territory of Orleans & Louisiana as fixed by Congress will cross this river about 32″ ½ seconds of a degree above the said point of Isle de Mallett.

About half past 3 p.m. passed the Bayu of the grand Marais or great swamp on the left; It has a very inconsiderable opening at the river, but extends some distance up nearly parralel with the Ouachita.—The lands are still low; the river generally gentle & deep averaging one hundred yards wide, sometimes more sometimes less, but when remarkably less we have generally noted it.—The timber pretty good, much as yesterday as well as the banks. Made this day 16 miles, 42

35. Dunbar may have had some type of intestinal ailment.

36. This may be Coffee Creek, which today forms Mossy Lake just below Lock and Dam No. 6 in Ashley County. Arkansas Quadrangle Maps.

perches, One of our men in the Canoe ahead killed a Deer pretty fat & a racoon [;[37] we] were presented with a quarter of the former. Plenty of Wild Geese & Ducks but very shy.

Friday 16th

DUNBAR

Thermomr. in air 38° in river water 54°.—Cloudy—Calm. Set out at 6h 58′ and continued our voyage, the wind rises northerly against us, nevertheless we make 7½ perches pr. ½ min: whereas with our former boat we should not have exceeded 4 per: still however our improved progress is short of the velocity which a boat for our purpose ought to attain; it should not fall short of 12 per: pr. ½ min: [12 perches per ½ minute] which would be about 4½ miles pr. hour. No observation to day the weather being cloudy, damp and disagreeable. Between 11 & 12 o'clock passed on the right the 'marais de la Saline" (Salt-lick marsh) There is here a small marshy lake, but it is not intended by its name to convey any idea of a property of brackishness in the lake or marsh, but merely that it is contiguous to some of the licks, which are sometimes termed "Saline' & sometimes 'glaise,' being generally found in compact clay which might serve for potter's ware; the bayou de la Tulipe forms a communication between the lake and the river: there is opposite to this place apoint of high land forming a promontory and advancing within a mile of the river, to which boats resort when the low grounds are under water: a short league after, we came to the mouth of the grand bayou de la Saline (Salt-lick Creek) on the right; this is a creek of considerable length & tollerably good navigation for small boats, the Hunters ascend it to an extent of a hundred of their leagues in pursuing their game. They all agree that none of the springs which feed this Creek are salt; it has obtained its name from many buffalo salt licks which have been discovered near to the Creek. Altho' most of those licks by digging will furnish water holding in solution more or less marine salt, yet we have reason to believe that many of them would produce Nitre.[38] we now begin to observe a stratum of a dirty white colored

37. The raccoon (*Procyon lotor*), 26–40 inches long, remains extremely abundant in Arkansas and Louisiana. Sealander and Heidt, *Arkansas Mammals*, 19, 22, 29, 210–211, 213–216; Sutton, *Arkansas Wildlife*, 10, 19–20, 37–38, 241–242.

38. The "Marais de la Saline" (today Marais Saline) enters the Ouachita on the right bank inside Felsenthal National Wildlife Refuge in Ashley County. Hunter, in his entry for this date, calls it "Marais de Saline." The "bayou de la Tulipe," or "Bayu called de La Tulipe" as identified by Hunter, probably received its name from the same trapper-hunter as referred to at the site "Cache la Tulipe" (Tulipe's Hiding Place). The "grand bayou de la Saline (Salt-lick Creek)" is the Saline River, one of the last free-flowing larger streams in southern Arkansas. It rises in west-central Arkansas from three forks in Saline

clay under the alluvial soil; this clay is similar to what we observed before we entered the alluvial tract; we have therefore reason to expect, that we are gradually emerging from this sunken tract & shall soon ascend into the high land country. Made this day 17 miles, 185 perches. In the evening it began to rain. Thermomr. at 8h p.m. 42°. Extremes 38°–51°.

HUNTER

Therm. 30° in air & in the water of river 54°. at 7 a.m. Cloudy calm.—at 3 p.m. 51° & at 7 p.m. 42.

Set out near seven A.m. The current is still so gentle as scarcely to be perceptible, Banks still low, with the appearance of having prairies & ponds behind them. In some places the hills or high grounds come to within a mile of the river, in others a league off, more or less according to local circumstances. The Timber, soil, plants &cc much the same as in the last two days course. At Breakfast time our pilot went off for a few minutes to a pond on the right & shot a pair of Ducks, called Duck & Mallet which proved poor.[39] About half past eleven came to a small Bayu called de La Tulipe, which leads to a small pond on the right shore called Marais de Saline, of about 1 mile in circumferance, a retreat for wild fowl. It is surrounded by Cypress trees. About half past twelve came to Bayu de Saline on the right; This is of considerable extent & is called one league from Bayu de la Tulipe. I went up this Bayu whilst dinner was preparing, for about a mile & found it considerably enlarged, tho the mouth seemed nearly choaked up with dreft & fallen timber. This afternoon the banks are just sensibly rising in height by very slow gradations. Passed several hunting camps during this days course; but the hunters were gone—Made no observation this day on account of the cloudy weather. Came 17 miles 158 perches this day. This afternoon about 4, it began to hail, which in time turned to rain which cont[in]ued with increased violence the greatest part of the night. The banks where we encamped still shew alluvial ground by being higher next the river & gradually decending from it.

---

and Garland Counties: the Alum Fork, the South Fork, and the Middle Fork. These join in western Saline County and the river flows south-southeasterly until it joins with the Ouachita at the juncture of Union, Ashley, and Bradley Counties in southern Arkansas. Arkansas Quadrangle Maps. "Nitre" is potassium nitrate, KNO_3 (also called saltpeter, niter, nitric acid, and potassium salt). It is used in the manufacturing of gunpowder and fertilizer.

39. The "Duck & Mallet" may be a mallard duck (*Anas platyrhynchos*), length 20–24 inches, which has a green head, a chestnut-colored chest, and a gray body. James and Neal, *Arkansas Birds*, 109–111.

Saturday 17th

DUNBAR

Thermomr. in air 40°. in river water 54°.—fog on the river—calm—river risen 2½ inches during the night.

Continued our voyage; the low lands are still alluvial, at least to a certain depth; an understratum of clay appears in many places, where the banks have been undermined & broken down: we remarked that since we entered the alluvial country about 32° 52′ Lat: we have seen no long moss (Tilandsia) altho' this low damp country seems in all respects well adapted to favor its production; upon enquiry of our pilot, he informs us, we shall see no more of it; probably its limit of vegetation northerly may be fixed by nature near to 33° Lat: Saw a great quantity of the long-leaf pine, which is frequently found in rich & even inundated lands as is the case here; the short leaf or pitch pine on the contrary is always found upon arid lands & generally in sandy & lofty situations; but our County furnishes it in a hard meagre clay.[40] In the forenoon saw the first swan which was shot by one of our hunters;[41] it was a solitary one whose mate had probably been killed: this is the season when the poor inhabitants of the settlement of the Washita turn out to make their annual hunt; they carry no provision with them but a little indian corn, depending on their guns and ammunition for the rest. The Deer is now fat & their skins in perfection; the Bear also is now in his prime with regard to the quality of his fur and the quantity of fat or oil which he yields, he has been feeding luxuriously for some time upon the autumnal fruits of the forest, such as pirsimmons, grapes, pawpaws, walnuts, packawns, hickory-nuts, chinquapins, beech-mast, a

40. "Tilandsia" refers to Spanish moss (*Tillandsia usneoides*) or, as Hunter called it, "Spanish Beard" or "Carolina moss," a member of the Bromeliaceae (pineapple) family. Dunbar is basically correct about the range of Spanish moss. Although it does appear in some sporadic places higher than this latitude, the primary boundary remains near this point. Smith, *Keys to the Flora of Arkansas*, 316. The longleaf pine (*Pinus palustris*), 75–120 feet tall, has needles 8–16 inches long and cones 6–10 inches long. This tree is an introduced species and is not native to Arkansas. Instead of the longleaf pine, Dunbar probably was seeing the loblolly pine (*Pinus taeda*), height 90–130 feet, needles 6–9 inches, cones 2–6 inches. The shortleaf pine (Pinus echinata) is 70–110 feet tall, with needles 3–5inches long and cones 2 inches long; the pitch pine (*Pinus rigida*) is an eastern species whose range is from New England through the Appalachian Mountains. Moore, *Trees of Arkansas*, 16–17; Little, *Field Guide to Trees*, 5, 472, 294.

41. The trumpeter swan (*Cygnus buccinator*, length 84–95 inches, wingspan 6–7½ inches), is the largest member of the swan family. The trumpeter swan was abundant in Arkansas prior to the 1820s. By the time of the Civil War, overhunting had caused the numbers to dwindle. During the 1990s a few pairs were returning to the central portion of the state near Heber Springs in Cleburne County. In 2003 two counts yielded between 52 and 55 birds in that area. Sutton, *Arkansas Wildlife*, 15–16, 256; James and Neal, *Arkansas Birds*, 100–101; *Arkansas Democrat Gazette*, February 1, 2004.

great variety of acorns &c &c; it is however well known (notwithstanding the fancies of some writers) that the Bear does not confine himself to vegetable food;[42] the planters have ample experience of his Carnivorous disposition. He is particularly fond of Hog's flesh, but no animal escapes him that he is able to conquer: Sheep & Calves are frequently his prey and he often destroys the fawn when he stumbles upon it; he cannot however discover it by the sense of smelling notwithstanding the excellence of his scent; Nature has protected the helpless young by denying it the property of leaving any effluvium upon its tract, which property is so powerful in the old Deer: perhaps it may not be generally known to Naturalists, that between the hoofs of Deer &c is found a sac with its mouth inclining upwards; this sac always contains more or less musk, which by escaping over the opening in proportion as it is secreted, gives to the foot the property of leaving on the ground a scent wherever it passes: during the rutting season the musk is most abundant particularly in old males, which may often be smelt at a considerable by the hunters.[43]

The Bear unlike to most other beasts of pray does not kill the animal immediately he has seized upon, but regardless of its struggles, cries and lamentations, fastens upon it and (if the expression may be allowed) devours it alive: the taste of Mr. Bruce & his Abyssinians may have been formed upon this excellent model.[44]—

42. The pawpaw (*Asimina triloba*) is 20–40 feet tall and has 8–10-inch leaves, broad at the end and tapering at the base; its 2–3-inch fruit is greenish-yellow to brown and rounded. The black walnut (*Juglans nigra*) is 60–100 feet tall; its compound leaves are 1–2 feet long with 15–23 leaflets, and its fruit is a nut 1–2 inches in size. The chinquapin oak (*Quercus muchlenbergii*) is 80–100 feet tall; its 3–6-inch leaves are oblong and largely toothed or notched. There are at least eight hickory species (genus *Carya*) in Arkansas. The American beech or Carolina beech (*Fagus grandifolia* or *Fagus caroliniana*) grows to a height of 70–100 feet, with 3–4-inch leaves that are lance-shaped and notched. Moore, *Trees of Arkansas*, 73, 26, 29–33, 47, 40. The American black bear (*Ursus americanus*), length 5–6 feet and weight 200–400 pounds, was once so abundant that Arkansas received the nickname "the Bear State." By the turn of the twentieth century, only small numbers of bears survived in remote areas such as along the lower White River.

43. The white-tailed deer (*Odocoileus virginianus*), length 4–5 feet, weight 75–300 pounds, was abundant in Arkansas until the beginning of the twentieth century. Restocking and management efforts have restored white-tailed deer to abundance. Dunbar was correct about the lack of scent or "effluvium" of fawns and the musk glands between an adult deer's hooves. Sutton, *Arkansas Wildlife*, 3, 15–16, 18–21, 28, 89–99, 206–211, 226–227; Sealander and Heidt, *Arkansas Mammals*, 11, 19–24, 29, 36, 39, 206–210, 243.

44. Dunbar refers to James Bruce (1730–1794). Born a Scottish laird in Stirlingshire, he became the first modern British explorer to search for the source of the Nile River. Bruce found the source of the Blue Nile on Lake Tana in Abyssinia (today Ethiopia, Eritrea, and southern Yemen) around 1770. In 1790 the stories of his many expeditions were published in five volumes under the title *Travels to Discover the Source of the Nile, in the Years 1768, 1769, 1770, 1771, 1772, and 1773*.

The hunters count much of their profits from the oil drawn from the Bear's fat, which at New-Orleans is always of ready sale, and is much esteemed for its wholesomeness in cooking, being preferred to butter or hog's lard; it is found to keep longer than any other oil of the same nature, without turning rancid: they have a method of boiling it from time to time upon sweet-bay leaves which restores it or facilitates its conservation. At noon found our Latitude to be 33° 13′ 16″.5. In the afternoon saw a small Aligator,[45] which we did not expect in so northern a situation; passed a few rapids & saw cane brakes on both sides, the canes of a small size, which demonstrates that the water does not surmount the bank above a few feet: the river widens & a number of sand-beaches are seen. Thermr. at 8h p.m. 44°.— Extremes 40°–51°. Made this day 15 miles 308 perches.

HUNTER

Therm. at 7 a m 40 & in the river 54[,] at 3 p m 51° & at 7 p.m. 44° Cloudy calm fog

Set out at ¼ past seven. for two hours met with rather more current than usual, the river being in many places much narrowed by the slipping in of an acre or two of the bank at a time, carrying the trees with it, which falling prostrate into the river cause impediments to the navigation till time cause them to decay. When we thought we had lost sight entirely of the Alligators since two days journey beyond the Post were surprised to see a small one basking in the sun on the bank. Saw this morning for the first time a solitary tho stately Swan in the river which o[u]r guide said had probably lost its mate by the Hunters. Breakfasted at a small bayu on the right called Marais de Cannes, or Cane Swamp.[46] after which as we coursed along came to land still a little higher more & more elevated above the water as we ascend the river. The river now & then running pretty strong, & in places so many trees carried bodily into the middle of the river some times almost opposite in both sides at the same time by the slipping in of the bank that it was with some considerable exertion the boat could be forced over the branches in such interstices as afforded water sufficient to float it, for these trees with the bank that fell into the river with them often formed bars or shallows[,] rapids & difficult places; Tho it generally happens that the banks on both sides are very different opposite to each

45. Alligators (*Alligator mississippiensis*) were once abundant in southern and eastern Arkansas, but by the mid-twentieth century they remained in only a few areas. Restocking occurred between 1972 and 1984 from Louisiana. Sutton, *Arkansas Wildlife*, 18, 28, 165, 250, 254–255.

46. Caney Marais Bend is located in Bradley County on the east bank of the Ouachita above the mouth of the Saline River. Caney Bayou runs parallel to the west bank of the Ouachita in Union County and possibly once entered the river. Arkansas Quadrangle Maps.

other, thus where on one side the bank is high & steep close to the river, opposite it is generally the reverse viz, rising to its natural height by slow & irregular slopes, & where on one side the banks are undermined by the current which is generally in the bends, on the other side there is a sand bar or beach projecting out & the land is gaining ground. It is observable that for this day or two past in these low grounds lately passed there is not to be seen on the trees as usual the long moss called Spanish Beard, or in Philada. called Carolina moss; our Guide tells us that we will see no more of it. Perhaps we are now too far North for it. The lands now bear amongst other trees the long leafed Pine, & we begin to observe here & there on the edges of the banks in high places small scrubby canes, a mark that the inundation is not of long duration on those parts. The river this day of various breadth averaging about 50 or 60 yard wide. The muscle shells are still strewed along the beaches & banks in places. In the evening after we had encamped our canoe came up with us, having been behind this forenoon & brought the swan that we saw this afternoon. It was about 4 ½ feet high from the tail to the bill.—The skin was preserved to make a muff.—Came this day 15 miles 308 perches.

Latitude by observation Lat. 33°.12′.00″

Sunday 18th

DUNBAR

Thermomr. in air 32°—in river water 52°—Serene—Calm—river seems rather on the rise. Set out at 7h.20′ and continued our voyage; passed along a narrow passage this morning, about 70 feet wide; the whole of the water of the river runs thro' this passage; on the left the old channel of the usual breadth leaves an interval which becomes an Island when the water passes along the old bed of the river during freshes: Came up to a place at the hour of breakfast where there is an appearance of some clearing called 'Cache la Tulipe' (Tulip's hiding place) this is the name of a french hunter who concealed his property in this place. It continues to be a practize of both white and red hunters, to deposit their skins &c often suspended to poles or laid over a pole placed upon two forked posts in sight of the river, untill their return from hunting; these deposits are considered as sacred and few examples exist of their being plundered.

The banks of the river have now the appearance of the high land soil, with a stratum of 3 or 4 feet of alluvion deposited thereon by the river, this superstratum is greyish and very sandy with a small admixture of loam, which indicates the poverty of the mountains and uplands where the sources of the river take their rise. At noon we found our Latitude to be 33° 17′ 13″—In the afternoon passed on the right, the entrance of a bay, which within must form a great lake during the inundation. We now see a considerable number of the long-leaf pine tree; the canes along the bank have a better appearance being much larger in size, this indicates a better or more elevated soil: Canes subject to be inundated, i.e. the land to be inundated, 3, 4 or 5 feet, are always small and tough; they grow much finer where there is little or no inundation, provided the soil be rich & loose. Passed a high hill (300 feet) on the left clothed with lofty pine trees. Thermomr. at 8h. p.m. 57° cloudy weather threatens rain. made this day 18 miles 75 perches. Having been much indisposed for some days past, the number of remarks are probably fewer than might have been made—I still remain in the same situation.[47]

HUNTER

Therm. 32°. at 7. a m in air & in the river 52° Hoarfrost. at 7 p.m. 57°.

Set out after seven a m & about 20 minutes afterwards came to an Island in the general state of the waters, tho now only a peninsula. There the whole of the river is compressed to about the width of 20 yards at breakfast time came to a small cleared place on the left called Cache de Tulipe, or where a person of that assumed name concealed his game when he went from there to hunt. Here it is said a few Indians now encamp, observed the Indian corn growing.

Saw a canoe with two french hunters belonging to a larger party ashore. We heard their dogs & were informed they had just killed a bear of which they had the skin & part of the meat in their boat. about 11. a.m. came to another rapid. about ¼ past two p.m. passed on the right the Bayu Moreau. The lands on both sides are now percepti[b]ly rising, tho the strata of sand & clay & the general appearance of the bank where it is bare, are the same as before described. White or long leaved pines are now very common along the banks & Cypress, Oaks, Hickory Persommon, gum,[48] &cc with Willow & Chenier to the waters edge. after dinner came to lands above the inundation on both sides of the river, observed on one point on the left the appearance of half formed stone, coloured by Iron of a brown color. The gravel of which the sand & gravel Beaches & rapids are composed consists of small stones about the size of a Goosses egg more & less composed of a sort of white freestone before described, which is on the outside become black by the water & air, the edges worn off by rolling & by the current. This day found for the first time

47. The "entrance of a bay" that Dunbar notes may be Moro Bay (formed by Moro Creek) at the conjunction of Bradley, Union, and Calhoun Counties. Hunter, in his entry of this date, refers to the site as "Bayu Moreau." The bay empties into the Ouachita at Moro Bay State Park. Arkansas Quadrangle Maps.

48. The sweetgum (*Liquidambar stryaciflua*) is 40–120 feet tall, with star-shaped 4–6-inch leaves; its fruit is a 1-inch-diameter prickly ball containing seeds. Moore, *Trees of Arkansas*, 75.

a piece of white flint on one of the beaches.—Latitude 33°.16′.47.6″ Made this day 18 miles, 75 perches.—

Monday 19th

DUNBAR

Thermr. in air 54°.—in river water 54°—Cloudy—Calm—river at a stand. Set out at 6h 56′ and continued our voyage—The banks present still more the appearance of the high land soil, the under stratum being a pale yellowish clay and the alluvial soil of a dirty white surmounted by a thin covering of a brownish vegetable earth: the trees begin to have a better appearance, growing to a considerable size and height, tho' much inferior to those of the alluvial banks of the Missisippi: passed the 'bayou de hachis' on the left this morning; points of highland not subject to be overflowed frequently touch the river, the valley is said to be league or more in breadth on each side of the river: passed some pine hills on the left called 'Cote de Champignole',[49] the river has been narrow during the courses of this day's voyage, not exceeding on the average from 50 to 60 yards. Thermometer at 8h p.m. 62°. Extremes 54°–67°. Made this day 18 miles 120 perches.

HUNTER

Therm at. 7. a m. 54° & the same in the river[.] at 7 p.m. 68°. at ½ past 7. 62° Calm & cloudy.

Set out about 7 a.m. & went on rowing as usual. The Canoe started 2 hours before with the hunters to try their luck as to killing game. About half past 7. a m came to Bayu de Hachè, or the bayu of Hashed meat. We still see canes on each side of the river, which may overflow here at high freshes about 3, or 4 feet. Have observed for these several days past small flocks of wild ducks swimming in the river about 2 gun shots off, which as we approach them, fly before us still keeping at a cautious distance, tho once in a while a straggling pair lag behind & get shot by us. Sometimes a point of high ground comes near the river, but generally the Valley of overflowed ground in freshes extends 1 or more leagues of [on] each side befor we come to the High grounds which are cheifly pine hills, light sandy soil.—

49. The confluence that Dunbar mentions must have been near Calion, Arkansas, in Union County and may have been either Salt Creek, which today runs into Calion Lake, or Chapelle Slough, just north of the town of Calion. Both are in northern Union County. Champagnolle Creek meets the Ouachita on the east bank in Calhoun County north of the town of Calion. The Ouachita forms the border between Union and Calhoun Counties, as well as the border between Calhoun and Ouachita Counties, in this area. In the "Hunter Official Report," November 19, 1804, Hunter referred to the site as "Cabane Chapignole." Arkansas Quadrangle Maps.

The water is now very clear so that we can see the gravelly bottom in most places, where it is not very deep. The Soil where the banks are laid bare by the washing of the river is as before viz. a Stratum of clay greyish intermixed with sand of from ten to twenty feet thickness above the water, above that two or four feet of sand more or less, then about a foot of alluvial ground mixed with the decayed vegetation & sand forming the surfaces Now & then springs Issue out about the waters edge in small streams: I tasted many of them, they were insipid.—[50]

About the middle of the day passed the hills of Champignolle on the left. They shew as gentle rising grounds about 60, or 80 feet above the river. They continue but for a short distance on the river, the banks again become lower & are during the freshes overflowed a few feet, bearing cane &cc—Being cloudy made no observation this day, but came 18 miles 120 perches, where it came on an heavy rain a little before sunset when we encamped, leaving the canoe with the hunters ahead. These several days Mr Dunbar has been indisposed by a diarhoea & one of the men with the same complaint, which I ascribe to the effects of exposure to damps & cold.

Tuesday 20th

DUNBAR

Thermr. in air 59°. in river water 54°—Cloudy—Calm. No change in the river. Set off at 6h 48′—The banks of the river appear to be higher and the river wider, we meet with a number of sand beaches and some rapids but good deep water between them. At 7 ½h a.m. passed a creek which forms a deep ravine in the high lands and has been called 'Chemin Couvert'—a little past 8h. we ascended a rapid where the water was confined to a breadth of 40 yards, a little farther we had to quit the great channel on account of its shallowness and rapidity, & passed along a narrow channel 60 feet wide: without a guide a Stranger would have taken this passage for a Creek.[51] Between 11 and 12h saw an aligator, which surprised us much at this late season and so far north. The Banks (exclusive of the large timber) are covered by Cane or thick underbrush, frequently so interwoven with thorns and briars, as to be impenetrable, untill the way is cut with an edge tool: we see also some species of timber not common below; such as Birch, Maple, holly & two kinds of timber to which no other name has yet been given but 'Bois du bord de

50. The term *insipid* means without distinctive qualities.

51. "Chemin couvert" (covered way or corridor) is Smackover Creek, which forms a portion of the northern border of Union County. The creek begins in Columbia County and runs past the town of Smackover, entering the Ouachita on the west bank. Arkansas Quadrangle Maps.

l'eau' (water side wood).[52] Pirsimmons and small black grapes are plenty in some situations; the first are often very large and excellent, the last a mixture of sweet and tart; those are also common on the Missisippi. The weather being cloudy we did not land to observe. In the afternoon observed some feruginous earth on the right: the margin is frequently fringed with a variety of plants & vines, of the latter several species of the convolvulus, which no doubt in their season ornament this river with their elegant flowers.[53] Thermomr. at 8h p.m. 54°. Extremes 54°–62°. Made this day 18 miles 308 perches.

HUNTER

Cloudy, calm. Thermometer at daylight 59° & in the river 54°. at 3 p.m. 62° & at 7 p.m. 54

Set off a little after day light, & before breakfast came to a small creek on the left, said to be pretty deep[,] called chemin couvert. In general the water is about 100 to 150 yards wide altho in places it is narrowed very much by gravel beaches, fallen Timber, &cc, one of which we passed this morning when the water was only 40 yards wide or there abouts, which we went thro, it appearing for above a mile, more like a narrow bayu or creek than the main river. At breakfast time our canoe came down the river to us; they having supposed we had passed them whilst they were ashore after game, had pushed on, up the river & by that means passed a very dissagreable night in the rain Came up with a canoe having a consumptive person on board on his way to the hot Springs for the recovery of his health The canoe belonged to & was conducted by a Mr Cambel an house carpenter by trade & cotton engine maker.[54] They accompanied us the rest of this day & encamped near us at night.

52. The river birch or black birch (*Betula nigra*) is 20–90 feet tall, with leaves 2–3 inches long that are somewhat oval and double-toothed. The maple may have been either red maple (*Acer rubrum*, height 50–100 feet, leaves five-lobed and coarsely toothed, 4–8 inches long) or silver maple (*Acer saccharinum*, height 60–90 feet, leaves three- or five-lobed and coarsely toothed with a silvery-white underside). The American holly (*Ilex opaca*) grows to a height of 20–50 feet and has 2–4-inch leaves and red berries for fruit. Moore, *Trees of Arkansas*, 92–93, 88. Dunbar's "Bois du bord de l'eau" may be a bois d'arc or osage orange (*Maclura pomifera* or *Toxylon pomifera*, height 20–60 feet, leaves smooth and spear-shaped measuring 3–5 inches, with large globular fruit 3–6 inches in diameter and greenish-yellow to yellow). The inside of the fruit is milky and sticky. Native Americans used the wood of this tree to make tools, especially bows. They also used the roots as a yellow dye. Its original range was very narrow, extending from southern Arkansas and northern Louisiana west to central Texas and Oklahoma. Ibid., 39, 67; Moerman, *Native American Ethnobotany*, 327.

53. The "small black grapes" are of the *Vitaceae* family or of the genus *Vitis*. The "convolulus" (*Convolvulaceae*) is in the morning glory family. Smith, *Keys to the Flora of Arkansas*, 146.

54. For more on Mr. Cambel, see Hunter's entry for November 21.

As the day was overcast we did not go ashore to observe. Saw slight appearances of iron, in places on both sides of the river near the waters edge—The clay as well as some half formed stones being tinctured of a rusty iron colour. The place where we encamped this night is about 40 feet perpendicular above the water & entirely above all inundations; Here are growing Beach, Maple, very large Hollys[,] oak[,] Hickory & Pines &cc. This day saw another Alligator after we were by information far beyond the region where they choose to live. Mr Dunbar is now much recovered of his complaint, & only feells now some debility. The soldier is also recovered.—Made this day 18 miles 308 perches.—The trees & shrubs & plants are now about half stripped of their foliage, the approaching winter had coloured the leaves that are still left, of various tints, which Mr Dunbar thinks designate the colours that might with proper management be extracted from those trees &cc, for the use of the Dyer. some are of a pale Lemon colour, others of a brighter & deeper yellow, some orange, some red, & many contain various shades of brown.—It would be desireable to be able to fix the beautiful crimson of the Pokeberry.[55]—At 5 p.m. passed a small Island by a very narrow passage to the right. A little above which we encamped for the night. Made this day 18 miles. 308 perches.

Wednesday 21st

DUNBAR

Thermr. in air 43°. in river water 54°—a little fog—calm. Set out & passed a hill and Cliff 100 feet perpendicular crowned with lofty pines called 'Cote de Finn' (Finn's hill) a chain of high land continues some distance on the left; the cliff presents the appearance of an ash colored clay; passed a strong rapid, and a little farther a Creek on the right called Bayou d'Acassia (Locust Creek): The river varies here from 80 to 100 yards wide; we frequently see indications of iron along the

55. The word *tinctured* refers to a slight infusion or trace of an element. The "Beach" is American beech (*Fagus grandifolia*); "Maple" is red maple (*Acer rubrum*) or silver maple (*Acer saccharinum*); "Hollys" are either American holly (*Ilex opaca*) or deciduous holly (*Ilex decidua*). The "oak" could be any of more than fifteen species of oak that are found in Arkansas (genus *Quercus*). Eight varieties of hickories are native to Arkansas (genus *Carya*). "Pines" may be longleaf pine (*Pinus palustris*) or loblolly pine (*Pinus taeda*) or shortleaf pine (*Pinus echinata*).

This was the last alligator that the group noted during their ascent.

Pokeberry (*Phytolacca amercana* or *Phytolacca decandra*), also known as pigeon berry, red inkberry, garget, Virginia poke, and pokeweed, is a shrub about 3–9 feet tall with smooth leaves that are 2–3 inches in length. It produces round, deep-purple berries. The short young shoots of this plant are edible and were served as food during this period. The berries were also used for purging. Saunders, *Edible and Useful Wild Plants*, 119, 244.

banks and some thin strata of ore from ½ inch to 3 inches thick, but no other metalic appearance, nor indeed any thing uncommon in the fossil kingdom; a little cloudy this morning, but cleared up before noon & got ashore hastily at a steep inconvenient place among trees and brush, and had a tollerably good observation notwithstanding: Latitude found 33° 29′ 29″. The day proves mild, warm and agreeable, which acted as a restorative to myself and others who had been indisposed for some days past: Thermr. at 3h. p.m. 72°. Altho' Ducks, Geese and Turkeys are often seen, yet we cannot say they are in that abundance which from report we expected, and they are so shy, that we seldom can get a shot from our large boat; but by sending the canoe a head some game is procured; it is probable that higher up, we shall be more successful.[56] Thermr. at 8h. p.m. 58°—Extremes 43°–72°. Made this day 18 miles 36 perches.

HUNTER

Therm at day light 43°. river 54°. at 3 p.m. 72 & at 7 p.m. 58°. Fog on the river, tho clear overhead, calm.

Set off about 7. A M. & shortly afterwards passed some land on the left bank about 100 feet perpendicular in height above the present water, This high land which is called Fin's, (or Beach) hill clifts, continues bordering on the river for about one mile & then gradually becomes like the rest. It is surmounted with pines white & yellow & other forrest trees before mentioned. The soil as appears by being laid bare by the waters consists of a quantity of loose whitish sand heapped up above the common level of clay, with this difference, that in general where the high land comes to the edge of the river, it shows some half formed shelly masses of stones composed of gravel cemented together & coloured of a brown rusty iron colour.—About breakfast time passed a Bayu to the right, called Bayu Accacia (Locust) The average width of the river this days journey might be about 100 to 1200 yards—altho in some places not half that distance. In the course of this day killed 4 pair wild ducks & a Turkey. The bottom of the river consists for these two or three days of clear gravel. The banks as before described of clay & sand rising very im-

56. "Cote de Finn (Finn's hill)" is possibly near the community of Crossroads in southern Ouachita County. Hunter records the place as "Fin's (or Beach) hill clifts." Locust Bayou enters the Ouachita on the east bank in Calhoun County. Arkansas Quadrangle Maps. The wild turkey (*Meleagris gallopavo*, length 37–46 inches) suffered drastic reductions in population during the nineteenth and early twentieth centuries, but this game bird can now be found throughout Arkansas and much of Louisiana. The French explorer Bénard de LaHarpe saw an abundance of these birds as early as 1721. James and Neal, *Arkansas Birds*, 6, 44, 49, 52–53, 154–156; Sutton, *Arkansas Wildlife*, 3–4, 16, 42, 80–81, 90–95; Smith, "Exploration of the Arkansas River by Bénard de LaHarpe."

perceptibly above the rise of the river. The current in general is still gentle, except at streights & rapids as before mentioned, yet the current is upon the whole sensibly more swift. & the rapids & shoals become more frequent than lower down. Observed where we encamped at Breakfast some Indian Hieroglyphics[57] on a tree at the waters edge, at a place called Ross's station, where a person of that name formerly resided, now dead.

The bark was taken off a cypress tree about breast high, for about 18 Inches, & two thirds round it, & on the bare place was painted black in a rude manner, the figure of a person on horseback with one hand extended to the water & the other towards the woods, two other persons whose figures were a little defaced seemed to be shaking hands, one of whom had a round hat on: on both sides of these persons were the figures of about a dozen of large & small four footed animals apparently feeding, some thing like deer without horns. One of our people went up the bank in the direction of where the principal figure pointed, & found an old encampment where a fire had lately been made, & the remains of a temporary shelter from the weather composed of 4 Sticks with crotches at the ends drove in the ground & the whole probably had been covered with skins to defend them from the weather; near which was found an hoe & upon a pole elevated about 12 feet & leaning against a tree were found 14 deerskins tied up in a bundle, At this instant Campbell (who had left his canoe behind for the sick man to bring up & had come this way by hunting along by land,) made his appearance & claimed the bundle of skins, which he asserted he had fixed there a year ago for an Indian cheif of the name of Habitant, so saying he seized the pole & took it down, untied the bundle & hastily counted them over & said they were right for that he had left there 13. upon which I had them counted again & found them in number 14, upon which we insisted that they were not his. We then concluded to take them into the boat & if we should come up with the Owner, who we concluded must be a Choctaw Indian, we should deliver them to him or some of his nation. & in case we were dissapointed in that, to take them back with us & leave them with the commandant at the Post or Garrison to be delivered to the true owner for we had no doubt that if we should leave them on the pole now Campbell was informed they were there, that he would take them & lay the blame on our boat. In a short time the sick man also made his appearance, he had left the canoe at a strong rapid below where he could not alone bring it over. We presently set out & left them behind,

57. Dunbar does not mention the "Hieroglyphics." This is curious, given the fascination with Indian languages that he and Jefferson shared. Jefferson to Dunbar, Washington, January 12, 1801; Philadelphia, January 16, 1810, Jefferson Papers; Rowland, *William Dunbar,* 111–112, 209–214.

for we did not chuse (if we could prevent it) to alow them to go a head & frighten the game from before us.

Found our Latitude by observation to be 33°. 29′.1.7″ N which is about 27″ less than by Mr Dunbars Observations. I have for these last three observations made my Latitude from one to one half minute less than him. Made this day 18 miles, 36 perches.

Thursday 22nd

DUNBAR

Thermr. in air 40°. in river water 53°—Light clouds—calm.—No change this morning in the general appearance of the country, the timber such as had been mentioned, with an increasing proportion of holly, birch, maple and beautiful pine-trees; at 10½ h a.m. came to the road of the Cadadoquis Indian Nation leading to the Arcansa Nation; a little beyond this is the Ecor à Fabri (Fabri's Cliffs 80 to 100 feet high: it is reported that a line of demarkation run between the french and spanish provinces, when the former possessed Louisiana, crossed the river at this place; and it is said that Fabri a french-man & perhaps the supposed Engineer deposited lead near the cliff in the direction of the line: we could not however obtain any authenticated account of this matter, and it is not generally believed: a little farther is a smaller cliff called 'le petite ecor à Fabri' (the little cliff of Fabri); those cliffs appear to be composed chiefly of ash-colored sand with a stratum of clay at the base, such as reigns all along under the bank of this river.[58] The day being hazy and cloudy we made no observation for the Latitude at noon. In the afternoon we encountered a great many difficult rapids, the current of the river being frequently confined to a very small space, where the depth of water is but barely sufficient for the passage of the boat; the additional rapidity of the current indi-

58. The "road of the Cadadoquis Indian Nation" is the Caddo Trace, a trail that stretched from the great bend of the Red River and crossed the Ouachita at present-day Camden. It continued on to the Quapaw villages along the Arkansas River. Dickinson, "Historic Tribes"; Whayne, *Cultural Encounters in the Early South,* 41, 81–84, 146. "Ecore à Fabri" (Fabri's Cliffs) is the location of present-day Camden, Arkansas, in Ouachita County. McDermott credits the naming of this site to André Fabry de La Bruyere. This man, identified as a French engineer, supposedly attempted to establish a trade route to Santa Fe under the orders of Governor Bienville. He also supposedly surveyed the Arkansas River as far as the "plains of the Osages." McDermott, "Western Journals of Hunter," 95–96; Lafron to William Dunbar, New Orleans, August 19, 1805, in Rowland, *William Dunbar,* 179. Don Juan (Jean Baptiste) Filhiol, the Spanish commandant of the Ouachita District, established an outpost at Ecore à Fabri as his seat of authority in 1783. Two years later he moved his operations to Ouachita Post (Fort Miró). For more on the establishment of Ecore à Fabri as an outpost, see Margry, *Découvertes et établissements des Français,* vol. 6; also see Dickinson, "Don Juan Filhiol."

cates that we are ascending into a higher country. The water of the river now become extremely clear and is equal to any in its very agreeable taste as a drinking water. The general breadth of the river to day has been about 80 yards, altho' in certain places not above one half of this quantity. We now find immense beaches of gravel and sand, or which the river passes, in the season of its floods with the rapidity of a torrent, carrying with it vast quantities of drift wood which are in many places piled up in prodigious masses, lying 20 feet above the danger of ascending or descending this river in certain degrees of its flood: accidents nevertheless are rare with the canoes of the Country; ours is the first barge of so large a size that ever ascended this river: passed a very intricate rapid in the evening, which we could not get up untill we had carried a rope ashore. Encamped upon an elevated gravel beach: Thermr. at 8h p.m. 54°. Extremes 40°–68°. Made this day 14 miles 317 perches.

This day an unlucky accident happened, which was very nigh being extremely serious. Doctor Hunter was employed in the Cabin of the boat loading one of his pistols; he held it between his legs upon a bench with his head almost over the muzzel: while in the act of ramming down the ball, the pommel slipt from the bench & the cock of the lock came with force against it, which giving ways discharged the pistol, the rammer and ball passed thro' the fingers & thumb of the right hand & also thro' the brim of the hat within little more than an inch of the Doctor's forehead; his thumb & fingers were much torn, but no bone was broken, the concussion of the head was most severely felt: the bottom of a new powder horn (not well secured) which lay upon the table was forced outwards & the powder partly spilt upon the table, which providentially did not take fire altho' the wadding was found smoking upon the table: the circumstance of the bottom of the powder-horn being forced outwards, points out a curious effect of the elastic power of the air, viz after sustaining a considerable compression the returning vibration causes a partial rarefaction & at the same instant the common air confined within bodies involved by the sphere of rarefaction, exerting its spring to restore the equilibrium, forces outwards all obstacles not sufficiently secured to resist its action.[59] The Doctor's wounds were dressed; he suffered great pain and debility, but after some repose felt better in the evening.

59. Ever the scientist, Dunbar even took this opportunity to make scientific observations concerning the air concussion. As would be expected, Hunter presents a much more vivid description of the accident. For the next two weeks Hunter relied on Dunbar's journals and/or memory to reconstruct his journal entries.

HUNTER

Therm. in air at day light 40° in water of the river 53° & at 3 p.m. in air [blank in MS.] & at 7 pm 54°. Calm.

Set off about 7 a m. The trees on the banks as before with the addition of Birch, Maple, holly, iron wood, dogwood, Ash[,] Sweet Gum ecc. Observed several new species of Hawthorn shrubs particularly some with fruit of a fine scarlet red, others of a red inclining to orange.[60] Both farinaceous & pleasantly sweet to the taste, free of worms, & having but few thorns. The banks are still composed of clay under a body of white sand, more [or] less thick according to circumstances[,] without stones, except where a ridge of hills comes to the edge of the river, which as yet very seldom happens, At about half past ten a.m. came to a sort of road or path called the trace of the Cadeaus, leading from that nation to the nation of Arkansas. In a quarter of an hour more came to some sand hills or cliffs on the left hand about 100 feet high, called Ecor de Fabri, (an old Hunter) who it is said by some buried some lead there, as a land mark to designate where the Line should be, between the French & Spaniards. This Fabri is dead & no person can tell when, or where this was done or whether it was ever done. Our pilot who we find not remarkable either for his judgment or veracity, also relates another tale of the same probability, viz that the only person who was said to be with Fabri when he buried the lead disapeared or died near the spot, or was killed, & that Hunters who encamp there have frequently heard him in the woods in the night call out aloud, Fabri! Fabri!—About ¼ past 11. a.m. passed another rapid where the river is narrowed to the space of 40 yds altho it is in general twice that width. & in an hour more passed the small cliffs of Fabri. The various places on this river named after different persons, have been so called, not from the land being granted to them, but because Those persons were generally hunters, who used to fix their camp at such places & make excursions round & bring the game they killed to such camps.—

Came to a rapid a quarter before 3 & another a quarter after 4. p m & between them the river was again narrowed to 30 yards. We are now entering gradually into another sort of ground, where as it rises a little more than the ground below, above the water of the river, it consequently causes the current to become more swift.

Being cloudy made no observation for the Latitude.

60. The ironwood or eastern hophornbeam (*Ostrya virginana*) has a height of 20–30 feet, and its leaves are alternate, oblong, and toothed. Five varieties of ash occur in Arkansas, including white ash (*Fraxinus americana*), green ash (*Fraxinus pennsylvanica*), Carolina ash (*Fraxinus caroliniana*), pumpkin ash (*Fraxinus tomentosa*), and blue ash (*Fraxinus quadrangulata*). More than sixty varieties of hawthorn (genus *Crataegus*) are native to Arkansas. The hawthorn is called an "under tree" by Dwight Moore. Moore, *Trees of Arkansas*, 38, 107–111, 81.

This day I met with an accident that had nearly cost me my Life. As I thought it prudent to be prepared against any event that might happen from Indians or from any other quarter, I was in the act of loading my pistols & whilst ramming down the ball, I was sitting on a trunk with the pistol between my knees resting its but on the trunk. by the motion of the boat or otherwise it sliped & immediately went off in my face The whole charge with ball & ramrod went thro between my right thumb & two principal fingers, which were thereby lacerated considerably, & then passed along my face, burning my eye lashes & eye brows entirely off & the skin round my eyes & nose. the charge bruised my forehead & caused two black eyes, & then passed thro my hat within an inch of my right temple & finally thro the roof of the boat. The strike or concussion of the air was such as burst my powder horn lying on the table by forcing out the bottom & scattering the powder on the table, it contained about lbs ¾ which if it had taken fire, (& it was a miracle it did not) the cabbin & all the people in it would have been destroyed. This Accident deprived me of the use of my hand for two weeks so that I could not write; which must appolagise for my not being able to take down for that period in writing the particular description of the river & the coasts of it, except from the notes taken on the spot of the courses &cc by Mr Dunbar & from memory. About 5 p.m. passed another rapid & shoal. Made this day 14 miles 317 perches.—

Friday 23rd

DUNBAR

Thermr. in air 48°. in river water 54°—light clouds—calm. River upon the fall. Set off and continued our navigation thro' difficult passages; the river is broken into a number of small streams by Islands, short turning rapids, sunken logs, shoals, bars, and every impediment to be expected in our situation, and this continued at short intervals during the whole of the day, so that our courses and distances cannot be expected to be perfect; every allowance which could be judged necessary at the moment was made: I fortunately obtained a good observation of the Sun's mer: altitude in the interval of some shifting clouds: Latitude deduced 33° 41′ 35″.[61] The banks of the river as we ascend are less elevated, being now only

61. These locations are near present-day Poison Spring State Park and near the confluence of Tulip Creek and the Ouachita River in Ouachita County. Toney Old River, Old River Island, Pine Lake, and Lower Old River are old river lakes that are adjacent to the Ouachita in this area. Hunter refers to passing a series of islands called "Drunken Islands." In his "Journal of a Geometrical survey," Dunbar named them "Drunkards Islands." They were probably in what is today northern Ouachita County. Arkansas Quadrangle Maps.

from 9 to 12 feet, and probably the freshes surmount them some feet; we passed a great number of high & low gravel and sand-beaches; on those were to be seen fragments of stone of all forms & of a great variety of colors; some highly polished and rounded by friction, and may have belonged to the mountains, rivers and oceans of a World, from the ruins of which the Globe we inhabit may have been formed. The banks of the river in this upper Country suffer greatly from abrasion, one side and sometimes both being broken down by every flood. We saw nothing to day worth noticing, no change being observable in the appearance of the lands and timber along the hills and banks of the river: we found on a gravel beach some fragments of the same kind of matter we found lower down resembling pit-coal; it burns without blaze to a white ash, but will not consume (in common temperature) without other fuel: under the burning glass, it emits smoke & consumes, yielding a faint smell of sealing wax; it is light and friable, & affords very little evidence of being penetrated by bituminous matter.[62] Thermr. at 8h p.m. 54°. Extremes 48°–72°. Made 13 miles 28 perches.

HUNTER

Therm. at daylight in Air 48° river 54°[.] at 3 p.m. 72° & at 7 p.m. 54°. Light clouds, calm.

Set off about 7. a.m. This morning find the river which had been raised about 3 feet or more by the late rains, begin to fall again. This addition of water in the river was of considerable use to us, by enabling us to bring our boat over many places otherways impassable. The banks of the river continue to rise irregularly, & the soil now assumes the appearance of better land, the sand being now more intermixed with black vegetable mold; producing timber of a larger growth & greater variety. the current becomes more swift the rapids & shoals more frequent. About 10 a m. came to a number of gravel barrs or small islands forming rapid currents with breakers caused by the water being precipitated over the shallow places, with impetuosity, from which I suppose these islands are called Drunken Islands.

About 11 a.m. passed a small hill called cote de Sofrion About 2 p.m. passed a place called point coupie or cut point where the river had formed a new channel leaving the old one on the east.[63] Observed by Mr Dunbar (my hands & eyes be-

62. He is referring to "pit-coal" lignite or a similar lower-grade or anthracite coal. There are many lignite deposits in northern Louisiana and southern Arkansas and in the vicinity of the Ouachita River. Several mining operations exist today in northern Louisiana within or near the Ouachita River valley. Troy, French, and Ales, *Coal and Lignite in Louisiana*, 1–4.

63. Dunbar made no reference to the "cote de Sofrion" in either the "Dunbar Trip Journal, Vol. I"

ing dissabled) & found the __Double mer. appt. altitude to be 70°, 59′, 13″ Index error 0°, 14′, 8″ Lat. 33°, 41′, 35″

Made this day 13 miles, 28 perches.

Saturday 24th

DUNBAR

Thermr. in air 48°. in river water 54°—light clouds—calm—river at a stand. Set off & continued our voyage thro' a country in all respects similar to that thro' which we passed yesterday, excepting that our obstacles from strong rapids are considerably augmented: at a place on the left called 'Auges d' Arclon' (Arclon's troughs) we observed some laminated iron ore, and a stratum of tenacious black sand shining with minute chrystals. The general breadth of the river is now 80 yards, tho' in many places greatly enlarged by Islands & shallows, and at other places contracted to 80 or 100 feet. The river is now in many places rocky of a greyish color & rather friable. Observed some willow very different from what is found below and on the banks of the Missisippi, the last is very brittle, this on the contrary is extremely pliant & resembles the osier, of which it is probably a species, I propose on our return to take some plants along with us; its foliage is now of a golden yellow & falling: we also found some of the larger Whortle-berry in fruit, the berry is of a Sub-acid agreeable taste, the leaves not yet fallen of a beautiful crimson.[64]

The weather being cloudy we had no observation at noon & went on to dine at the forks of the Washita and Missouri the lesser; the latter comes in from the left hand and is a considerable branch, perhaps about ¼ of the Washita:[65] Hunters of-

---

or the "Dunbar Report Journal" (reproduced here). The place called "point coupee or cut point" is possibly near a present-day site named Old River Island, which is part of the old channel and forms a C-shaped or oxbow lake at this location on the west bank. Arkansas Quadrangle Maps.

64. "Auges d'Arclon" would have been located in what is today the northeast corner of Ouachita County. Ibid. Osiers are willows that have tough, flexible twigs or branches that are used in wicker or weaving. Whortleberry, of the *Vaccinium* genus, is a shrub also known as black whortleberry, bilberry, blueberry, huckleberry, dyeberry, wineberry, and hurleberry. Moerman, *Native American Ethnobotany,* 500–509, 584; Grimm and Kartesz, *Illustrated Book of Wildflowers and Shrubs,* 558; Smith, *Keys to the Flora of Arkansas,* 94–95.

65. Dunbar was mistaken about the extent of the Little Missouri. The river originates in the Ouachita Mountains near the community of Athens in Howard County. It flows south-southeast and forms the borders between Clark and Pike, Clark and Nevada, and Clark and Howard Counties. Arkansas Quadrangle Maps; Hunter journal entry for this date. Dunbar also recorded in his "Journal of a Geometrical survey" a branch he called the "Petit-Washita." This is a branch of the Ouachita that curves southward and enters the Little Missouri just above the confluence. In "Western Journals of Hunter,"

ten ascend the little missouri, but they are not inclined to penetrate far up, because this branch reaches near to the great planes or prairies upon the red river, which are often visited by the lesser Osage Tribe settled on the river Arcansa:[66] These last frequently carry war into the Cadadoquis tribe who are settled on the red river about W.S.W. from this place, and indeed they are reported not to spare any nation or people.[67] They do not come upon the head waters of the Washita, because they are surrounded by a number of mountains or steep hills rising behind each other, and so extremely difficult to travel over, that those savages perceiving no desirable object, do not attempt to penetrate to the river, & it is supposed to be unknown to the nation: The Cadadoquis (or Cadaux as the french who are fond of abbreviations generally pronounce the word) may be considered as Spanish Indians; They boast, I am told with truth, that they never have imbrued their hands in the blood of a white man: it is reported (perhaps falsely) that they are excited to enmity by the Spanish officers at Nacocdoches against the Americans.

We are told there is a mine up the little Missouri, it is said that the stream runs over a bright splendid bed of mineral of a yellowish and whitish color, it is most probably martial pyrites:[68] some 30 years ago, several of the inhabitants hunters worked upon this mine and sent a quantity of the ore to the Government at New

---

97, McDermott wrongly identified this as Terre Noire Creek. Terre Noire Creek is a long running stream that enters the Little Missouri approximately 1½ miles from the confluence with the Ouachita. Arkansas Quadrangle Maps.

66. By "great planes or prairies" Dunbar may be noting an area of black-land prairies that are located today in Howard and Hempstead Counties in southwest Arkansas. These prairies, drained by a tributary of the Little Missouri River called Ozan Creek, near the site of Mound Prairie, an early settlement. Mound Prairie was located just west of Washington, Arkansas, in Hempstead County and along a trail called the Fort Towson road. Today the Arkansas Game and Fish Commission operates the Rick Evans Grandview Prairie Conservation Education Center and Wildlife Management Area on the site. Trey Berry, Ray Granade, and Tom Greer, "Grandview Plantation and Mound Prairie History," unpublished manuscript prepared for the Arkansas Game and Fish Commission in 1998 and available in Special Collections, Riley-Hickingbotham Library, Ouachita Baptist University.

67. The Osage long harassed the Caddo Indians, and these raids may have been one of the reasons the Caddo relinquished their lands in southwest Arkansas. For more on the Osage and their warfare campaigns, particularly those against the Caddo, see Din and Nasatir, *Imperial Osages*, 46–47, 93, 102, 137, 148, 156–157; Arnold, *Rumble of a Distant Drum*, 4, 20, 60, 73–75, 92, 100, 114–115, 153; Swanton, *History and Ethnology of the Caddo Indians*, 71, 74, 78–79, 85; Smith, *Caddo Indians*, 14, 56, 66, 70, 75–78, 87, 110; Carter, *Caddo Indians*, 43–45, 185, 192, 205, 206–208, 220–221, 241; Bolton, *Arkansas, 1800–1860*, 67–69, 75–77. Dunbar probably refers to the Caddo as "Spanish Indians" because of their known relations with the Spanish. Smith, *Caddo Indians*, 26, 31–32, 59, 64–67, 71, 73, 89, 106.

68. Dunbar is probably describing an iron sulfide (FeS) or sulfate of iron. It usually appears with a reddish or yellowish color. Bates and Jackson, *Dictionary of Geological Terms*, 504–505.

Orleans, but they were prohibited from working any more. Thermr. at 3h. p.m. 59°. Extremes 48°–72°. Made this day by a very uncertain reckoning 11 miles 152 Perches.

HUNTER

Therm. in air at daybreak 48° in the river 54° & at 3 p.m. [blank in MS.] at 8 p.m. 59° Light clouds, calm.

Got under way about 7 a.m. & presently perceived on the banks near the waters edge poor Iron ore, unfit for use, accompanied with black sand on the bottom, at a place called Auges d'Arclon. (Arclon's Troughs)[69] The land rises on both sides more & more, forming small stony hills sometimes on one side & sometimes on the other but seldom on both at the same time. This stone is now hard, & well formed freestone in irregular Strata inclining from 20° to 30° down the river, & those strata being split crossways form fragments of from one to six feet broad; These stones have a grit very sharp & fine, fit for grind stones of one sort. The current still is more swift. The rapids are now become small falls, & nearer to each other, The soil however improves upon the whole, having a greater proportion of black mold amongst the sand, & a greater proportion of the bank above all floods. The pine trees still appear to be the most numerous. These rapids or small falls, often impede our progress & cause frequent stops, obliging us to search for other channels to pass thro. The weather now growing thick & cloudy prevented taking an observation for the Latitude. Near noon passed an old corn field on the left & in half a hour came to Bayu Talien on the left. About ¼ past 1 p.m. passed the little Missouri to the left, which at its entrance appears to be nearly as large as the Ouachita, & is said to run thro much good land, in its course[.]About 4 p.m. came to little Ouachita on the left, which runs also into the little Missouri, both going then obliquely towards the waters of the red river. About 20 minutes past 4 p.m. passed a place called Belle Ance (handsome bend).[70] Our Guide or Pilot informs us that there are many small bodies of good land lying on both sides [of] the river

69. During the descent of the Ouachita, Hunter writes in his journal on January 10 that this point occurred about "3 leagues" (approximately 9 miles) below the Little Missouri River confluence. In that entry Hunter also says that the name origin of the location came from "one Arclon [who] made troughs to carry down his bears Oil to market." Arkansas Quadrangle Maps.

70. What Hunter calls "the little Missouri" may be French Creek, which meets the Ouachita just below Tate's Bluff in northeastern Ouachita County. "Belle Ance" is a bend that forms a beak-shaped land feature just below the Dallas-Ouachita-Clark County lines. On this day, William Dunbar does not mention either "Bayu Talien" or "Belle Ance." Arkansas Quadrangle Maps; "Dunbar Trip Journal, Vol. I," November 24, 1804.

at a distance amongst the pine ridges & broken hills which compose a great part of the interior of this part of the country. Made this day 11 miles 152 perches.—

Sunday 25th

DUNBAR

This morning proved very rainy, having commenced raining before day, we were therefore constrained to continue encamped: a cessation took place after breakfast, which gave us some hopes of being able to proceed, but this was not of long duration; the rain recommenced and we remained all day in our tents. we have the consolation however to expect that the river will rise a little in consequence of the rain, which will facilitate our ascent over the shoals that are to be expected above. Thermtr. at 8h. p.m. 62°. Extremes 54°–70°.

HUNTER

This day being wet having rained the greater part of the preceeding night, we remained in our tents & did not move until the next day.

Monday 26th

DUNBAR

Thermtr. 50°—river water 57°—Clear above. calm—river risen 3½ inches in the night. Contrary to expectation the morning proved not only fine and serene, but of a mild, agreeable temperature. In general after the winter season sets in, the changes in the weather are made by extremes. A day or two of rain is commonly succeeded by a cold and blowing north wester, and the day following a frost of some severity, which has not been the course upon this last occasion, it appears also that the rain has raised the temperature of the river 3°. The water is now remarkably clear and fine, and it does not seem to have been discoloured by the last rain. There is still a great sameness in the appearance of the river banks, the Islands are skirted with osier, and immediately within on the bank grows a range of birch trees & some willows; the more elevated banks of the River are clothed by a thick growth of Cane & the timber which rises above the Cane is such as has been already mentioned Viz. oak, white, black, and red; many species of each: black Maple, white maple, Sycamore, Elm, several species, Ash, hicory many Species. Dog wood, Holly, Iron wood &c—[71]

71. Black maple (*Acer nigrum*) is not indigenous to Arkansas. Dunbar is probably confusing this tree with the red maple (*Acer rubrum*). Dunbar's "white maple" may be a silver maple (*Acer saccharinum*). The American sycamore (*Platanus occidentalis*) grows to a height of 70–120 feet and has 4–8-inch leaves; the bark is white on large trees. Four elm species are indigenous to Arkansas: American

Saw a number of yellow butterflies fluttering about the banks of the River. We continue to encounter the same obstacles from the Shoals & rapids; the valley of the river, in its present low state is filled with Islands, which dividing the current reduces the depth of the Channel; We find no great difficulty where the water is collected into a single channel. Our Pilot informs us that there is a body of excellent land upon the little Missouri & more especially on the Creek called the 'Bayou á terre noire,' which falls into the little Missouri; this land reaches within a few miles of the Washita, and is said to extend to the Red River being connected with the great prairies above the Cadaux nation & in the proximity of the red River; this rich tract of Country is said to be of very considerable extent perhaps a square of 30 miles & is connected with the great prairies which are the hunting grounds of the Cadaux Nation, consisting of about 200 warriors, they are warlike, but frequently unable to defend themselves against the tribe of Osages who are settled upon the Arcansa river, who passing round the mountains which give birth to the Washita, along the prairies which enclose those mountains on the West separate them from the main Chain of mountains which furnish the waters of the red & arcansa river, pass down in the Cadaux Country & rob & plunder them of their horses and other effects, & not unfrequently take a few scalps; for it seems that this detached tribe of the Osages is a lawless gang of robbers, making war with the whole world.[72]

Thermtr. at 8h p.m. 62°—Extremes 50°–68°. Made 12 miles 21 perches.

HUNTER

Therm. at daylight in air 50° river 57°. at 3 p.m. 68°, & at 8 p.m. 62°.

Set off about 7 a m. The river has risen 3½ Inches last night About breakfast time came to an old encampment called Bears head camp which is overflowed at high waters. About 11¼ a.m. passed a small Island called pettite cote (small Hill.)[73]

---

elm (*Ulmus americana*), slippery elm (*Ulmus rubra*), winged elm (*Ulumus alata*), and cedar elm (*Ulmus crassifolia*). Moore, *Trees of Arkansas*, 93–94, 76, 59–62.

72. The "yellow butterflies" are probably one of several species of butterflies called sulphurs: subfamily *coliadinae*, genus *Colias*. Opler and Malikul, *Field Guide to Eastern Butterflies*, 71–90. "Bayou a terre noire" is Terre Noire Creek, which rises south of the community of Alpine in Clark County and flows southwest through the county. It meets the Little Missouri River about 1½ miles from the Little Missouri's confluence with the Ouachita. Arkansas Quadrangle Maps. Four branches, the Prairie Dog Town Fork, Salt Fork, Elm Fork, and North Fork, meet along the border of northern Texas and southwestern Oklahoma to make the Red River. The main branch, Prairie Dog Town Fork, originates in northeastern New Mexico. The Arkansas River begins near Leadville, Colorado, in the Sawatch Range of the Rocky Mountains and flows southeasterly through Colorado, Kansas, Oklahoma, and Arkansas. *National Geographic Atlas of the World*, 39–41.

73. Hunter identified several sites on this day: "Bears head camp" and "pettite cote (small Hill)"

Still find shoals, rapids & small falls in the river with a gradual increase of current. At ½ past 3. p.m. passed a Bayu on the left called Bayu dc Cypre ⊙ d. Appt. mer. alt 69°, 37′, 45″ Index error 0′, 44″. Lat 33°, 54′, 21.8″

Made this day 12 miles 21. perches

Tuesday 27th

DUNBAR

Thermtr. 54°—river water 58°—Cloudy—River risen above the mark which was 12 inches out of water: set off at 7h. 1′. and continued our Voyage with the same obstacles from rapids, which were very violent at particular points from the encreased body of water descending from the higher position; but we obtained at the same time the advantage of approaching the willows & even passing thro' them, to avoid the most difficult passes. During the hour of breakfast the river rose 1½ inches perpendicular. The general height of the main bank is now from 6 to 12 feet above the level of the water, and the land is rather of a better quality, the Cane &c shewing a more luxuriant vegetation: the superficial soil subject inundation is of brownish appearance greatly mixed with Sand; At noon arrived at 'cache á Macon' (Masons hiding place) on the right, stopped here for dinner. Having been informed of some pit coal reported to be in the neighborhood, we determined to explore its position. Doctor Hunter with the Pilot set out for this purpose, & at about 1½ mile N.W. of the Boat found in the bed of a Creek a substance similar to what we had formerly seen under the name of coal; some pieces of it were very black, solid, & of a homogenous appearance greatly resembling pit Coal, but it was deficient in ponderosity, & did not seem to be penetrated by bituminous matter in a sufficient degree to constitute Coal; We may perhaps therefore be permitted to consider it as vegitable matter in a certain stage of its progress of transmutation into Coal, we were the more confirmed in this opinion by discovering other fragments, which still retained very evidently the fibrous texture of wood, one piece in particular seemed to have been a large chip taken out by the felling ax.[74] [T]hose last pieces were not so far advanced in the transmuting progress as the first mentioned; although black it was not so perfect, being rather a very dark brown black,

---

were both possibly on a section of the river near Beaver Creek on the east bank in Dallas County; "Bayu de Cypre" is Deceiper Creek (also called Caney Creek), which rises in eastern Clark County and runs parallel with the Ouachita River, emptying into the west bank about 9–10 miles south of Arkadelphia. Arkansas Quadrangle Maps.

74. What Dunbar calls "cache a Macon" is probably Casa Massa Creek, which enters the east bank of the Ouachita about 4 miles southwest of the community of Dalark in Dallas County. The "pit Coal" was probably some type of bituminous lignite, or low-grade coal, which was also referred to at this time as pitch coal or peat coal. Parker, *Dictionary of Geology and Mineralogy*, 37. A "felling ax" is a broad ax.

retaining the exact form & shape of the wood as it had been separated from the log: as this incipient or imperfect coal was found imbedded among clay & gravel, which appeared to have been washed down by the torrent, no clue could be found to lead to a discovery of the process by which nature effects so extraordinary a change, an ingenious enquirer placed in favorable circumstances, will probably have the good fortune to make this discovery: The time may arrive when the Planter who shall be clearing his Plantation or farm of useless timber, will be enabled from the instructions of the Chemist to place the whole in a situation to be transmuted into an usefull article capable of long preservation. This is no doubt the carbonated wood described by Kirwan & other Chemists. We found along the banks a species of the white thorn loaded with abundance of ripe fruit, being a small oval berry of a cornelian colour & agreeable sweetish taste; the whortle berry was also found in the same situation.[75] The white maple has now a beautiful appearance, its leaves before their fall first assume a pale yellow, but this soon fades, and they change into a splendid white and present at some distance the appearance of clusters of elegant flowers. Being cloudy at noon we made no observation for the Latitude.

We suppose the river to have risen at least 30 inches and it now flows with great rapidity, which obliges us to pass sometimes among the willows to avoid its impetuosity: this afternoon we passed some reaches of the river, which were very handsome, being of considerable length, and at least 150 yards wide, and flowing with a full current from bank to bank. We found a considerable number of unknown (to us) plants some of them very handsome, but our very limited knowledge in practical botany, did not enable us to discover what they were, particularly as they were not in flower. Made this day 13 miles 39 perches. Thermr. at 8h p.m. 66°. Extremes 54°–71°.

HUNTER

Therm. at daybreak in air 54°. in river 58°. at 3 p.m. 71° & at 8 p.m. 66°. Cloudy.

Set off at 7. a m. When we observed that the river had risen one foot perpendicular during the night. About half past 7. a m. came to another rapid which continued with more or less force for ¾ hour & at breakfast found the river to rise 1 ½ Inch during the hour. at half past 11 a.m. passed a large Island to the left. The clouds prevented an observation this day. About noon passed Cache a Macon & Bayu Macon on the right. I went with our Pilot by land on the right hand side of the river in a direction nearly NNW mostly up the river & slanting obliquely in-

75. For more on Kirwan, see Dunbar's entry for November 3. The "white thorn" is possibly of the Hawthorn genus *Crataegus*. Moore, *Trees of Arkansas*, 81. Also see Dunbar's December 1 journal entry.

land about 1 ½ miles distance till we came to the banks of a creek called coal mine creek to search for a coal mine said to be there.[76] At length found it but it proved to be simular in quality & form to what we had found below the Post of Ouachita, altho somewhat more advanced in its progress towards perfect coal: took some samples of it, to bring home[.] there was a narrow stratum of it laid bare on both sides of a small Bayu about 6 Inches thick with but little consistence being very friable between the fingers & split where it was exposed to the air into small peices of about ¼ Inch thick. In the bottom of the river in a part now dry I found several masses of about lbs 100 to lbs 50 weight detached from each other, closs to which lay several pieces of wood turned into the same sort of coal. some in part & some entirely. The Stratum of coal in the Bayu lay over a bed of yellowish clay & was covered by about 1 foot thick of gravel; next above that was 18 Inches of yellow loam & a few Inches of vegetable mold. The coal lay in the bank about 3 feet perpendicular above the present state of the river. The average of the land a little back might be about 20 to 30 feet above this. This sort of coal is considerably lighter than our Virginia coal[,] is difficult of Ignition, burns without flame into a white light ashes, It smells not unlike other coal when ignited. It appears as yet applicable to no useful purpose. This Stratum of coal appears manifastly to have been once the surface of the ground[.]the Stratum of gravel & yelow loam, have been carried there by the strength of the current.—The river rose whilst we were at dinner 4 Inches in two hours. The river continues to run with gradual increase of swiftness[.] at 4 p.m. is 150 yds wide. Various shoals & ripplings cause frequent short stops & delays. Made this day 13 miles 39 perches.

Wednesday 28th

DUNBAR

Thermr. 68°—river water 60°. fallen 4 inches in the night—Cloudy—calm. Set off at 7h 5′ and continued our voyage, meeting the same species of obstacles as yesterday—the river appears to increase in width being sometimes 170 yards broad, flowing at this time with a full tide from shore to shore, the Current is in some places extremely rapid, that is where the depth of the Channel is diminished and the bed contracted, in such situations we are under the necessity of catching hold of the willows &c, & hauling up along shore, oars and poles being insufficient to stem the violence of the torrent; in other situations for miles together the current in inconsiderable, in fact it is nothing under the shelter of the points, this ad-

76. This may have been Mill Creek in northwestern Clark County. Mill Creek enters the Ouachita about 1 mile south of the community of Griffithtown. Arkansas Quadrangle Maps.

vantage is the result of the enlargement and encreased depth of the river. Being cloudy we had no observation for the Latitude. Some of our people who walked out with their guns at the hour of dinner discovered some buffalo tracts we are therefore in hopes soon of getting some fresh beef. We past some beautifull Pine Forests. The Lands in many places appeared of a pretty good quality producing trees and a variety of vegetable subjects indicating good soil. Encamped in the evening after making by our reckoning 12 miles 255 perches. Here we found an old dutch Hunter with his party consisting in all of 5 persons.[77] This man has resided 40 years on the Washita and before that period has been up the arcansa river, the white river and the river of St. Francis; the two last he informed us are small rivers of difficult navigation similar to that we are now upon, but the Arcansa river is a river of great magnitude, a large and broad channel, and when the river is low with long and great sand beaches like to the Missisippi. So far as he has been up, the navigation is safe and commodious, without any impediment from rapids or shoals, upon all those rivers, the soil is of the first rate quality, the countries are of easy access, being lofty open forests, unembarrassed by canes & other under growth: the lands on the Arcansa are generally level and not subject to inundation, with here and there gently rising hills. The river is not embarrassed with rocks so far as this informant has ascended, but its bed is composed of mud and sand: the water of the river is extremely bad to drink, being of a disagreeable red colour and very brackish when low, a multitude of creeks which flow into the river furnish sweet water, which the voyager is obliged to carry in vessels on board to supply his immediate wants, hence this inconvenience is not of much moment. This man confirms the frequent reports given of silver being abundant up this river; he has not been so high as to see it himself, but says he has received a silver pin from a hunter who assured him that he himself collected the virgin silver from the rock, out of which he made the Epinglete by hammering it out; The tribe of Ozages live higher up than this position, but the hunters rarely go so high, being afraid of those savages who are at war with the world and destroy all strangers they can meet with. It is reported that the arcansa nation with a part of the Chactaws, Chicasaws, Shawnese &c. have formed a league and are actually gone or going 800 strong against those depredators, with a view to destroy or drive thm entirely off and possess themselves of their fine prairies which are most abundant hunting grounds, being plentifully stocked in Buffalo, Elk, Deer, Bear and every other beast of the

77. Hunter stated in the "Hunter Official Report," 39, that Paltz was of "German extraction" and that his home was Ouachita Post, or Fort Miró. He also described his party as consisting of five people, including "his three Sons men grown & an hired hand with a gang of half a doz n dogs."

chase, common to those Latitudes in America.[78] Our old Dutch Hunter informs us of a saline or salt spring from which he has frequently supplied himself with salt by evaporation, we shall visit it in the morning, being only half a league distant. Made 12 miles 255 perches. Thermr. at 8. p.m. 73°, Extremes 68°–78°.

HUNTER

Therm. at daylight in Air 68° river 66[.] At 3 p.m 78° & at 8 p.m. 73° Cloudy & calm weather

Set off about 7 A.M. in half an hour passed an hill to the right called Ecor a paux de bois (Wood Tick Hill) or cliff. About 9. a m breakfasted opposite a large number of tall Pine trees on the right. at half past 10 A.M. passed another ripple & Bayu de l'eau froide (cold water bayu) on the right.[79] The water for these six days past has become more clear than formerly so that the rocks & gravel on the bottom is seen more distinctly than before The soil on the banks improves in quality, having a greater proportion of black mold intermixed with the sand. The land is still as we ascend the river, getting higher & we come nearer to the hills This day being overcast made no observation.

Here we met with a Delaware Indian painted with Vermilion round the eyes. He called himself Capt. Jacobs & informed us that a large number of Chickasaw & Choctaw Indians had gone to hunt on the waters of the Arkansa. This day passed as usual a number of small rapids, shoals & ripples causing short delays[.] about 1/8 past 4 p.m. came to Grand Glaise[80] opposite to a Bayu called de Cypre from hav-

78. An "Epinglete," from the French *épingle*, is a decorative pin. The regional tribes such as the Caddo, the Choctaw, the Cherokee, and the Coushatta engaged in war with the Osage by 1817, and some of these tribes fought much earlier as well. Smith, *Caddo Indians*, 110–111.

79. The "Wood Tick Hill" may have been on the east bank about 3 miles southwest of the community of Griffithtown. "Bayu de l'eau froide" is L'Eau Frais Creek, which rises in north-central Dallas County and joins the Ouachita about 2 miles due west of the community of Griffithtown. Both sites are in present-day Dallas County. Arkansas Quadrangle Maps.

80. Beginning in the late eighteenth century, the Delaware and other eastern tribes began to cross into Spanish Louisiana, where they lived in what is now northeastern Arkansas and southeastern Missouri. By the 1780s the Delaware were settling along the White River in northern Arkansas. Some Spanish officials attempted to keep them out of Arkansas, and the Spanish encouraged Don Juan (Jean-Baptiste) Filhiol at Fort Miró to do the same. Other eastern tribes that started crossing the Mississippi during this time and shortly after included the Miami, the Cherokee, the Shawnee, the Choctaw, and the Chickasaw. Whayne, *Cultural Encounters in the Early South*, 57–60, 190. In the "Hunter Official Report," 39, Hunter added that when the explorers approached the Indian, he retorted, "O! Canoe damned big." Dunbar failed to mention the Delaware Indian in his daily entry. See also "Dunbar Trip Journal, Vol. I," November 28, 1804. The area Hunter identified as "Grand Glaise" was probably just south of Arkadelphia in Clark County.

ing a number of Cypress trees growing round it, remarkable because those trees terminate hereabouts, being found very seldom much farther north of this place on this river, which is here 170 yds. wide. Made 12 miles 255 perches this day.—

Thursday 29th

DUNBAR

Thermr. 72°. river water 62°—Cloudy—wind South, blew strong all night—This morning Doctor Hunter went with a party and the old dutch hunter to visit the saline, which was found in the bottom of the bed of a dry gully near a Creek;[81] after digging a few feet found the water which proved very brackish to the taste; the saline lies about 1½ mile northerly from our encampment, a creek falls into the river a little above our encampment, being the same which communicates with the saline, a quantity of the water was brought into camp whose specific gravity was carefully ascertained by comparison with the river water and found to be as 1.02116+ to 1. Evaporated 10 quarts of the water which produced a saline mass weighing when dry 8 ounces. It began to rain about 9h a.m. which obliged us to remain in camp untill after dinner, when it cleared up, and we set out at 1h. 27′ p.m., the water of the river has now become whitish and less transparent in consequence of the rain and appears to be rising again altho' it seemed to have stopped since last night: the water was tollerably favorable in the afternoon having met with only one rapid of difficulty and considerable length: since we have had so much difficulty to encounter from the shoals and violence of the current, the Soldiers have exerted themselves with a considerable degree of vigor and perseverence and seem desireous that we should accomplish the end of our voyage. Thermr. at 8h p.m. 52°. Extremes 52°–76°. Made this day 8 miles 2 perches. The weather clears up and begins to grow cold, we expect a north-wester in the morning.

81. This stream is Saline Bayou, on the east side of the Ouachita River about 1½ miles east of Arkadelphia in Clark County. It is apparent that Caddoan peoples lived and made salt at this site for centuries. They possibly used the salt for regional and interregional trade. (Some of the sites in the area are known as "Bayou Sel" and "Hardman Site" by archaeologists.) Around 1811–1812, John Hemphill began manufacturing salt in the Saline Bayou area and continued for many years. The Confederate Army in Arkansas also used the site for a short period during the Civil War. The salt apparently seeps into the stream at many locations. For more on this site, Indian salt manufacturing, and John Hemphill's operations, see Early, *Caddoan Saltmakers in the Ouachita Valley;* and Wendy Richter, ed., *Clark County Arkansas: Past and Present* (Arkadelphia, AR: Clark County Historical Association, 1992), 10, 14–15.

HUNTER

Therm. 62° in river water & in air at 7. a.m. 72° At 3 p m 76° & at 8 p.m. 52° Wind South.

It rained the greater part of last night & this forenoon which prevented us from moving from the camp till after dinner. At this place were also encamped an Old Hunter named Paltz with his three Sons men grown & an hired hand with a gang of half a dozn. Dogs with a Veiw of Bear hunting[.]As yet they had only killed a few deer; their provisions consisted of only a few bushels of Indian corn which they pounded as they waited in a hollow block of wood with a short cut of an hicory saplin by way of pestle. depending on what they could kill & providing for the rest. This old man who did not speak a word of english was well acquainted with this part of the country, having lived & hunted in it & on the Arkansa river for upwards of thirty years. He informed us that there was a party of Chickasaws, Choctaws & other neighboring Indian nations about eight hundred in number now on their way to the Arkansa Waters to drive off those 400 Warriors of the Osages who had lately come to that country whose hands were against every other description of persons white or red.[82] He also informed us of a Saline about half a league to the northward of where we were, & agreed to accompany me to it. I set out with him & two men with Spades & a kettle to bring back some of the water for examination We travelled on the right side of the river upwards & nearly in the direction thereof only edging inland withal for near two miles, Over land which in times of high freshes is generally about 3, or 4 feet under water, until we came to a Bayo or Rivulet which contained about as much fresh water as might turn two or three grist Mills, in the present low state of the waters, the bottom was yellow gravel, & no place very near it except an old Indian Mount or Cemintary with a base of about 80 feet diamitor, but what was overflowed in high freshes. At the foot of this Mount, in the bottom of the Creek we saw a few stones & some black marks on the gravel where fire had been made to boil the salt water. The ground was manifestly salt, & bitter, to the taste, & the water oosing from a kind of puddle brakish. I set the men to dig two holes one about three & the other six feet deep which they did, & after passing about three feet in the first hole of a blue clay came to quicksand & gravel from which Issued the salt water very strong to the taste In the second hole we dug 6 feet to come to the quicksand & water in doing which found several pieces of Indian earthen pots, probably used by them in making salt. This 2nd wa-

82. For more information concerning Osage relations with other tribes and Europeans, see Dunbar's journal entries for November 24 and 26, December 13, and January 10.

ter tasted also very strong, & to the taste scarcely weaker than the first, being both judging from the taste about as strong as sea water.—Brought to the camp two & an half Gallons of the Strongest & about a pint of the other. In the first [Baumè]s pese Liqueur [hydrometer] marked 4 degrees below o. & the 2nd shewed 3° degrees Found the Specific gravity by a glass ball suspended to a nice delicate balance by an hair. This ball weighed in air 31 dwt 2 ¾ gr & in water of the river

6 dwt 15.3/16 gr	
& in the Salt water	6″3—
the other Salt water weighed	6″2 5/8

The specific gravity found by dividing the difference between the weight of the bulb in Air & salt water, by the difference between its weight in Air & fresh water proves to be

of the strongest water in the shallowest hole	1.0272
That of the other water	1.02104

I afterwards evaporated ten quarts of the strongest water to dryness[;] it yielded ten Ounces Averdupois of salt which proved deliquessent & develloped to the taste besides Sea Salt Muriat of Lime or Magnesia,[83] perhaps both.—

The weather clearing up, we set out about half past one p.m. & passing a few rapids & shoals came at 3 p m. to Ecor a Chicots,[84] continued our course till sun set & encamped, having made 8 miles & 2 perches this day.

83. In the "Hunter Official Report," 40, Hunter wrote that the water of the spring "is about as strong as the famous Salt Lake of Kentucky called Bullits Lick & Mann's Lick from which so much salt is made, yet I have no doubt, but that if the ground was perforated to a sufficient depth, water of much greater specific gravity would be obtained." Hunter would have known of the "Salt Lake" in Kentucky because of his early travels in that area. Arkansas Quadrangle Maps; "Dunbar Trip Journal, Vol. I," November 29, 1804; Dunbar journal entry of this date; McDermott, "Western Journals of Hunter," 19–57. "Magnesia" is Magnesium sulfate, $MgSO_4 \cdot 7H_2O$. This is also called bitter salt, epsom salt, sal amarum, salt anglicum, muriate of magnesia, and epsomite. Dunbar used the term "muriated magnesia."

84. The translation of the French name of this place is "Stump Cliff." In the official map of the expedition (drawn by Nicholas King of Philadelphia), this site is on the east bank of the Ouachita about 1 mile north of Arkadelphia in Clark County. This could have been, however, a reference to the DeSoto Bluff, a 300-foot cliff about 1 mile above Arkadelphia that would have been a significant landmark. The DeSoto Bluff is on the west bank, though. It is possible that the river has changed course in this area and that the 1804 channel may have taken the explorers around DeSoto Bluff, thus preventing them

Friday 30th

DUNBAR

Thermr. in air 38°. in river water 60°—river risen 19 inches—clear calm. Set off & continued our voyage against a strong current during the greatest part of the day, altho' frequently we found favorable eddies or little or no Current where the bed of the river became enlarged, which sometimes extended to 150 and even 170 yards in breadth. Saw great flocks of Turkeys to day, two of which were killed. At 10½h a.m. arrived at the large branch on the left called 'Fourche des Cadaux' (Cadadoquis fork) about 100 yards wide at its entrance into the Washita; immediately beyond which on the same side the land is considerable elevated (abt. 300 feet.) The wind from North and N.W. opposed us most of the day, so that our progress was not very rapid. At noon landed & observed the Sun's altitude in a difficult place, in some measure thro' the branches of trees, the Latitude deduced was 34° 11′ 37″. As we advance to the north we perceive more of the effects of winter; the trees are now nearly stripped of their foliage, which a week below seemed to be nearly entire, altho' changed in color: Being informed of a saline or salt-lick, we landed before 3h. p.m. and the Doctor with a party went to view it, thermr. at 3h. 57°. The Doctor returned in the evening with a quantity of water from the saline, which from taste appeared to be less impregnated than the former, and on trial its specific gravity was found to be when compared with the river water, which at that time was principally rain water, 1.017647. This salt pit was found in a low flat place subject to be overflowed from the river, it was wet and muddy, the earth on the surface yellowish, but on digging into the stratum which yielded the salt water, it was found to be a bluish clay; probably the water was fresher in consequence of the rain of the day before, which had not fallen when the first water was collected. Ten quarts of this last water produced by evaporation six ounces of a saline mass, which from taste was principally marine salt, it was however evident that it contained besides marine salt, some soda and a bitter salt, which last no doubt was muriated magnesia, but the marine salt greatly predominated. Made 7 miles 28 perches.[85]

---

from seeing and recording what is the largest bluff before the "Narrows" near Hot Springs. King, "Map of the Washita River in Louisiana"; Arkansas Quadrangle Maps.

85. The Caddo River begins in Montgomery County in the Ouachita National Forest and runs southeasterly through northwestern Pike County and northern Clark County. It meets the Ouachita about 4 miles north of Arkadelphia. From site examinations today, it does not seem possible that the mouth of the Caddo River could have been "100 yards wide." Hunter called the Caddo River the "'Fourche de Cadeau' (forks of the Cadeau nation)." Arkansas Quadrangle Maps; Hunter entry for November 30.

HUNTER

Therm. in air at 7.a.m. 38° in river water 60° & at 3 p m. 57° Calm Clear Weather

We found this morning that the river had risen about 1½feet during the night Set off about 7 A.M. The Banks now in many places exhibit to veiw rocky shaggy banks, these rocks are composed of hard free stone, suitable for Whetestones or grindstones of a fine grit. The river continues to increase its velocity, the land on each side grows higher[,] Rapids, small cataracts & shoals more & more frequent. about half past ten A M. passed the Fourche de Cadeau (forks of the Cadeau nation) on the left hand & an hill 300 feet high the river being 100 yds wide & at ¼ past two p.m. passed Bayu de Roches (Rock Creek) on the left.[86] Mr D.'s observation for the latitude was ⊙ ap. mer. d. alt 67°, 25′, 30″ Ind. error +13′,42″ Latitude found 34°.11′.37″ My eyes & fingers still preventing me from taking an observation. distance made this day 7 miles & 28 perches. Our Pilot having informed us of another Saline, or salt spring called Saline de Bayu de Roches we stopped about ¼ before three p.m. when I accompanied him to veiw it, taking a couple of men with spades & kettles for the water & travelled on the right bank of the river in a S.W. direction going up the river edging inland for about 1 mile, when we came to an almost dry Bayu, in the bottom of which was a large clay flat, the clay in one place tasted brakish & we began to dig[,] in about the depth of 4 feet cheifly thro blue clay[,] came to quicksand & water of which we carried to the camp the kettle full[.] it tasted salt & strong, tho not so much quite as the former By the pese liqueur of Baumè it shewed 2½ degrees below 0. Ten quarts being excicated to dryness yeilded about 7 ounces salt which besides sea salt shewed bittern of Lime or Magnesia, It proved deliquescent, tho not quite so much as the former water. Its specific gravity was found as before viz by weighing the same bulb in it: Its weight then was in the salt water 5 dwt 4.13/16 gr Specific gravity 1.0176

86. DeRoche Creek rises in Hot Spring County and runs through northeastern Clark County. It joins the Ouachita approximately 1 mile north of the Caddo-Ouachita confluence. DeRoche Creek forms one of the borders between Clark County and Hot Spring County. The hill beside the Caddo-Ouachita confluence is approximately 60–70 feet high. Recording this in his journal at the end of the day, Hunter may not have correctly remembered the height of this hill. Arkansas Quadrangle Maps.

December 1804

Saturday 1st

DUNBAR

Thermr. in air 32°. in river water 54°. Clear—calm—river fallen 18 inches. The morning was cold & damp; we passed a considerable Island on the right about ¾ of a mile in length, called 'Isle du bayou des roches' (rocky creek Island)—we were greatly impeded this day by rapids, it was with much difficulty, some hazard, & great exertion of the men, that we ascended some of the rapids: we passed several points of high land full of rocks and stones, much harder and more solid than we have yet seen; the rocks were all silicious, and we began to observe, that their fissures were penetrated by sparry matter: indications of iron were frequent, & even fragments of poor ore, but no rich ores of that or any other metal have presented themselves to view. Some of the hills appear to be well adapted to the cultivation of the vine, the soil being a sandy loam with a considerable proportion of gravel & stone and a superficial covering of good vegetable black earth: the natural productions were sufficiently luxuriant, consisting of several varieties of oak, Pine, Dogwood, Holly &c with a scattering underwood of Whortleberry, Hawthorn, China-briar and a variety of small vines. It is probable that a skilful Vigneron, who shall undertake the establishment of a Vineyard in a well-chosen position in this neighbourhood, will find his labors amply compensated; the market of New Orleans is at hand, where his wines (if good) may be immediately sold and paid for at a high price.[1] At noon we were detained upon a very bad rapid & shoal, by which

1. "Silicious" refers to the silicates, which make up one of the largest groups of minerals. They are very difficult to identify without detailed training. Dunbar's "sparry matter," as in feldspar or rocks, contains flakes or small crystals of silica. The material can also be described as substances containing silicate minerals; such substances include microcline orthoclase or potash. Chesterman and Lowe, *Field Guide to Rocks and Minerals*, 789; Parker, *Dictionary of Geology and Mineralogy*, 276–277. The "Hawthorn" could have been one of sixty species that are native to Arkansas. Hawthorns are small trees or bushes and are often called "under trees." They include the barberry hawthorn (*Crataegus berberifolia*), the blueberry hawthorn (*Crataegus brachyacantha*), the pear hawthorn (*Crataegus calpodendron*), the cockspur hawthorn (*Crataegus crus-galli*), and the bittmore hawthorn (*Crataegus intricata*). Moore, *Trees of Arkansas*, 81; Little, *Field Guide to Trees*, 461–488. A "Vigneron" is a vine-dresser or vine-grower.

we lost the opportunity of making a meridian observation: In the evening also we landed a little earlier than usual at the foot of a long and difficult rapid, which we did not think it prudent to encounter so late, from the danger of getting fast upon it all night: we are now encamped upon the declivity of one of those hills about 150 feet high, commanding a fine prospect both up and down the river, & will at a future day become a rich Vineyard. Thermr. at 8h. p.m. 35°. Extremes 32°–58°. Made this day 7 miles 148 perches.

HUNTER

Therm at 7. a.m. in air 32°. river water 54°[.] at 3 p.m. 58° & at 8 p.m. 35°. Clear Calm. Set of[f] a little past 7. a.m. when we found the river had fallen 1½ feet perpendicular during the night. We immediately passed on the right an Island & rapid about ¾ of a mile long called Isle de roches, rocky Island.[2] The river now presents a series of shoals small cataracts & rapids with peices of level water between each, of more or less length according to circumstances; Sometimes these are occasioned by accumulations of gravel which in many places are forced to the sides of the river ten or twelve feet in heighth, & in others the river is pricipitated over ledges of Rocks thro which it has wore small channels, Upon the whole as far as we have come, the navigation for boats made of a proper construction capable of carrying five or six tons, cannot be called difficult: & with a very moderate expense might be improved in one season to admit boats of twice or thrice that burthen. The gravel on the beaches now becomes larger or rather moderately small stones amongst which are found many peices resembling hones, Turkey Oil stones, & whetstones of various colours & degrees of fineness in the grit. After various interruptions from the above mentioned impediments came about ¼ past noon to the Bayu de lisle de Mallon on the right.[3] Made no observation for the latitude this day.

2. The island noted in Hunter's journal is probably Morrison Island (also called Watermelon Island) in Hot Spring County between the communities of Friendship and Donaldson. Dunbar in his entry of this date calls the island "'Isle du bayou des roches' (rocky creek island)." It is important to note that most of what Hunter recorded in his journal since his accident was either reconstructed from memory at a later date or partially copied from Dunbar's entries. Arkansas Quadrangle Maps; "Dunbar Trip Journal, Vol. I," December 1, 1804.

3. "Turkey Oil stones" is a nineteenth-century term for certain whetstones. In this case the stone is novaculite, which was and is used mainly as a sharpening stone. Novaculite, also known as oilstone or whetstone, is a sedimentary stone, a recrystallized chert that is composed mainly of microcrystalline quartz. According to the Arkansas Geological Commission, there are two categories of these stones in the state: "Arkansas" stone and "Ouachita" stone. Novaculite is mined today in moderate quantities. McFarland and Bush, *Geology of Arkansas*, 119–134; Griswold, *Whetstones and the Novaculites of Ar-*

At half past 3 p m came opposite to a saline which lay about 2 miles off from the river on the left. but as the winter had set in & our stock of provisions were near a close, we did not go to see it being eager to push on up the river as fast as convenient. Made this day 7 miles, 148 perches; The land, Timber &c much the same as yesterday, everything shewing that we are approaching a more elevated country.—

Sunday 2nd

DUNBAR

Thermr. in air 30°. in river water 50°. Clear—calm—river fallen 4 inches. Continued our voyage and passed over a series of strong rapids, which opposed us untill the hour of breakfast. The Country appears now to wear a new aspect; high lands and rocks frequently approach the river; the rocks are extremely hard, and altho' the grain resembles that of free-stone, yet the stone is hard enough to be used for the purpose of hand-mill stones, to which object it has been applied; the river beaches also exhibit a great variety of fragments of flint and other stone of the most solid kinds; the quality of the land seems to improve, the superficial stratum of Vegetable earth being of considerable thickness (from 6 to 12 inches) and of a dark brown color mixed with loam and some sand; at 2½ h p.m. passed a rock on the margin of the river consisting of blue slate, which we shall probably find time to examine on our way down; more of the same is to be seen higher up. About a league from the river a little above the slate quarry is a considerable plane called 'prairie de Champignole,' often frequented by Buffalo; some salt licks are to be found near it, and in many situations on both sides of this river at small distances from it, we are informed that Salines or salt-licks exist which may be rendered very productive; when this river comes to be settled, so necessary an article as marine salt will therefore be in sufficient abundance for the consumption of a full population.[4] We

---

kansas, 443; Parker, *Dictionary of Geology and Mineralogy*, 179; Bates and Jackson, *Dictionary of Geological Terms*, 351. The "Bayu de lisle de Mallon" may have been the confluence of Ten Mile Creek and Black Branch on the east bank in Hot Spring County. Dunbar does not mention the bayou, but in the official maps drawn by Nicholas King (and of which Dunbar approved), the stream is identified as "Bayou de Isle de Millon." Arkansas Quadrangle Maps; King, "Map of the Washita River in Louisiana."

4. The "freestone" is a late-eighteenth-century and early-nineteenth-century name for a thick-bedded or even-textured sandstone. It is usually light gray or tan in color and is easily carved. Freestone was used extensively for building during this period. The White House was constructed of a similar stone. Flint or chert may range in color from white to black, yellow, gray, red, brown, and green. Blue slate is a metamorphic rock that is bluish-gray in color. Bates and Jackson, *Dictionary of Geological Terms*, 196, 85, 471; Parker, *Dictionary of Geology and Mineralogy*, 102, 54, 279. The "prairie de Champignole" is possibly an area near present-day Social Hill, Arkansas, in Hot Spring County. Dunbar labeled the

are greatly impeded today by rapids and were unable to get ourselves landed in a situation favorable enough to make an observation for the Latitude before it was too late. We encamped just below some rapids which we are to encounter in the morning, upon excellent level and rich land, being almost entirely an Oak forest; it is not improbable that this land is sometimes subject to inundation, having the appearance of alluvial Land which has acquired permanency & stability, it is now at least 20 feet above the level of the river water. Thermr. at 8h p.m. 38°. Extremes 30°–59°.

HUNTER

Therm in Air at day light 30°[.] in river 50°[.] at 3 p.m. 59° & at 8 p.m. 38°. Calm weather & clear

Set off about half past 7 a.m. Find the water of the river fallen 4 Inches during the night. Our course all this day was interrupted by a series of rapids, ripplings, & shoals which occupied us so much that we did not take time to make an observation for the latitude, besides when it came near the time, both banks were so steep as not to afford a proper place. At about ¼ past 2. p m. passed a quarry of imperfect Slate near the waters edge under a stratum of stone of the nature of freestone fifty feet thick. The layers of slate were perpendicular near which was a creek on the left, in few minutes more passed Isle de Chevreuil (Deer Island)[.] next passed Bayu de prairie (Medeow Creek[)].[5] The Hills of rocks now come to the river's edge except here & there a stripe of low land enterveen This hills are cheifly covered with Pines & the low land with a variety of all sorts of timber formerly mentioned with many others unknown; between these hills & broken grounds are many fine tracts of land, tho not in extensive bodies contiguous. The soil on the low lands is tolerably good for cotton, wheat Corn &cc.

This day made only 6 miles 118 perches.

Monday 3rd

DUNBAR

Thermr. in air 38°—in river water 48°—clear—calm—river fallen 8 inches. Continued our voyage with favorable water until breakfast, after which we encountered a great many very bad rapids during the remainder of the day; some

---

area "Bayu de prairie de Champignole" in his "Journal of a Geometrical survey." It seems apparent that Dunbar received much of the information concerning place names and adjacent land features from the expedition's pilot, Samuel Blazier. Arkansas Quadrangle Maps.

5. Prairie Bayou enters on the west bank 1 mile south of the community of Social Hill in Hot Spring County. Arkansas Quadrangle Maps.

were so difficult, that it was impossible to ascend without sending the greatest part of our people ashore with a good rope, & sometimes they were obliged to walk in the water; the exertions of the Soldiers on some very difficult and trying occasions were equal to every thing which could be expected, and exceeded greatly my expectations: at noon we had a good observation about 4 miles below the 'Chutes' (falls) Latitude deduced 34° 21′ 25″.5 we were now anxious to see the famous Chutes, which it was supposed at the Post, we should never be able to pass with so large a boat. The land on either hand continues to improve in quality; there appears to be in general a superficial stratum of good earth of a dark brown color, upon which vegetation is sufficiently luxuriant; hills frequently arose out of the level country, full of rocks & stones, generally of an extremely hard flinty kind, often resembling the Turkey oil stone, of this kind was a promontory which came in from the right hand, a little before we arrived at the Chutes: this promontory presented some appearance at a distance, of the ancient ruined fortifications & Castles so frequent in Europe, the effect was greatly heightened by a flock of swans which had taken their stations under the Walls which rose out of the Water; as we approached the Birds floated about magestically upon the glassy surface, and in tremulous melancholy accents seemed to consult each other upon measures of safety, the ensemble produced a truly sublime picture: several masses of the same hard rock insulated by the river conveyed the idea of redoubts and out-works; we expect to visit this place in our descent. A little after 4h p.m. we arrived at the Chutes.[6] We found these falls to be occasioned by a chain of rocks of the same hard nature with those we had just seen below, here they extended quite across the river, the water making its way over the chain thro' a number of breaches, which by the impetuosity of the torrent had been worn out of the rock: this chain seemed to proceed from a lofty rocky hill on the left side the appearance of which conveyed the idea, of its having been cut down by the abrasion of the waters to its present level: the various breaches thro' which the water poured, were so many cascades, thro' one of which it was necessary to pass; otherwise the Barge must remain below the Chutes: it was quite uncertain which of the Cataracts ought to be preferred; it was also doubtful whether our barge (9 feet wide) could find sufficient breadth & depth

6. Two decades later, this site became a major river crossing for the Southwest Trail (later called the Military Road). In 1832 the community of Rockport was founded at the Chutes by Samuel A. Emerson, and by 1844 it had become the county seat of Hot Spring County. A ferry was operated at the site for many years. The first river bridge in Arkansas was constructed across the Ouachita River at Rockport; however, the bridge was destroyed one year later by flood waters. The Chutes are located today near the town of Malvern, where Interstate 30 crosses the Ouachita River. "Early Roads of Hot Spring County," 34; "History of Hot Spring County," 3.

of water clear of pointed rocks to pass over the Chutes. We came up to the rocks & stoped between two of the Cascades, & sent a couple of Men with a small Canoe, who crept along shore & got above the Falls, they made fast a rope to a tree, and letting themselves gradually down by the same rope, came on board in great safety; having now got a number of hands ready to haul in upon the rope, we employed the remainder with poles to give a proper position to the Barge & to guide her into the best passage: we accordingly entered one of the Cascades, but after many fruitless attempts we found there was a deficiency of water; with some pointed rocks which opposed our passage; we therefore dropped down a little way, and moved laterally by poling to a second Cataract much more considerable than the one we had just attempted: the rolling impetuosity of the water is not easy to describe, above and below the fall there was a rapid descent, but just at the fall there seemed to be a step of nearly one foot perpendicular; difficult & dangerous as this place appeared for a frail bark like ours, we were determined to make the attempt & we lost no time in entering the strait, in which our Barge soon stuck fast at the bows, we then concluded it would be impossible to pass; it seemed that an inch or two were just wanting to our success; we however continued our efforts by moving from side to side by the stern, while great efforts were making upon the rope; we perceived a small advancement by every new exertion, our hopes revived, the Barge was in this manner forced half way thro' the Cascade, & now she seemed so completely wedged into the narrow passage, that every effort to stir her in any direction proved ineffectual; the water tho' extremely rapid was not deep & we got four of our boldest men into the water at her bows, as far as possible from the suction of the fall, who by feeling for rocks on which she rested, & raising her sides with all their might, enabled us to advance a step or two farther, beyond which it seemed impossible to move: it was now night, the stars were visible, the water was cold, and altho' the weather was not freezing, it was far from being mild, the thermr. being a 45°; we now repented that we had make the attempt to pass so late in the evening, & wished we had delayed until the morning; at the same time the river was falling, & it seemed not proper to defer the attempt, lest we should not get above the Chutes until another swell of the river: in this situation we determined to lighten the Barge, by sending all the men, except four, ashore to haul upon the rope, while the 4 who remained were with hand levers to endeavor to raise up & lighten the bows of the vessel: the first man who went out discovered, that by the violence of our exertions the rope was beginning to give way & that one of the three strands of which the rope was composed, had actually parted; we were now in a perilous situation, for if the rope had separated, no force on board could have prevented our being dashed to pieces upon the rocks: we immediately ordered every man on board to his pole to support the boat; in the mean time a man was dispatched thro' the water with the end of a rope from on board, which being made fast to the same tree, we were again placed in a state of security; we now sent the other men on shore as had been intended, who gaining a firm footing and exerting themselves with great vigor soon extricated us and drew us safely ashore, greatly rejoicing to find ourselves without accident above the 'Chutes': we are encamped under the incessant roar of the cataracts, which resembles nothing so much that I have heretofore witnessed, as the horrid din of a hurricane at New Orleans in the year 1779: the course of the chain of rocks across the river is nearly S.W. and N.E.—Made this day 7 miles 218 perches—Thermr. at 8h. p.m. 44°—Extremes 38°–59°.[7]

HUNTER

Therm in Air at 7. a.m. 38°. river water 48° At 3 p.m 59° & at 8 p m 44°. Weather Clear & calm

Set off about ¼ past 7 A.m. The river continues to fall. Last night it fell 3 Inches. Passed on the left Bayu de L'eau froide (cold water creek) about half past 8 a.m.[8] The current is generally very strong, passing rapids & shoals & nearly level ground in succession, as yesterday. Landed to observe & found the sun's double Apt. mer. alt to be 66°.27′.20″ Ind. er.—1′.0.5″. Lat. 34.21.11.2 About 4 p m. passed an high steep rocky hill on the right with a continued ledge of Rocks crossing the river forming small cataracts or little falls trenching about S W & N.E. thro clifts of which the water had wore several small passages, thro one of which we hauled our boat with some difficulty The river at this place is about 200 yds. wide

About 4 p m came to the great falls or Chuttes & hauled our boat over it with considerable exertions, as the water in the whole might have a fall of about 4 feet in ¼ of a mile The ledge of Rocks crossed the river nearly in the same direction as

7. The group acquired the canoe mentioned in this entry during their stop at Fort Miró. The hurricane in New Orleans that Dunbar is referring to occurred on August 18, 1779. The Spanish governor Don Benardo de Galvez assessed the damage by saying, "There are but few houses that have not been destroyed, and there are many wrecked to pieces; the fields have been leveled; the houses of the near villages, which are the only ones from which I have heard at this time, are all on the ground." Dunbar conducted a series of experiments during this hurricane to attempt to discover the true nature of such storms. Because of his extensive study, the hurricane is known as "William Dunbar's Hurricane." Roth, "Louisiana Hurricane History"; Garvey and Widmer, *Beautiful Crescent*, 49.

8. "Bayu de L'eau froide (cold water creek)" could be Blakley Creek, which meets the Ouachita on the west bank just north of Social Hill in Hot Springs County. Hunter resumes taking observations on this day, stating, "my hands & eyes were much recovered from the accident with the Pistol." Arkansas Quadrangle Maps; "Hunter Official Report," 43; McDermott, "Western Journals of Hunter," 100.

at the smaller falls above described in the same manner differing only in degree. Came this day 7 miles 218 perches.

Tuesday 4th

DUNBAR

Thermomr. in air 36°. in river water 48°—clear—calm—river fallen 2 inches. Immediately above the Chutes, the water possesses little or no Current, owing no doubt to its depth & breadth & we went on without opposition untill after breakfast; about 8h a.m. passed a ledge of very hard freestone rocks with moderate current: this reach is spacious being not less than 200 yards wide & is terminated by a high rocky hill (about 350 feet perpendicular) crowned with beautiful pine woods, a fine situation for building: at 10½ h passed a bald hill on the left being chiefly uncovered rock, and arrived at the foot of a most tremendous rapid full of breakers, the passage being studded with pointed rocks of all magnitudes, which raising their rough heads above water, seemed to threaten with destruction the unwary voyager who should presume to attempt their passage; this place appeared to me much more difficult and dangerous than the Chutes, the water descended along a plane of considerable inclination with a most impetuous velocity, the spray & white foam dashing over the rocks, occasioned a very perceptible mist or vapor which spread about at a small elevation, it is probable it might ascend into the atmosphere at a higher temperature.[9] We stopped to contemplate this embarrassment & ordered out a rope, which was carried along shore by a certain part of the people, the rest using their poles on board; we made many fruitless essays to pass upwards by several openings near the shore; at length we attempted the center of the Cataract where the current was the most violent, but the water deeper, & by very great exertions we got over into moderate water, having consumed 1 ½ hour in making about ½ mile; 300 yards of this distance is difficult & perilous, the greatest prudence with unceasing exertion being indispensibly necessary to the safety of such a barge as ours. We landed above this rapid & by a good observation found the latitude to be 34° 25′ 48″; on our right stood a high rocky hill crowned with very handsome Pine-woods; the strata of this rock were inclined 30° to the Horizon in the direction of the river descending; this hill may be from 300 to 350 feet high: we have now frequently the hills touching the river on both sides; a border or list of green Cane skirts the margin of the river, growing out of the alluvial soil, beyond is generally a high & sometimes barren hill. At 2h p.m. we passed a hill on

9. This "bald hill" and the rapids were possibly near Narrow Mountain and in the vicinity of Jones Mills in Hot Spring County. Arkansas Quadrangle Maps.

the left containing a great body of blue slate, in some places hanging over the river; a little farther came to another rapid or cataract, which appeared if possible more terrible than the last, the descent of the water was extremely precipitate; from the very irregularly undulating surface, it was evident that the bottom was composed of innumerable fragments of rock, many of which just shewed their heads out of water; we halted on the right shore & sent up our rope, but after many fruitless & some dangerous attempts, in which we were always repelled by the rocks, we were obliged to give up the expectation of passing up on that shore; we therefore had recourse to the expedient of swinging the barge into the middle of the river & by the aid of the rudder and the exertions of poling, we with some difficulty got hold on the opposite shore, notwithstanding that the rope was caught under a rock in the middle of the river. We hauled the rope on board and sent it up the shore, and passed up the most violent part of the rapid: we ascended a second rapid of less importance and encamped, our people being almost exhausted with fatigue; on the right is a creek called 'bayou de la saline';[10] about a league up the Creek is a salt-lick, which by digging yields salt water resembling what we have already seen; there is also blue slate near the same situation. This afternoon our hunters shot twice at a Buffalo & wounded him severely, the blood flowing as he run, but he escaped. Our tents were pitched on a stony and gravelly beach, they were completely paved with stones of a great variety in kind, color and size. Thermr. at 8h. p.m. 36°—Extremes 36°–50°. Made only 4 miles 164 perches.

HUNTER

Therm. at 7 A.m. 36° river water 48°. at 3 p.m. 50° at 8 p.m. 36°.

Set off after 7 A.M. The river has fallen 2 Inches more during the last night, The weather still clear & calm. About ¼ past 8 a.m passed another ledge of hard freestone rocks crossing the river, simular to those of yesterday, but not so difficult to get over, & as usual high rocky hills with scattering pine trees to the very top; one in front 350 feet high. About ¼ past 10 a m passed a Bare hill of stones on the left afterwards came to a very voilent rapid. & at ¾ past one p.m. came to another rocky pine hill on the right about 300 feet high—about 2 p.m. came to hills of imperfect slate bluish & in about ¼ of an hour came to a considerable rapid which took much time & exertion to overcome; Indeed we are now into the hilly country where the river seems to rise like the steps of a stair making small falls over the

10. This may have been Cove Creek, which joins the Ouachita below Remmel Dam (which forms Lake Catherine) and near the community of Jones Mills in Hot Spring County. The expedition was now entering an area near the present site of Remmel Dam called "the Narrows." The river below the dam is still dotted with rapids in this area. Ibid.

rocks of from a few Inches to two or more feet in a very short space. These rocks also display considerable variety;[11] in some parts they are very hard having a flinty appearance, some white, others milky opake, & others again cream coloured & some black, all with a very fine grit, hardly to be distinguished from the Turkey Oil stones, so much used to sharpen edge toolls; Some varieties of hard & soft, coarse & fine grit freestones, & others of an hard blue excellent cutting whetstones. ⊙ d. mer. Apt. Alt 66°, 2°,45 Ind er.—1′,4″ Lat. 34°, 25′,21.7″

We encamped near a bayu on the right called Bayu de Saline Made this day 4 miles, 164 perches.—

Wednesday 5th

DUNBAR

Thermr. in air 23°—in water of the river 47°—very serene—calm—river fallen 2 inches. The morning tho' cold was agreeable, the air being very dry: all night we hear'd the roaring of a Cataract, which we were to encounter this morning; we were presently at the foot of it; the violence of the rapid was about 100 yards in length, & as I sat in the cabin of the barge with my eye lowered to the level of the still water of the reach above the rapid, I found there was a fall of 4½ feet; we sent our rope ahead as usual; but made very little progress for some time, the rope being entangled among sharp rocks which endangered its cutting, the consequence of which might have been fatal to all on board the barge, with the entire destruction of the boat and every thing contained in it; the passage was full of breakers and studded all over with pointed rocks, so that it was necessary to guide with the utmost care, to be able to pass clear of those unfriendly obstacles: the men on shore exerted themselves greatly, but were frequently obliged to rest, & the boat was often at an entire stand, at length the rope escaped from the rock which bent it out of its course, and we began to move up very slowly, frequent rests were necessary & in about an hour and a half we ascended above the rapid which was only about 150 yards in length; a small island here divided the river into two channels, we took the shortest tho' the most rapid, because it was most favorable for the use of the rope: The french hunters have denominated this place 'La Cascade' on account of the rapidity & great fall of the water within so small a space: below the Cascade, we had rocky hills on both sides, the quality very hard freestone, but that found in the bed of the river which was rolled down by the floods from the upper countries, was very frequently of the hardest flint, sometimes resembling the Turkey stone.

11. For more information on "Turkey Oil stones" (novaculite here), see Hunter entry for December 1.

Being embarrassed upon the rapids we could not land to observe at noon. we were obliged to use the rope a second time to ascend a very impetuous rapid, altho' much inferior to that of the morning: at 1h 45′ p.m. passed a creek on the right called 'fourche au Tigre' (Tiger creek) 4 computed leagues from the Chutes; it would seem that the Early Hunters have calculated their leagues by the time required to ascend the stream, & not by distance, as it appears from our calculation, that the distances passed over are frequently not above half those by computation: we now carry the rocky hills with us very often on both sides; rich bottoms nevertheless are not infrequent, & the upland is sometimes of moderate elevation & tollerably level: we are informed that up the fourche au Tigre, & other Creeks there are many extensive tracts of rich level land. The stones and rocks we now meet with are chiefly penetrated along their fissures by sparry and chrystaline matter. Last night a band of Wolves howled in our neighborhood a great part of the night.[12] Turkeys become now much more abundant & less difficult of approach than below, our hunters generally kill some every day. The opposition on the river was to day so great, that we made only 3 miles 128 perches, altho' by the old computation our days voyage was little short of 3 leagues. Thermr. at 8h. p.m. 38°. Extremes 23°–56°.

HUNTER

Therm at daylight in Air 23°. in river water 47°. at 3. p.m. 56° & at 8 p.m. 38°

Set off about half past 7. a.m. Find the river fallen 2 Inches during the night. The weather clear & calm.—We have now high hills on both sides consisting of those sorts of stones described yesterday; before breakfast passed not without some extra exertion a very smart small cascade, with a fall of about 4 feet in about 80 yards. The rapids, cataracts in minature, peices of level water, & shoals, with

12. "La Cascade" and the series of rapids within are known as "the Narrows"; they are near Jones Mills and Remmel Dam in Hot Spring County. Dunbar's "'fourche au Tigre' (Tiger Creek)" is today Tigre Creek in northern Hot Spring County. In Hunter's journal entry for this date, he labeled the stream "Fourche a Tigre (Tiger Creek)." Tigre Creek's original confluence is now under Lake Catherine in northern Hot Spring County. Arkansas Quadrangle Maps. Plentiful populations of red wolves (*Canis rufus*) inhabited Arkansas and Louisiana forests prior to the 1830s. LaHarpe had seen and heard wolves during his Arkansas expedition in 1721–1722. Beginning in the late 1830s, bounties began to be placed on wolf pelts in Arkansas, and the number of wolves began to decline. The rapid drop in the red wolf populations occurred during the 1920s and 1930s. By the 1930s, coyote-wolf interbreeding began to threaten the animal, and the pure species was thought to be extinct in Arkansas by the middle to late 1970s. Smith, "Exploration of the Arkansas River by Bénard de LaHarpe"; T. H. Holder, *A Survey of Arkansas Game*, Federal Aid Publication Project, 11-R (Little Rock: Arkansas Game and Fish Commission, 1951); Sutton, *Arkansas Wildlife*, 110; Sealander and Heidt, *Arkansas Mammals*, 14, 22–23, 25, 36, 192, 194, 198–200.

rocky hills on each side, now succeed each other so fast that it seems unnecessary to particularize every one. The driftwood & brush caught on the banks shew that the freshes at high water rise about 25 feet perpendicular above the present level of the river.

About noon passed Fourche a Tigre (Tiger Creek) which our Guide says has a considerable quantity of good land on its banks. Got no observation for the latitude this day, we were so engaged with surmounting the difficulties in the river, yet we made only three miles 128 perches.—

Thursday 6th

DUNBAR

Thermr. in air 45° in river water 48°—cloudy—light wind at S.W. river fallen 2 inches. We were encamped last night upon excellent land, tollerably level, and of a good dark brown or blackish soil at the surface, about 12 inches deep, lying upon a yellowish loam; the growth of timber is large and handsome, chiefly a forest of Oak with an admixture of ash, hickory, elm &c, a field of corn has been formerly cultivated here by one of the hunters during the summer recess from hunting. This morning the Weather being cloudy we apprehended rain, but hoped to reach the 'fourche of Calfat' (Caulker's creek) the point which is to terminate our navigation, & encamp before bad weather;[13] we accordingly proceeded on without material interruption until the hour of breakfast, carrying with us high hills on the left and good level lands on the right, subject perhaps to be inundated: at 9h a.m. arrived at the foot of a very long precipitous rapid, it seemed to be divided into four steps, one of which was at least 15 inches perpendicular exclusive of the inclined plane above and below, the whole could not be less than 5½ feet perpendicular from the beginning to the end, which was about 400 yards, altho' the swift water continued half a mile: the rope was carried along the bank as usual, and many stops were made upon the rocks before coming to the great fall; at last the barge entered between two high rocks, the men exerted themselves vigorously both on shore and aboard; the barge appeared to be ascending an inclined plane of 12 or 15 degrees; great exertions were necessary, she however passed without touching any other obstacle but the impetuous torrent and in a few seconds was drawn into moderate water to the infinite joy of the whole party; upon another part of the rapid higher up, we got upon a rock, which seemed to serve as a pivot, upon which the boat turned as a Center; after reiterated exertions, we could neither advance nor re-

13. The "fourche au Calfat (Caulker's creek)" is today Gulpha Creek in Garland County. Gulpha Creek flows into the northern shore of Lake Catherine, and the site where the group later makes their base camp on the Ouachita is undoubtedly beneath the lake today.

treat, we therefore unloaded about one quarter of the cargo which enabled her to pass up without difficulty: we immediately re-loaded having spent three hours in getting over this rapid, and proceeded a quarter of a mile farther to Ellis' Camp a little below the 'fourche au Calfat' (Caulker's creek): Here terminates our voyage upon the river upwards, for the present. Our pilot considers this the most convenient landing, from whence to transport by land our necessary baggage to the hot-springs, the distance being about three leagues. There is a creek about 2 leagues higher up, called 'bayou des sources chaudes' (hot-spring Creek) upon the banks of which the hot springs are situated, about 2 leagues only from its mouth, but the road is very hilly and therefore less eligible than the path from this camp or landing, which is almost a level road.[14] Upon ascending the hill to encamp we found the land extremely level and very good, with some plants in flower & a great many evergreen vines; the forest is chiefly oak with an admixture of other timber as before mentioned: soon after we arrived it began to rain, we were however tented before it commenced. Thermr. at 8h p.m. 56°. Extremes 54°–67°. Our short voyage this day was only 2 miles 32 perches.

HUNTER

Therm in Air at day light 45°. in river water, 48°. at 3 p.m. 67 & at 8 p.m. 56.

Set off at 40 minutes past 7 a m The weather being cloudy Wind S.W. In about an hour passed some hills to the left & tollerably good Land to the right, The current is strong & shoals so rocky & intricate that we were about three hours in passing a sort of cascade. & in about ¼ past one p m arrived at Ellis's camp a few hundred yards below the fourche a Calfat (Calkers fork) where we encamped. The course up the river is here S W. Came this day 2 miles & 32 perches.—

14. In reference to "Ellis' Camp," many places along the Ouachita, and indeed along many rivers in the lower Louisiana Purchase, were named for the trapper-hunters who deposited their wares and skins at these sites. In his "Western Journals of Hunter," John Francis McDermott said he believed that Ellis may have been an A. Ellis whom Thomas Rodney referred to in a letter dated October 20, 1804. This letter stated that an A. Ellis had traveled to Hot Springs with several other planters for medicinal purposes. Alternatively, the Ellis of Ellis' Camp could be John Ellis, who described the Ouachita and Hot Springs region to Joseph Macrery of Natchez in "Description of the Hot Springs and Volcanic Appearances in the Country adjoining the River Ouachita, in Louisiana." John Ellis's account of active volcanic features in the Hot Springs region makes his descriptions highly suspect among most scholars. A John Ellis also served as the president of the Legislative Council for the Mississippi Territory. In a letter dated June 1798, Dunbar referred to John Ellis as "the wheelwright." William Dunbar to Dinah Dunbar, Camp, June 1798, Dunbar Papers, Jackson; McLemore, *History of Mississippi*, 1:174. The "bayou des sources chaudes" is Hot Springs Creek. Today, the creek runs under "Bathhouse Row" and Central Avenue in downtown Hot Springs, Arkansas.

Friday 7th

DUNBAR

Thermr. before sun-rise 38°. in river water 47°. Cloudy—Wind N.W. river risen 4 inches. In the morning Doctor Hunter with the Pilot &c went to view a salt-lick about a mile to the West of our camp but found no salt water; the clay was extremely stiff and difficult to dig: after breakfast dispatched the Pilot with the greatest part of our people with their own baggage & some provisions to encamp at the hot-springs, hoping to find Cabins there sufficient to hut our party with orders to return early next morning so as to take out a load of more baggage and instruments.[15] Took the sun's meridian altitude: Latitude deduced 34° 27′ 31″.5—Thermr. at 3h. p.m. 50°—the weather cleared up about 9h. p.m. and became very serene and cool with wind at N.W. some venison and turkey were procured by the hunters: altho' we have frequently seen the tracks and other marks of buffalo, we are hitherto disappointed in killing any of them.

HUNTER

Therm I Air 38°[.] in river 47°. at 3 p m 50° & at 8 p m 24°. Being now by estimation about 9 miles distant from the Hot Springs by land & as near as we could approach them by Water. The Pilot with a couple of men were detatched to visit them, to examin the path &cc. In the mean time the following observations were made to ascertain the Latitude viz ⊙ d. Apt. Alt 65°.16′.40 Ind. error—0.1′,15″—Lat found 34°,26′ 45″. An accident happening to the pedestal of the sextant at the moment the sun was on the meridian may have caused a small error of some seconds—[16]

The ground whereon we are now encamped is elevated above all freshes, a tol-

15. The fact that Dunbar mentions cabins in this entry seems to show that he had heard of the possibility of such dwellings at the site. He might have gained the information from his pilot, Blazier, or from others. The buffalo (*Bison bison*) was plentiful in Arkansas until the late eighteenth century. As trapping and hunting increased along the major Arkansas rivers, these animals were some of the first to experience decline. The last buffalo herd in southern Arkansas was killed in the Saline River bottoms around 1809. Only smaller herds survived in remote areas of eastern Arkansas until they were destroyed during or just before the Civil War. The track of the buffalo is 5½ inches long and 4 inches wide. Sutton, *Arkansas Wildlife*, 3–4, 8, 10, 13–16, 27–28, 33, 35–36, 57; Sealander and Heidt, *Arkansas Mammals*, 21, 23, 37, 242–243, 250; Murie, *Field Guide to Animal Tracks*, 254–255, 309.

16. According to Dunbar's entry for December 6, Blazier had calculated the distance at "three leagues," or approximately 9 miles.

Notice that Dunbar's reading for December 7 was different from Hunter's. Dunbar later took additional observations and recorded what he considered as the fixed latitude for the Ellis's Camp site (see Dunbar entry for December 8).

erably level spot[,] soil but thin, stones of various colours tho cheifly Silicious; many of them give very good fire with steell; we found many fragments nearly in the shape of & very suitable for gun flints.[17]

Our Hunters killed a few wild Turkies & one Deer, neither of which were in good order.

Saturday 8th

DUNBAR

Thermr. in air 10°. in river water 43°—very serene—light wind at N.W. river risen 4 inches. We found the weather this morning extremely cold, the thermr having fallen lower, than we expected in this latitude, particularly at the present early period of the winter season; it is perhaps to be ascribed to the elevation of the country and neighbourhood of mountains: as we have no barometer with us to indicate the pressure of the atmosphere, we shall when we get to the hot springs, ascertain the degree of the thermometer at which water boils, from which scientific men may draw their own conclusions respecting the elevation of the land.

At 10h. a.m. our people returned from the hot-springs, each giving his own account of the wonderful things he had seen: they were unable to keep the finger a moment in the Water as it issued from the rock, they drank of it after cooling a little and found it very agreeable; some of them thinking that it tasted like Spicewood tea. The people after refreshment were dispatched with another load of necessary baggage.[18]

Took the Sun's meridian altitude again to day & found the latitude to be 34° 27′ 27″ being 4″ less than yesterday; should no more observations for the Latitude be made here, we may consider it as fixed at 34° 27′ 29″. The Thermr. at 3h. p.m. 47°. We may prepare for another cold night: a flock of swans passed us to day: we have had an abundance of venison & turkey since we landed here, sufficient to supply the whole party with fresh provisions. The bank or hill upon which we are en-

17. From Hunter's description, it appears that this may have been the site of a prehistoric campsite where ancient people fashioned novaculite tools. For more on novaculite quarries, see Early and Limp, "Fancy Hill"; Baker, "Aboriginal Novaculite Exploration"; Holmes, "Aboriginal Novaculite Quarries"; Griswold, "Indian Quarries in Arkansas," 25–26; Etchieson, "Prehistoric Use of Geological Resources"; Etchieson, "Prehistoric Novaculite Quarries."

18. Spicewood, also called wild allspice or feverbush (*Lindera benzion*), is a fragrant shrub with yellow flowers and is usually found in bottomlands. French Canadians referred to the plant as a pepper plant or "poivrier." Spicewood tea, made from the plant's young twigs, was popular during this period in the South. Indians used the plant for a variety of medicinal purposes including relief of aches, colds, and even syphilis; they boiled the leaves in water. Saunders, *Edible and Useful Wild Plants,* 144–146; Moerman, *Native American Ethnobotany,* 308–309.

camped is a least 50 feet perpendicular above the present level of the river, and therefore I presume 30 feet clear of inundation. Some hills of considerable height are in view, clothed with pine trees, but the lands around us extending far beyond our view, lie very handsomely for cultivation; the superstratum is of blackish brown color from 8 to 12 inches deep, lying upon a yellowish basis, the whole intermixed more or less with stone & gravel & fragments of blue schistus, which is frequently found so far decomposed as to have a strong aluminous taste.[19] The thermr. at 8h p.m. 26°; very serene and calm, the stars shone with uncommon lustre: in an hour more the face of the heavens was changed, a general cloud produced an intense darkness; the thermr. rose to 36°. and we expected snow or rain; after midnight notwithstanding, the clouds were dissipated, the face of heaven recovered its brightness & the Stars shone with undiminished splendor. Extremes of the thermr. 10°–47°.

HUNTER

Thermometer before sunrise 40° & in the river 43° & at 3 p.m 47°. Wind N.W. clear.

Took an observation for the latitude again this day by which ⊙ d. Apt. Mer. Alt. 65°.2′.50″. Ind. err.—1′,12″ Lat. 34°.27′.15″.

Found the river has risen last night 4 Inches.

Our people returned this forenoon from visiting the hot Springs & set out again after dinner with what they could carry to the same place, bringing information that there was a log Cabbin built there & several sheds of split boards, bark &cc, so that it was not necessary to carry our Tents; This saved much time, for even the little baggage & provisions that were absolutely necessary took them till the evening of the 12th Inst to bring out to the Springs a distance by computation of about 9 miles[20]—

Sunday 9th

DUNBAR

Thermr. in air 19°. in river water 41°. very serene—Wind moderate at N.W. river risen 2 inches. The people returned from the springs between 9h & 10h. a.m. and after some time given for repose and refreshment, the party set out again with such baggage as was immediately wanted, and Doctor Hunter and myself accom-

19. The "blue shistus" is probably alum shale or alum schist, a shale that has been decomposed by weathering. It forms a sulfuric acid similar to alumina or potash and may also contain pyrite. Parker, *Dictionary of Geology and Mineralogy*, 10.

20. This entry is a clear indication that Hunter was completing his daily entries after the fact.

panied them; the people complained of the length of the road and weight of the loads, we therefore diminished the latter: The Sergeant and one private remained in care of the Barge and her stores. We left the river camp about noon and with many delays and haults for resting we arrived at the hot springs at 4½ h p.m.—the distance is computed to be 9 miles, which we shall verify by actual measurement, probably on our return: the first six miles were in a general westerly direction with many sinuosities and the last three northerly, which courses were necessary to avoid crossing some very steep hills. We found on the way three principal salt-licks & some inferior, which are all frequented by buffalo, deer &c the soil around consisted of a white tenacious clay, probably fit for Potter's ware; hence the name 'Glaise' which the french hunters have bestowed upon most of the licks which are frequented by the beasts of the forest, altho' salt is not always to be found in such places so as to merit attention: we saw on the way recent tracts of the Buffalo and several Deer skipped along before us; we did not follow the game, being desireous of arriving at our destination before evening. The people were much fatigued with this days labor, altho' the road is by no means bad or hilly, but there is no doubt that a heavy load constantly bearing a man down must be very fatiguing upon the best of roads: the time and difficulties of moving our small baggage and provisions, altho' nothing but what is essentially necessary, to so small a distance, naturally suggests the inconveniencies which must arise in transporting over unknown mountains between the sources of the red and Arcansa rivers, baggage & provisions indispensibly necessary, with tools and implements for the construction of a boat or boats to descend the 2d river.[21] Soldiers accustomed to carry moderate loads only, would find it intollerable to transport burthens which would be thought light by a Canadian or other woodsman enured to such hardships: a little calculation will shew what ideas we ought to form upon this subject. The provisions, instruments, arms & other baggage which may be deemed indispensible for 15 persons engaged on such an expedition, i. e. what must be transported from the head of one river to the commencement of navigation on the other, are certainly not over-rated at 3000 lib; of the whole party 10 carriers are the highest number we

21. The "white tenacious clay" may be kaolin, a clay composed of kaolinite minerals and also known as white clay or bolus alba. It is used mostly in ceramics and paper. Parker, *Dictionary of Geology and Mineralogy*, 137; Bates and Jackson, *Dictionary of Geological Terms*, 279. Dunbar's reference to "the road" may mean that a well-worn path ran between Ellis Camp and the Hot Springs. He had also begun to give much thought to the problems and necessities of portage. He later expressed these reservations to Secretary of War Henry Dearborn concerning the Grand Expedition on the Arkansas and Red Rivers. Dunbar to Dearborn, the Forest, May 4, 1805, Henry Dearborn Papers, Library of Congress; Dunbar Papers, Jackson.

can calculate upon, some being necessary to guard the two camps while the scientific persons unattended would explore the environs: those 10 carriers from what we have seen could not be expected to carry for a number of days successively more than 50 pounds each (several of our people were incapable of doing so much) and ten miles to go loaded & return empty day after day even on a tollerably level road, is perhaps beyond what we can flatter ourselves with accomplishing; thus it would require at least six days to transport the baggage 10 miles, and the seventh would be demanded as a day of repose: now if the heads of navigation should be only 50 miles apart, & the passage not rugged or mountainous, it would require at the least 35 days to pass along the unknown region; and if allowance be made for such difficulties as ought to be expected including bad weather, we shall perhaps still flatter ourselves, if we expect to complete this portage in 50 days: on due consideration therefore it may be more advantageous (if the expedition is to be carried on by soldiers who cannot travel without their rations, tents, baggage & above all their execrable whisky) to explore one river only at a time. When arrived at the head of Navigation which will constitute a kind of head quarters and point of departure, the scientific men with a sufficient party may make with tollerable convenience excursions of 30, 40 or 50 miles in all directions, prolonging the time according to the fortune of procuring game, which will enable the party to reserve the provisions taken from Camp for their return: an advantage resulting from this plan would be the facility of transporting specimens of natural history meriting attention; it is evident that this benefit must, upon the other plan, be nearly given up excepting on the descent of the second river. I am not ignorant that the plan originally proposed may be carried into effect, but this must be done by persons chosen for the object, in order that it may be done with economy & in a reasonable time: Two young men of science of robust constitutions attended by four Canadian or other woodsmen inured to fatigue and who can depend altogether on their guns for subsistence may accomplish this object; they will be able to transport at once, their blankets, their arms and amunition, a little parched meal, very light instruments, such as a 3 inch sextant which may be graduated to 20″ of a degree, a pocket case with a few re-agents for mineralogical assays, and 3 or 4 days provisions in case of disappointment in finding game; (spirituous liquors must be out of the question:)[22] Such a party, each carrying a light ax for the purpose of building Canoes &c may accomplish the object proposed, upon supposition that no hostility is to be apprehended from the natives.

22. Following the conclusion of the Ouachita River expedition, Dunbar continued to question the feasibility of an expedition that must traverse overland between the headwaters of the Arkansas and Red Rivers. Dunbar to Dearborn, May 4, 1805, in Roland, *William Dunbar,* 147–149.

From the river camp for about two miles, the lands are level and of second rate quality, the timber chiefly oak intermixed with others common to the climate and a few scattering pine-trees; further on, the lands on either hand arose into gently swelling hills, clothed chiefly with handsome pine-woods: the road passed along a valley frequently wet, by numerous rills and springs of excellent water which broke from the foot of the hills: as we approached the hot-springs the hills became more elevated and of steep ascent & generally rocky; those hills are here dignified by the name of mountains, altho' none of those yet in view exceed 4 or 500 feet; it is said that mountains of more than five times the elevation of these hills are to be seen in the North-west towards the sources of the Washita river; one of those has been called the glass, Chrystal or Shining mountain, on its surface is to be found vast numbers of large hexagonal prisms of very transparent colorless chrystal, generally surmounted by pyramids at one end, rarely at both; they do not produce a double refraction: many searches have been made over those mountains for the precious mettals, but hitherto without success, so far as I can learn.[23]

We found at the Hot-springs an Open Log-Cabin and a few huts of split boards, all calculated for summer encampment, & which have been erected by persons resorting to the Springs for the recovery of their health; we shall endeavour to render our temporary lodging comfortable for the people and ourselves during the short time we expect to stay here: we are a little discouraged by the dilatory ways of the Soldiers; it is evident that to promote the advancement of an object similar to ours, they ought to be commanded by a commissioned officer, whose manners and disposition would render him an agreeable companion to his fellow laborers: it cannot be said that the Soldiers are disobedient, on the contrary they are to me uniformly respectful, but it sometimes appears that a spur is wanting, & there is no person here who treats them otherwise than with civility; there is also some appearance of design to prolong their return to new-orleans, the present service being much more agreeable to them than the duty of a garrison under the eye of their officer.[24]

23. The term *rills* refers to small brooks. Dunbar's phrase "those mountains" undoubtedly makes reference to the Ouachita Mountains, a range that consists of a series of parallel ridges that run from eastern Oklahoma to east-central Arkansas. In Arkansas they run east to Pulaski County, south to Polk, Montgomery, and Garland Counties, and north to Yell and Logan Counties. The Ouachitas are composed primarily of shale, sandstone, chert, and Ordovician limestone. Roberts, *Field Guide to Geology,* 266–269. The hexagonal prisms, or quartz crystal prisms, are found within the Ouachita range. Commercial mining and private crystal hunting continues in the Ouachitas today.

24. It is difficult to precisely calculate the location of the dwellings mentioned, but they were very near the springs, possibly in the vicinity of "Bathhouse Row" and Central Avenue today. In the "Dun-

On our arrival we immediately tasted of the hot-spring water, that is, after a few minutes cooling, for it was impossible to approach it with the lips when first taken up, without scalding: having arrived here without prejudice for or against the springs I did not discover any other taste except that of very good water rendered hot by culinary fire; some of our people pretended to have discovered cathartic properties, which must be feeble, as I have been unable to detect the existence of such a quality in the waters. Thermr. at 8h p.m. 28°. Extremes 19°–42°.[25]

HUNTER

This day set out a foot for the Springs accompanied by the Pilot & 8 of the soldiers each carrying a load of necessary articles; We traversed a level country for about a mile in a northwesterly direction, & then passed a deer Lick, which is only a place in the ground about 18 Inches deep, & about 40 feet irregularly in diameter, it is said to be licked away by the wild animals deer & buffalo, & in several places was hollowed out seemingly by their means, It consisted of a yellowish clay soil, but was not sensibly salt to my taste. In pursuing our course in the same direction about a couple miles further passed Bayu Califat[,] a small creek not boatable now here[,] & then in about 2 miles further passed three more such Licks, & came to the Bayu of the Springs, when after having crossed it we altered our course more to the northward. The soil which before was tolerable wheat land tho rather thin & cold, now becomes more elevated & stoney, the stones in this Bayu (which is only now a mill stream with many falls & unboatable) are like those last described in the Ouashita, viz some a sort of blue slate pretty hard,[26] fiter for whetestones than any other purpose, some white, some cream coloured & some blackish, silicious, resembling the various kinds of Turkey Oil stones, & some an hard species of freestone reddish or brown on the outside & white within.—This last species

---

bar Trip Journal, Vol. I," December 13, 1804, Dunbar calculated the distance from the cabin to the Hot Springs as one-half mile. The presence of the cabin and the huts is also a clear indication that the springs were used (probably infrequently) by people venturing there for medicinal reasons in the years before the Louisiana Purchase. The explorers recorded no other sign of a sustained population in the area.

Dunbar continued to emphasize the need for a commissioned officer to organize the soldiers. In connection with the later Freeman-Custis Red River expedition, Dunbar wrote to Jefferson saying that a commissioned officer ought to accompany future trips. Dunbar to Jefferson, Natchez, March 16, 1805, Jefferson Papers; Rowland, *William Dunbar,* 147–148.

25. The term *cathartic properties* refers to purgative properties for the bowels.

26. Deer and other hoofed animals commonly find dirt or clay with salt deposits and will rut out depressions after prolonged use. Presumably, the "Bayu of the Springs" is Hot Springs Creek. The "blue slate" may be the "schistus" or shale that Dunbar writes about on many occasions.

composes the upper parts of these Hills some of which are about 500 feet high[.] In about 4 miles more in the latter direction near sun set came to the high Hills which give birth to the Hot springs 4 of which Issue at the foot of an hill to the right & one on the left of the Bayu or creek. some near the waters edge, & others about from three to 8 feet above it. The water is so hot as to make it impossible for a person to hold his hand half a minute in it, causing a considerable vapor round each. To the taste it differs very little from other warm water. As soon as we get ourselves fixed & get the medicine chest from the boat, (which with all the rest of our heavy baggage we left in charge of the Sergeant) we intend to examine these waters & such other matters as shall present themselves—[27]

Monday 10th

DUNBAR

Thermr. 26°—very serene. Wind moderate at N.W.—We spent a cold night in our new lodgings, not being able to keep up a large fire in the Cabin, which is only 12 feet square without a chimney. From the complaints of great fatigue by the people, we found it necessary to allow some repose, and ordered the people to go into the river camp, there to remain during the night and return the day following with more of our baggage, directing the loads to be made still lighter: the day proved serene and fine, but as we had been obliged to leave our instruments yesterday at the river-camp, no astronomical observations could be made this day. We visited all the hot springs; they issue from the sides and foot of a hill placed on the east side of the narrow valley where we are hutted, one small spring only rises out of the face of the west bank of the creek; from the quantity of calcareous matter deposited by it it does not appear to be of long standing; a natural conduit probably passes under the bed of the creek to supply it. There are four principal springs arising immediately on the east bank of the Creek, one of which may rather be said to spring out of the gravel bed of run; a fifth smaller one is that just mentioned rising on the west side of the creek; a sixth of the same magnitude is the highest or most northerly one rising near the bank of the Creek; those are all the sources which merit the name of springs near to our huts; but there is a considerable one some distance below, & all along the creek at intervals the water oozes out or drips from under the bank into the creek, which during the present cool season in very evident from the condensed vapor which floats along the margin of the Creek, where those drippings are visible & even where none is to be seen; a statement will hereafter be given of the temperatures of the respective springs with the quantity

27. "Sergeant" is Sergeant Bundy. McDermott, "Western Journals of Hunter," 65.

of water delivered and references to their respective positions; from some slight trials, it appears that the highest temperature is about 148° to 150° of Farheneit's thermometer.[28]

In the afternoon we ascended the hill of the hot springs. It is of a conical form terminating at top with a few loose fragments of rocks covering a flat space of 25 feet diameter.[29] altho' we have said the hill is conical, yet it is not entirely insulated, for it is connected by a very narrow ridge with the neighbouring hills.

The primitive rock of this hill above the base is chiefly silicious, some part of it being of the hardest flint, others of the nature of freestone extremely compact & Solid, and of a great variety of colors; the base of the hill, & indeed for a considerable extent, is composed of blackish blue schistus, which divides into perpendicular laminae like blue slate, The water of the hot springs is therfore delivered from the siliceous rock, but this is generally invisible at the surface, being encrusted by or rather buried in the mass of calcareous matter, perpetually precipitated from the water, iron in small proportion was also deposited in form of a red calx, the colour of which was frequently distinguishable in the lime.

Under the hotest water we observed a lively green appearance, which at first induced us to suppose that Copper might be present, but on closer inspection, we found it to be a soft tender matter, perhaps a feculum deposited by the water, it may possibly be of the same nature with the green matter found in conduits or even in well buckets under pure water at common temperature respecting which a dispute arose /I think/ between Doctor Priestly and other Philosophers, whether this green matter is a perfect vegetable or only a feculum; the question is perhaps now decided (if we suppose the green matter of the hot springs to be of the same kind) for by reasoning from analogy, no vegetable can be supposed to exist in the temperature of 150°; but we must beware of presuming to set bounds to the power of Nature: we shall thereafter examine this matter with due attention; we shall only now observe, that this substance seems to be deposited by successive thin laminae.[30]

28. Hunter recorded the same temperatures. See Hunter's entry for December 10. There are forty-seven springs that emerge from Hot Springs Mountain from a geological formation called Hot Springs Sandstone. In a 2000–2001 test, the water ranged in temperature from 130° to 145°. Petersen and Mott, *Hot Springs National Park*, 10–13.

29. The hill that Dunbar ascended was Hot Springs Mountain. Arkansas Quadrangle Maps.

30. Joseph Priestly (1733–1804) was a renowned English scientist and clergyman. After moving to Pennsylvania in 1794, he became a fellow member of the American Philosophical Society with Dunbar and Jefferson. Priestley was best known for his discovery of oxygen, or the isolation of dephlogisticated air. Between 1773 and 1780 Priestly published two major works entitled *Experiments and Observations* and *Disquisitions Relating to Matter and Spirit*. For more information on Priestly, see A. Holt, *A*

As we advanced up the calcareous region of the hill, we discovered several patches of rich black earth, which appears to be formed by the decomposition of the calcareous matter: in other situations appeared an incrustation of limestone, i.e. the superficial earth was penetrated, indurated and encrusted by lime with fine laminae or minute fragments of iron ore; we entertained no doubt that the water of the hot springs had here issued formerly from the hill and run over the surface, and that the entire mass of the calcareous rock to the height of one hundred feet perpendicular has been created by the incessant depositions of the hot springs; in this high situation we found a spring whose temperature is 140°.

After passing the calcareous region, we found the primitive hill covered by a forest, whose trees were not of the largest size, they consisted chiefly of Oak, Pine, Cedar, Holly, hawthorn with many others common to the climate, with a great variety of vines, some said to produce black & some yellow grapes, both excellent in their kinds:[31] the soil is extremely rocky, interspersed with gravel, sand & fine black vegetable mold. When we had advanced about 250 feet perpendicular up the hill, we found a change in the soil; it was equally stony & gravelly as below with a superficial coat of black mold but immediately under the last was found a basis of fat, tenacious, soapy, red clay, inclining to the colour of bright spanish snuff; it seemed to be very homogeneous with scarcely any admixture of sand and no saline taste, but rather soft and agreeable; the same timber continues but diminishing in size as we ascend the hill, and rocks increasing to the top: we estimate the whole height of the hill to be about 300 feet above the level of the valley where we are hutted. Thermr. at 8h.p.m. 28°. Extremes 26°–50°.[32]

Life of Joseph Priestly (London: Oxford University Press, 1931); D. J. Rhees, *Joseph Priestly, Enlightened Chemist* (n.p;: American Chemical Society, Center for History of Chemistry, 1983); Joseph Priestly, *Autobiography of Joseph Priestly* (Cranbury, NJ: Associated University Presses, 1970). What Dunbar calls "feculum" is a siltlike matter, possibly algae or decomposed algae. The term "laminae," or lamina, refers to a thin sedimentary rock or sediment. These lamina layers are usually less than 1 centimeter thick. Parker, *Dictionary of Geology and Mineralogy*, 143, 277.

31. Calcareous rocks are those that are composed chiefly of calcium carbonate. Bates and Jackson, *Dictionary of Geological Terms*, 69. The grapes could have been from the genus *Ribes*. Included in this genus are the buffalo currant (*Ribes ordoratum*), which produces black and yellow fruits, and the Missouri gooseberry (*Ribes missouriense*), which produces a dark purple fruit. Grimm and Kartesz, *Illustrated Book of Wildflowers and Shrubs*, 437–441.

32. On this day Dunbar wrote in the "Dunbar Trip Journal, Vol. I":

> As we advanced up the side of the hill we found several patches of rich black earth, which Dor. Hunter says is such as is formed by the decomposition of limestone—in other situations appeared an incrustation of limestone, i. e. the superficial earth was penetrated, indurated and encrusted by lime with detached laminae or fragments of iron ore; we had no doubt that the water of the hot springs had here formerly issued from the hill & run over the surface caus-

HUNTER

Therm. at sunrise 26°. at 3 p m 50°. & at 8 p.m. 28

This day whilst waiting for our Baggage took a cursory veiw of the hot springs, again, & observed several more than presented themselves yesterday. One of them that seemed to give out the most water by one thermometer indicated 150° & by another 148°.—This gave about as much water as might with a considerable pressure pass thro a 4 Inch pipe. The others of various different sizes & also shewed different temperatures viz 145, (140 further up the hill) 136, & 132. As it is probable the whole originated from the same source, The difference of temperature may arrise from the difference in the length or other circumstances depending on their passages before coming to the Air, for they all issue within 130 yards not excepting some up the hill perhaps 100 yards perpendicular above the others.

The whole perhaps might be made to pass thro a 9 Inch tube with the pressure of about 20 feet perpendicular. This might be ascertained in the sumer when the creek is dry above & receives no water but from hot springs. Now when there is water in the creek to turn several Mills, & as the hot water is seen to rise in some parts even in the bed of it, & from several irregular shaggy rugged openings, it is not practicable to measure their contents with any kind of certainty.

The hot water appears to generally in former times to have issued out much higher up the hill than now; for it has deposited an immense mass of a porous limestone on the side of the hill for upwards of an hundred feet high, envelloping in its progress various other flinty stones of which the hill is composed. This porous limestone is in appearance like the rough shapeless masses of limestone which form the Bahama Islands & of which the filtering stones are made & not unlike externally the french bur mill Stones.[33] These stones are here found of all degrees of

---

ing the depositions just mentioned, about 200 feet perpr. up the hill we found a hot spring whose temperature is 140°—The timber here is not large consisting of oak, pine, Cedar, Holly Hawthorn, with many other common to this climate, with a great variety of vines, of which some producing the black and the yellow grapes, the soil extremely rocky with good soil interspersed with stones, gravel, sand &c. When we had advanced about 250 feet perpr. up the hill we found some change in the soil; it was equally stoney or rocky with some black superficial earth immediately under which was found a basis of tenacious fat, soapy reddish clay, inclining to the colour of high colored Spanish snuff; it appeared by taking a bit into the mouth, to be very homogeneous, without sand, & no saline taste, but rather soft & agreeable, this soil with the same growth of timber, but diminishing in size, rocks increasing, continued to the top of the hill, which for the present we estimate at 300 feet per-pr. above the valley where we are hutted.

33. Also called buhrstone or burrstone. This is a fine-grained sandstone or silicified fossiliferous limestone used for millstones. The burrstone can also be a flinty-type rock. Bates and Jackson, *Dictionary of Geological Terms*, 67; Parker, *Dictionary of Geology and Mineralogy*, 45.

hardness according to the date of their formation, some are as soft as clay, others friable between the fingers, & scarce adherent, others again, which have been long exposed to the air are pretty hard; & some peices which have been detached from the rest & fallen in the creek, have a superior degree of hardness resembling the corral rocks of the Bahamas, & other parts of the West Indies. In this situation many of them have atracted iron from the water, so that they resemble rich iron ore, being very heavy, & entirely penetrated by the iron, but still retaining their original rough shaggy appearance. Except at this place there is no appearance of limestone to be seen in any part of our course as yet, nor have I heard of any such in any part of the country hereabout. This limestone is confined exclusively to the small space either now or formerly overrun by the hot water, from its external appearance I suspect that besides lime it contains a considerable proportion of silicious matter that in some places it contains mush iron is very evident & to me it appears in some places to resemble the puzzelano or dutch terrace & capable of making with proper proportion of quicklime, a cement to harden under water. In one of the hot springs tem[p]erature 150° of Farenheit There appears a rusty brown sediment adhering to the rocks, stones & pieces of wood it meets in its course to the creek, & in several of these springs even at the temperature of 140° there are deep & lively mossy plants in full vigor, a large one of which I took up from where I could not bear my hand half a minute. The plant seemed to be all connected together like green shaggy plush or velvet, about the size & nearly the shape of my open hand, had branches about half an inch long, each of which produced many traverse branches about 1/32 Inch long not unlike feathers.[34] The whole plant seemed to have been produced at the bottom of a small recervoir of the hot water about half a foot deep & at [as] it increased in size it became specifically lighter & rose up about half way from the bottom, still bringing up with it some of the lime in a soft state adhering to its lower part, The plant itself seemed to be produced & supported by a tube 2 Inches long about the size of a Goos quill which tube joined an hollow sphere of the same soft white substance like a small bladder of the size of a pistol ball, with which little bladder the whole plant was connected.—

Our Hunters killed 4 deer this day in their way to & from the boat as they went to bring the rest of the baggage. so that we had plenty of Venison. No observation. cloudy

Tuesday 11th

DUNBAR

Thermometer before sun-rise 48°. Wind S.E. The weather changed very much in the night; it became much warmer & the heavens were overcast with one gen-

34. This could be any of a wide variety of mosses.

eral cloud, the air was still damp and penetrating, and our mansion pervious to the chilling blast, but we made good fires & comforted ourselves in expectation of good weather to enable us to complete our observations and researches. The People arrived about 1 o'clock in the afternoon with a few things including the instruments. At 3h p.m. the thermometer rose to 59° and in the evening at 8h fell to 50°, the weather being still cloudy & disagreeable. Some more venison was brought in after dinner[.] The people five in number went back to the river to fetch tools & necessaries, while others were occupied in building a log-chimney at the end of our Cabin which we proposed to line with stone as a security against fire. No change in the appearance of the weather at bed-time. Extremes of the thermr. 48°–59°.[35]

HUNTER

Them. at sunrise in Air 48°, at 3 p.m 59° & at 8 p.m. 50. Clear weather, The people being employed bring[ing] the rest of the baggage from the boat, we made a small excursion up the Hill or Mountain from whence the hot water issues,[36] & having crossed the rivulet began to ascend thro briars & shrubs for perhaps 150 feet high over the before mentioned shapless mass of stone deposited by the hot waters formerly, & found in our rout two or three more hot springs one of which upwards of 100 feet perpendicular above the creek, continued to discharge a considerable quantity of water at the temperature of 140°. of Farenheit; The stones above the deposited matter are simular to those at the bottom in the creek, before described. The soil as we ascend except a thin stratum of vegetable mold is a fine red clay free from grit

In several places found small masses of from one pound to about lbs 100 of shelly Iron Ore, as if it had been formed by attracting Iron from the hot water in its passage down forming a sort of crust or shell or the surface of the deposited limestone before mentioned, At the top of this mountain which is about 400 or 500 feet high, we saw another rather higher behind & connected with this by a neck which reached to within about 200 feet of the top. The stones on the very top are a flinty hard sort of freestone whitish within of a very sharp grit & a reddish tinge many of them on the outside.—These hills produce small yellow pines[,]several sorts of oaks white & red[,] red cedars & some vines[,] myrtle male & female, Cassina or tea shrub with a red berry[,]but all of a diminutive stature.[37] In taking

35. The "Dunbar Trip Journal, Vol. I" ends on December 10, 1804, and the "Dunbar Trip Journal, Vol. II" begins on December 11, 1804. For more information on the two volumes, see "Notes on Sources and Editorial Process."

36. Hot Springs Mountain.

37. Both the shortleaf pine (*Pinus echinata*) and the loblolly pine (*Pinus taeda*) are referred to as

another veiw of the springs as we decended, observed some of them to discharge a moderate quantity of gas. & some of them have deposited a reddish brown ochry sediment[,] others not: The thick overcast weather prevented an Observation for the Latitude.—

Wednesday 12th

DUNBAR

Thermometer before sun-rise 36°; the weather has become colder, but still continues overcast, damp and disagreeable the wind being about north, a few drops of rain fell last evening & during the night. As it still continues cloudy, no astronomical observations could be made. I therefore occupied myself in the forenoon in bringing up & completing my journals, and in the afternoon went to examine all the hot springs with the thermometer: four principal springs seemed only to merit attention; those which yielded the greatest quantity of water were of the highest temperature and are in the following order, No. 1–150° No. 2–145° No. 3–136° and No. 4–132°. The last in order is the only one on the west side of the Creek, & I did not perceive any sign of hot water any where else on that side of the Creek. I therefore conceived that the spring No. 4 is supplied by a channel under the creek from the general reservoir in the hill on the East; at the spring No. 3 was a small bason of some little depth, in which was a considerable quantity of the green matter in temperature 134°, it had much the appearance of a vegetating body, being detached from the bottom yet connected by something like a stem which rested in Calcareous matter, the body of one of those pseudo plants was about 4 to 5 inches diameter, the bottom a smooth film of some tenacity & the upper surface divided into ascending fibres of ½ to ¾ of an inch long resembling the gills of a fish formed into a kind of transverse rows; not being then prepared for a more minute investigation, a future examination will be made with the microscope. Should it prove

---

"yellow pines." The other trees and vines are the white oak (*Quercus alba*), the northern red oak (*Quercus borealis*), the southern red oak (*Quercus rubra*), the eastern red cedar (*Juniperus virginiana*), and the wax myrtle (*Myrica cerifera*). Moore, *Trees of Arkansas,* 16–17, 43, 49–50, 21; Grimm and Kartesz, *Illustrated Book of Wildflowers and Shrubs,* 396–397. The "Cassina" or yaupon holly (*Ilex vomitoria*) is a small tree, also called the cassine or Christmas-berry, which attains a height of 3–20 feet. Its leaves are one-half inch to 1½ inches thick, oval, and usually with blunt teeth, and it bears red fruit one-fourth inch in diameter. Indians used the plant to make ceremonial drinks. In the "Dunbar Trip Journal, Vol. II" Dunbar lists the plant as "Cassina Yassou," having apparently refined his spelling while later copying his journal. Dunbar may have been familiar with this species from specimens around Natchez. It is also called the emetic holly; various Indian tribes used this plant to induce vomiting. The dried younger leaves of the tree have been used for medicinal purposes and as a tea. Moore, *Trees of Arkansas,* 89; Moerman, *Native American Ethnobotany,* 273.

that this is a vegetable production and not an accumulation caused by precipitation, it will be a new proof of the wonderful powers of nature in the production of animal & vegetable life in temperatures which have been hitherto thought sufficient to extinguish the vital principle: Should this green matter prove to be vegetable, I shall confidently expect the discovery of animal life; for no plant I believe upon due research will be found without its animal inhabitants. A little farther on we came to another small muddy bason, in which a vermes about ½ an inch long, was moving with a serpentine or vermicular motion, the water was found a little warm to the finger: I observed invariably that the green matter forming on stones & leaves covered a stratum of Calcareous earth,[38] sometimes a little hard & brittle, but at other times soft and imperfect, but whether the lime favors the production of the green matter or vice versa, we probably shall not have time to ascertain. Thermr. at 8h. p.m. 36°. Extremes 36°–50°.

HUNTER

Therm. at daylight 36°. at 3 p m. 44°, & at 8 p. m 32° Hazy weather, No observation. Like for rain. Whilst some of the people were employed bringing the rest of the baggage, others were fixing up the Cabin & sheds so as to make them a sufficient shelter from the bad weather expected at this season. Our Guide went ahunting & brought in a Deer.

Thursday 13th

DUNBAR

Thermr. before sunrise 26°. Wind north. The weather still continues cloudy, dark & disagreeable; finding no probability of making any astronomical observations this day, I determined to make an excursion upon the neighbouring western mountain, & having gained one of its summits about ½ mile from the Camp, took various courses of hills & points on the river & having gone to its extreme summit to the westward about a mile distant, I took courses to the same points in order to ascertain nearly their position:[39] We had several fine prospects from this hill, which we estimated to be 300 feet higher than the valley of the hot Springs where we first ascended and 400 feet at its western extremity; the valley of the Washita

38. Scientists now know of vegetation and lower organisms that are found in hot springs throughout the world and under the oceans near geothermal vents. The "vermes" was a worm-like organism. Calcareous earth or calcareous soil contains magnesium carbonate, or magnesite ($MgCO_3$), calcium carbonate ($CaCO_3$) and/or accumulated calcium deposits. Parker, *Dictionary of Geology and Mineralogy,* 49; Bates and Jackson, *Dictionary of Geological Terms,* 308. Also see Dunbar's entry for December 10.

39. Dunbar's "western mountain" is called today West Mountain.

river comprehended between the hills on either side, seemed a perfect flat and about 12 miles wide, on all hands we saw the hills called here mountains rising behind each other: in the direction of north the most distant were supposed to be 50 miles off & are considered to be those of the Arcansa river; the rugged mountains which divide the waters of the Arcansa from those of the Washita prevent the Osage Indians from visiting the Washita river, of whose existence they are in general ignorant; were it otherwise, their excursions here, would prevent this place being visited by White persons or even Indians of other tribes as they make no difficulty of traveling round the mountains which give birth to the Washita by the great prairies which lie east of the great dividing Ridge, and it is known that those robbers plunder indiscriminately all they can find.[40] In the direction of S. W. we saw at about 50 miles distance a ridge perfectly level which we supposed to be the high prairies or planes of the red river, so that we had under our Eye an horizon whose diameter was 100 miles, incomplete to the East and N. W.[41] Notwithstanding the late severity of the weather, we found along the ridge a considerable number & some variety of plants in flower, & others retaining their verdue. We found indeed the ridge much more temperate than the valley; When we left the valley it was extremely damp, cold and penetrating; upon ascending the ridge, the atmosphere became dry & mild, so that walking thereon was perfectly agreeable: a few of the plants in flower were collected for specimens, but what surprised us much was to find upon this ridge a species of Cabbage, the plants grew with expanded leaves spreading on the ground, of a deep green with a shade of purple, the taste of the cabbage was plainly predominant with an agreeable warmth incling to the raddish; several tap-roots penetrated into the soil, of a white colour, having the taste of horse raddish, but much milder, a quantity of them were brought to camp & when dressed proved palatable & mild; it is highly improbable that any Cabbage seed has ever been scattered upon this ridge, the hunters ascending this river have

40. The "great dividing Ridge" is the Rocky Mountains. The Osage did conduct raids on the Caddo in areas farther south than the Ouachita Mountains. French traders had established themselves illegally on the Arkansas River in the 1770s and had begun to trade weapons with the Osage. As a result the tribe gained additional power and exhibited a propensity to dominate the region between the Arkansas and Red River valleys, undoubtedly enhancing the ferocious reputation of the Osage among European and American travelers. The majority of the Osage traded with the French near St Louis. In fact, until the late 1700s, French traders at St Louis held a trade monopoly with the Osage. For more see Din and Nasatir, *Imperial Osages*, 25–5, 152; John Joseph Mathews, *The Osage: Children of the Middle Waters* (Norman: University of Oklahoma Press, 1961), 236–237; Smith, *Caddo Indians*, 66, 100; Carter, *Caddo Indians*, 185, 206–208; Whayne, *Cultural Encounters in the Early South*, 56.

41. At more than 100 miles away, Dunbar would not have been able to see the Red River valley from West Mountain.

always pursued far different objects: we must therefore consider this Cabbage / untill farther elucidation / as indigenous to this sequestered quarter & may be denominated the Cabbage raddish of the Washita, I shall preserve and take with me several living plants in hopes of procuring in due time seeds from which the curious may be furnished. we also found growing here a plant which is now green, called by the French 'racine rouge' (red root) which is said to be a specific in female obstructions; it has also been used combined with the china root to die red; which last, probably acts as a mordant; having understood that it has also been used with the bark or root of an aromatic vine, (which I shewed to Mr. Bartram at Baton Rouge) for the same purpose of fixing a red die.[42]

The top of this ridge is in a manner crowned by rocks of a flinty kind, so very hard as to be improper for gun flints; when applied to that purpose it very soon digs out cavities in the hammer of the lock. This hard stone is generally white but frequently clouded with red, brown, black and some other colours, and no doubt in the hands of a practical mineralogist, would receive a variety of denominations such as agate, jaspar, Calcedony, Cornelian and perhaps some of the adamantine genus. Notwithstanding the abundance of rock, a great deal of excellent black vegetable earth was found along the ridge, and generally an understratum of darkish or greyish brown earth, producing oak and hickory with other woods & great number of grape vines, said to yield excellent black grapes; there is no doubt that this soil upon the top & sides of these hills is well adapted to reward the labors of an expert Vigneron.[43] Here & there we met with fragments of iron stone & often

42. Hunter listed the "Cabbage raddish" as a dwarf cabbage and described it as a plant "found upon a high ridge near the hot-springs, growing in rich black mold amidst masses of flinty rock; it is not confined to a particular spot, but extended along the ridge at least half a league, it is not therefore probable that the seed of this plant was dropt there by any hunters; I conclude it to be indegenous." "Hunter Official Report," conclusion. This may have been an Arkansas cabbage (*Streptanthus obtusifolius*), which occurs in the Hot Springs area and throughout the Ouachita Mountain Range. It may have also been a spring cress (*Cardamine bulbosa*). However, the spring cress occurs throughout the state, and Dunbar and Hunter describe the species as being very localized. Smith, *Keys to the Flora of Arkansas*, 89, 93. The redroot (*Ceanothus herbaceus*) is a shrub 1–2 feet high that grows in sandy or rocky soils. The leaves have an oval shape and a fine-toothed margin, and the plant produces small white flowers and dark brown fruits. Indians used this plant's roots as a cough remedy and its leaves for making a tea. Dunbar continually showed interest in finding sources that could be exploited for commercial uses. Grimm and Kartesz, *Illustrated Book of Wildflowers and Shrubs*, 502–503; Moerman, *Native American Ethnobotany*, 145.

43. The explorers may have come upon a prehistoric novaculite quarry. Many plants within the grape genus *Vitis* are native to Arkansas, including peppervine (*Ampelopsis arborea*), heartleaf peppervine or heartleaf ampelopsis (*Ampelopsis cordata*), Virginia creeper (*Parthenocissus quinquefolia*), summer grape (*Vitis aestivalis*), post oak grape (*Vitis aestivalis* var. *lincecumii*), riverbank grape (*Vitis riparia*), frost grape (*Vitis vulpina*), winter grape or graybark grape (*Vitis cinerea*), possum grape (*Vitis cinerea* var.

where a tree had been overturned by the roots, some schistose stones were brought to view which were suffering decomposition by their exposure to the atmosphere; In returning we descended the hill obliquely & found for 200 feet perpendicular the same kind of stone, much broken into loose fragments: & slipping under foot frequently endangered our falling, the hill being in many places extremely precipitous: in this position we dug into the side of the hill and found the 2nd stratum to consist of a reddish Clay somewhat resembling that found near the top of the Conical hill to the East of our Camp, but not so highly coloured nor so argilacious, the proportion of silex being manifestly much greater. We continued to descend & found at b of the hill downward, the rock to alter considerably, & altho' it still continued siliceous, yet it was rather a very hard free stone mixed with fragments of flint which had probably rolled from above, descending still lower we found a blue schistus, in a state tending to decomposition wherever it was exposed to the atmosphere; more interiorly the schistus was hard resembling coarse slate: few other argilacious stones presented themselves to view, the silicious were always predominant; and we often found what had much the appearance of the Turkey oylstone.[44] Towards the base of the hill was a considerable expansion of tollerably good land, lying sufficiently level for cultivation, & is supposed a good soil for wheat. The Timber such as above described with a large proportion of Pine.

Thermr. at 8h. p.m. 30°. Extremes 26°–40°. Wind N.

HUNTER

Therm. at day light 26°. & at [8] p m 30° Overcast weather, This day finished what repairs were deemed necessary for our cabbin.

The Sextant I brought from Philada being originally made for using at sea, offhand & being no adept in the use of Astronimical Instruments, I found it very awkward to hold it sufficiently steady so as to take the suns double altitude by

---

baileyana), red grape or catbird grape (*Vitis palmata*), bush grape or sand grape (*Vitis rupestris*), and muscadine grape (*Vitis rotundifolia*). Grimm and Kartesz, *Illustrated Book of Wildflowers and Shrubs*, 502–509; Smith, *Keys to the Flora of Arkansas*, 146–147.

44. "Argilacious stones," or argillaceous stones, are mainly composed of clay minerals such as shale. "Silex" probably means silica or silicon dioxide, SiO_2. The substance that Dunbar is describing is most likely chert. Some of the minerals that compose the Ouachita Mountain Range are Stanley shale, Polk Creek shale, Collier shale, Womble shale, Mazarn shale, Johns Valley shale, Mcallister shale, Blakely sandstone, Crystal Mountain sandstone, Blaylock sandstone, Jackfork sandstone, Bigfork chert, Caballos novaculite, and Arkansas novaculite. Arkansas Geological Commission, www.state.ar.us/agc/agc.htm; Haley et al., "Geologic Map of Arkansas"; Aber, *Ouachita Mountains*, 1–2; Fay, "Geology of Paleozoic Strata," 183–188; John C. Nichols, *Minerals of the Ouachita National Forest*, 1999, www.fs.fed.us/oonf/minerals; Bates and Jackson, *Dictionary of Geological Terms*, 28, 467.

means of the artificial mercurial Horizon, with a tollerable degree of accuracy, especially as the cover was very narrow (1½ Inches by three) with all the joints so open as not to exclude the influence of the wind on the mercury, which being shaken thereby often prevented an exact observation being taken. The box to hold the mercury was made of tin soldered at the corners & Japaned with a veiw of preventing an amalgamation with the solder & mercury. However after using it a few times the cracks in the japan were penetrated by the mercury & the solder at the corners entirely destroyed, by which the box became useless.[45]

To remedy these inconveniencies, I took the Idea from Mr Dunbars Circle of reflection which is supported by a pedestal of brass with three feet & nicely contrived, so as at the same time, to be steadyly supported & capable of every kind of motion, so regulated, by a variety of different sorts of joints & screws, as enables the observer to be very exact & to have as great confidence in his results as any instrument can give. I imitated the principle of this pedestal, as near as circumstances would permit to be done with wood, balancing the weight of the Sextant with a peice of Lead cast in sand for the purpose, & in place of screws used wedges to lighten or slaken the several motions. After several trials & alterations, I finally brought it to that perfection as to be able to take all the various observations with much more precision than before; having also made a box of wood for the mercurial horizon in place of the tin one rendered useless.—no observation

Friday 14th

DUNBAR

Thermr. 28° Wind N.E. Cloudy, dark, cold and sleet—This morning has made no improvement upon the weather; rain & sleet fell in the night & the ground is hard frozen: Dor. Hunter had proposed an excursion into the mountains with a Party this day, but the appearance of the weather forbids it: the bad state of our mansion calling for further repairs in the present severe weather, we employed some of our people in shutting up the cracks and openings between the logs, which will render our dwelling more comfortable; placed some of the flowers collected between hortus-siccus-paper & had the roots of the new Cabbage planted so as to be preserved until our return.[46]

45. The word *japan* is a reference to the process of applying a black varnish or lacquer coating to metal (japanning).

46. The term *hortus siccus* means dried garden. This type of paper, usually consisting of stout, smooth sheets (later, wax paper), was used by botanists and naturalists to preserve plant specimens. Diane M. Birdson and Leonard Forman, eds., *The Herbarium Handbook*, 3rd ed. (Port Jarvis, NY: Lubrecht and Cramer, 1998).

The day continues to drip a little from time to time, being still dark, damp and disagreeably cold. Thermr. at 8h. p.m. 32°. Extremes 28°–40°. We have news from the Sergeant that the river has fallen 5 feet.

HUNTER

Therm. at day light 28°[.] at 3 p.m. 40° & at 8. p.m. 32°. Clouds, rain—

This day intended to make an excursion with a small party to explore the country, but the stormy & rainy weather prevented it. Therefore examined the water of the hot or boiling springs by such chemical tests as we had with us.—

In specific gravity it resembled rain or distilled water—

It slightly tinged Litmus paper a little red[47]

It discovered nothing particular to the smell or taste, different from other hot water, except that shortly after drinking it hot, it caused a slight eructation.

Nitrat of Silver, Nitrat of Barytes, Sulphuric, Nitric or Muriatic acides, prussiat of Potash,[48] alcohol of galls, Solution of Potash, of Caustic & mild Volatil[i]ty produced no effect on the water.—

from these, & several other trials it appears to be pure water containing only a small proportion of Carbonic Acid,[49] or fixed air.—

Tried at the same time the rough shaggy shapeless mass of lime stone, & found it to effervess with every acid. Those pieces picked up from the bottom of the creek penetrated by a rusty iron coloured matter shewed after solution in the Sulphuric, Nitric & muriatic acids plain signs of iron by the Prussiat of Potash & alcohol of Galls. It is a poor iron ore, but not in any great quantity here.

It is remarkable, that this water should be the same that has in former times deposited such immense masses of calcarious stone as compose the bottom of this side of this hill, & also considerable portions of calx or oxide of iron,[50] now visibly

47. This means that the water was slightly acidic.

48. "Eructation" means the act of belching. Silver nitrate, $AgNO_3$, is a colorless, crystalline, caustic substance and extremely soluble. "Nitrat of Barytes" is also called nitre of heavy earth, $BaNO_3$. Sulfuric acid, H_2SO_4, is sometimes referred to as oil of vitriol. Nitric acid is HNO_3. Muriatic acid, HCl, is also called hydrogen chloride or hydrochloric acid. Muriatic acid is a technical grade of the acid and is often yellow in color. It is the most corrosive of the acids. HCl is produced by dissolving hydrogen chloride in water. Potassium carbonate, K_2CO_3, is usually white and crystalline, and it creates a strong alkaline solution. This could also be potassium hydroxide or caustic potash, KOH. Potassium hydroxide is also usually white, and when dissolved in water it makes a caustic solution and gives off heat.

49. "Carbonic acid," H_2CO_3, is a weak dibasic acid that is formed when carbon dioxide dissolves in water.

50. "Calx," or Calc, is a prefix meaning a substance contains calcium carbonate. Bates and Jackson, *Dictionary of Geological Terms*, 69.

intermixed in places with it, & yet at this time to be so pure as to yield nothing but a little carbonic acid or fixed air.—

How is this water heated? From what cause has the change in it taken place by which it is now so pure, & yet so formerly must have contained so much lime & iron, as to cover the whole side of the hills with it, down to the very margin of the creek, where it overhangs in its whole breadth, the creek water having excavated a passage for itself under it whilst yet soft? There is none of this sort of limestone, nor any other limestone on the other side of the creek; nor for many miles round it.

To resolve these questions will require some reflection & perhaps a knowledge of more facts than we are at present possessed of.—There is abundance of scistus of various sorts & colours in the neighborhood of this hill even on the other side of the creek within a few feet of where the hot water issues, some blue others grey & between these, some as soft as clay, & others advancing by degrees from that to hard silate & beyond that to a sort of stone, (whetstone) Those scisti containing clay, iron, Bitumen silex & sulphur by the action of air & water generate heat, sometimes take fire, the sulphur being converted into sulphuric acid, seizes the clay & forms Alumn[51]—Is this water heated by some such chemical mixture or combination, & thereby forced in the state of vapor to the upper internal cavities or regions of this mountain, where it is condensed again into water as in the refrigitory of an Alembic, whence it falls again & issues throu such apertures as present themselves very hot yet still many degrees below boiling water? This is mere conjecture.

Saturday 15th

DUNBAR

Thermr. 26° Wind N. W. strong. The morning was cloudy, but less dark and disagreeable than the day before, the air became drier & the clouds were disipating by 9 and 10 o'clock: prepared for a meridian observation. The wind blew very strong down the valley: we are here placed as in a point of convergence for whether the wind blows directly or obliquely into the valley from above or below, it is reflected from the face of the hills on one hand, and by three lesser vallies on the other so as to have its force directed against this point as a Center; there will therfore be a breeze here when there is none upon the adjoining hills, perhaps the rar-

51. Hunter is probably referring to shale or slate. Dunbar called it "schistus." Bitumens are solid hydrocarbons. Silex is usually a finely ground quartz. Alum, $KAl(SO_4)_2 \cdot 12\ H_2O$, is a mineral that is basically colorless to white with a sweet-sour taste. Bates and Jackson, *Dictionary of Geological Terms,* 449, 57, 16–17; Parker, *Dictionary of Geology and Mineralogy,* 262, 276; Chesterman and Lowe, *Field Guide to Rocks and Minerals,* 733–735.

efaction produced by the hot springs may also contribute in some measure at this season.

At noon had an observation altho' much disturbed by the frequent recurrence of violent blasts of wind, which greatly agitated the mercury of the artificial horizon; it appears that the Lat: here will be about 34° 31′, but as I intend to make a short series of observations, with the face of the Instrument both East & West, the final result will then appear.[52]

Thermr. at 3h. p.m. 32° at 8h. p.m. 30°.

HUNTER

Therm. at day light 26°[.] at 3 p.m. 32° & at 8 p.m. 30°

The weather this day was clear but the runs being swelled & the roads wet, with the late rains, deferred the small excursion I Intended till tomorrow. In the mean time took the following observation for the latitude of the Hot Springs. viz

⊙ doub. Appt. mer. Alt. 63°.52′.25″ Ind. error.—48″ Lat. 34°.30′.42″

This differs 6″ from Mr Dunbars Observation—at the same time.

Sunday 16th

DUNBAR

Thermr. 21°. Wind moderate N. W. This morning is cold but promises fine weather; the wind nevertheless arose at 9 o'clock & continued to blow strong all day. Prepared for astronomical observations. Took corresponding equal altitudes of the Sun with corresponding azimuths before & afternoon, with the help of a common circumferenter, by which it appears that the magnetic variation is 8° 20′ East; this being about the expected variation, we may conclude, that the needle is not here influenced by any local attraction.[53] Took also equal altitudes for the regulation of the watch before & afternoon. Took also the Suns mer. alt. with the face of the Instrument reversed, and in the evening between 10 & 11 o'clock, the thermr. being at 22°, perfectly serene & calm, took 9 lunar distances between the moon's east limb & ? arietis; the evening was perfectly agreeable & not sensibly cold altho'

52. Hot Springs, Arkansas, is located in a narrow valley that could cause the same wind-tunnel effect experienced around skyscrapers in major cities.

53. An azimuth is the arc of horizon that is measured either from the south point or the north point and to a certain point where a vertical circle through a given heavenly body intersects the horizon. Such measurements are used in astronomy and navigation. A circumferentor measures the angular diameter of an object, thus determining the size of the object. It usually consists of a wooden block or box with a brass protractor, a surveyor's compass, and a sighting instrument mounted on the top surface. Department of Defense, *Map Reading and Land Navigation*, 62–67.

the thermr. was so low; I conclude these observations to have been made with great accuracy from the advantage of the circumstances, the Circle was mounted on its pedestal very firmly, the Star towards the west & the moon over head, so that when both were brought into the field of view & the Star made to move gently across the limb of the moon by a turn of the foot screw backwards & forwards, or by sliding the foot a little to right & left so as to discover the true point of contact on the moon's limb, the Star being left a little open, the observer had only to wait with his eye fixed upon a permanent steady object untill he was convinced of the contact being perfect:[54] I consider one observation made in this way superior to any number or set of observations made by an instrument supported upon the arms of the most experienced observer; I would therefore recommend to all persons using a Sextant or reflecting Circle by land, to adapt a pedestal of support with the three necessary motions; the superiority is so great that he who has accustomed himself to use the one mode cannot reconcile himself to the manifest imperfection of the other; the observation being made the angle is read off without stirring the Instrument so that every thing is ready fixed to the eye for the next observation; I perceive that when all things are favorable, a set of distances may be taken by the difference of 1′ of a degree precisely between observations; i.e. by moving the index before making the observation, exactly one minute in advance,

54. The passing of one celestial body in front of another, or occultation, is usually mentioned when the moon blocks the view of a star or a planet. On December 16, Dunbar wrote in the "Dunbar Trip Journal, Vol. II" the following passage, but then crossed it out:

> I think necessary also to observe that I suffer the star to be just entered within the bright limb of the moon and fully half entered upon the dark limb/ when visible/ before I conceive the contact to be perfect, which is conformable to the opinion of De la Lande & his best astronomers, it being admitted by the best skilled in optics that there is an overflowing of the light on the retina of the Eye from bright images; the Star being a mere point but luminous gives us the measure of the overflowing which is equal to half its diameter; the light of the moon being equally bright must also overflow an equal quantity; hence appears the reason why the Star is apparently just within her disk; in confirmation of the truth of this idea, some acute observers who have been viewing an occulation of a planet or bright star by the moon, when the light of the former has been sufficiently bright or different in color from that of the moon, have actually seen the star or planet entered upon the moon's disk before the absolute occulation took place.

Joseph Jerome le Francais de La Lande (1732–1807), a French astronomer and cartographer, concentrated most of his work on planetary theory. The data he collected made possible the correct calculations of the distance between the Earth and the Sun. La Lande was appointed to the chair of astronomy at the College de France in Paris in 1762 and held this position for almost fifty years. In 1781 he published a four-volume work, *Astronomie,* that helped popularize astronomy.

so that it may be written down by the assistant before the time of counting seconds; this will operate as a check also upon the negligences of young assistants. a mistake in minutes of time would thus be easily detected; this mode I shall follow in future, as being easier & more perfect://

Thermr. at 8h. p.m. 22°. Extremes 21°–34°.

HUNTER

Therm. at day light 21° [.] at 3 p m 34° & at 8 p m 22

Set out after breakfast with a couple of our people & the Guide, carrying a spade, a matlock & our rifles. We made semicircle round the Mountain of the hot Springs, leaving it on the left, & proceeded in a northeasterly direction thro a very mountainous country, & poor, thin stoney pine lands, interspersed with other small timber of various kinds, sometimes we passed over the tops of high cragy mountains, sometimes along the narrow stoney valleys between them, examining the soil, stones &cc which were laid bare by the rivulets.

I found no essential difference in this days journey between the stones I met with & those formerly described in the Ouachita & on its banks. I saw much slatey scistus, & as for the rest they were all scilicious of various colours & qualities, such as, white, cream-colored, orange, red & black, some tinctured with iron, here & there a few detatched peices of poor Iron Ore This is not the country for mettals, at least for as far as we have gone—Came in the afternoon to the waters of the Califat.[55] and there found the stones & rocks in the hills as before except that they had in them very generally many veins of various dimentions filled up with white spar or flinty matter. We now returned by another rout to our Camp, crossing a very high mountain which brought us to the waters of our creek, by following their course got home in the evening, bringing samples of the most remarkable stones with us.—

Monday 17th

DUNBAR

Thermr. before sun rise 26°. wind moderate N.W. the morning is bright & promises a fine day. Yesterday Doctr. Hunter made an excursion into the mountains, and to day he goes again. he discovered nothing of importance hitherto, the only metal of which we have seen any indications has been iron, the one of which is scattered about in small fragments upon the hills & in the water courses. Prepared for observation—took equal altitudes of the Sun before & after noon to cor-

55. The "Califat" is Gulpha Creek. Arkansas Quadrangle Maps.

rect the watch, which compared with the result of yesterday's equal altitudes will give the rate of the watch's going, by which the true time of the Lunar observations will be precisely ascertained: took the Sun's meridian altitude with the face of the Instrument again reversed; . prepared to observe the distance of the moon from Aldcbaran, expecting fine observations from so bright a Star, but we were disappointed, the evening became hazy, the stars frequently obscured, & a large halo with a broad white brim appeared around the moon.[56] The night became cloudy & some drops of rain or sleet fell. Appearance of bad weather for to morrow. Thermr. at 8h. p.m. 28°. Extremes 26°–42°.

HUNTER

Therm. at day light 26°[.] at 3 p.m. 42° & at 8 p m. 27° Clear

The weather being clear, set out again with another small party; & as we had gone yesterday to the northeast, we directed our course now to the northwest. We proceeded on in that direction as nearly as the Mountains & rivulets would permit, up one of the forks of the hot spring creek, until we came to its source, where it is divided by a ridge from the head of some of the waters of the Califat; we crossed this ridge & proceeded on in the same direction as before, but observed no essential difference between the soil, trees, mountains, stones &cc now & yesterday, except the stones appeared to have more the appearance of a hard freestone than before, having more of sand in their composition. We then crossed a steep mountain to the west, & continued that course till we came to a bayu or creek not named that runs into Ouashita, continued down that creek for a mile or two, & then began to edge off to the southward & eastward towards our Camp again; In our progress our pilot shot a Buck which he skined on the spot in a few minutes, & having made three parcels of the meat, carried the whole to the Camp, which we reached in the same evening after crossing another high mountain[.][57] as this days journey produced no new matters, thot it unnecessary to trouble the men to carry any samples to camp.—

56. Aldebaran, also called Alpha Tauri, is a part of the constellation Taurus; it is the bull's left eye in that group. Forty times larger than the Sun, Alpha Tauri is a red giant star that shines as one of the brightest in the northern hemisphere. Seeds, *Foundations of Astronomy*, 21, 259. The "Dunbar Trip Journal, Vol. II" included on this day the phrase "read off from the *three Indexes* & took the contacts in the same manner."

57. The "Mountains" may have been Blowout Mountain. The unnamed "bayu or creek" could be Bull Bayou, which runs south and enters what is now Lake Hamilton near the community of Piney in Garland County. "Another high mountain" may have been Music Mountain and West Mountain. Arkansas Quadrangle Maps.

[TABLE 1]

16th . The suns magnetic azimuth[1] before & after noon with the same altitude

		A.M.	
9h.50′.1	⊙ lower limb	magnetic azimuth	S.42°.20E
9″	dble Alt 47°.30′		
		P.M.	
Time missed	do	do	S.45 .40W
			16.40
		Varn. E. ½	8. 20

Correction for change of declination[2]

Notes (added by editors):

1. The magnetic azimuth was an arc of the horizon intercepted between the vertical circle passing through any object and the magnetic meridian (north toward the pole).
2. For more on the "change of declination" see "Explanation of Navigational Techniques.

[TABLE 2]

Equal Altitudes

⊙ appt. Dble alt 54°.27′ (Ind. Er. +15′.46″)

	A M		P M
contact upper limb	10h.18m.59″	Lower limb	1h.42m.12″
Center	10 .21 .56	Center	1 .4 .15
Lower limb	10 .24 .59	upper limb	1 .48 .12

[TABLE 3]

Lunar Observations

Contacts of the ☽'s east limb with Arietis (Ind. Error—16.16)

Times	Distances	Times	Distances	Times	Distances
10h.31m.50″	55°.38′.20″	10h.41m.53″	55°.42′.40″	10h.53.48	55°.47′.20″
" 33 .57	" 39 .10	" 44 .49	" 43 .45	"	"
" 36 .46	" 40 .10	" 47 .12	" 44 .50	"	"
" 39 .11	" 41 .20	" 50 .12	" 45 .55		

Whilst I was making these short excursions of the 16th & 17th. Mr Dunbar made the following observations [tables 1, 2, and 3]

Monday 17th.

[See table 4]

[TABLE 4]

Monday 17th Equal Altitudes
⊙ appt. dble Alt. 45°.40′.0″ (Ind. Err. +15′.48″)
mag. Az S 44°.30 E with the Lower limb.

A.M.		P.M.	
Contact upper limb	9h .44′.56 ½″	Lower limb	2h .27m.57 ½
Center	9 .47 .12 A.M	Center	2 .30 .13
Lower Limb	9 .49 .30	Upper limb.	2 .32 .31

Tuesday 18th

DUNBAR

Thermr. 34° wind north, Cold, damp, disagreeable. The appearance of the weather evening prevents Dor. Hunter from making another excursion to day, some rain fell in the night, but the aspect of this morning bespeaks snow or sleet— Having no better occupation in the present State of the weather, I brought up my journals and began to form a list of all the vegetables I had seen here & in the neighbourhood upon the River, which will be inserted in this Journal when made a little more complete. The day continues dark, cloudy & rainy; in the afternoon it began to hail, & in the evening it snowed pretty fast; about 8h. p.m. it was 3 inches thick; thermr. at the same hour 32°. Extremes 32°–36°.—

This evening Doctr. Hunter was very much indisposed, but was relieved before bed time.[58]

HUNTER

Therm. at day light 34°[.] at 3 p.m. 36°. & at 8 p m 32° Bad weather rain & sleet

58. Dunbar had scratched lines through the following passage in his journal on this day: "This Evening Dor. Hunter was attacked by a pain in the region of the Kidneys which caused him to suffer greatly the whole evening, but was relieved before bedtime." Perhaps Dunbar considered his entry too personal and thus he scaled back the reference. Hunter reports in his journal entry of December 24 that he has had an attack of what he calls "gravel," also known as kidney stones. His illness is no doubt the reason he records only the temperatures and weather conditions for December 18–23. Clayman, *Home Medical Encyclopedia,* 2:224–225.

The list of "vegetables" did not survive within either the "Dunbar Trip Journal," either volume, or the "Dunbar Report Journal." Dunbar must be referring to sleet instead of "hail."

Wednesday 19th

DUNBAR

Thermr. 30°—Wind in the valley West, but changeable; This morning we have a full prospect of a northern winter; the ground is covered 4 inches deep with snow—and it continues from time to time to fall, tho' not remarkably fast, the eves of our Cabin hang with beautiful icicles, which we have the pleasure of admiring thro' the logs as we sit by the fire side: outdoor business being out of the question, I continue to augment my list of vegetables from memory & with the help of the pilot, who proves to be tolerably intelligent. The Doctor has been unable to discover any thing in the water of the hot springs except some weak acid which is probably carbonic; the water has been from this cause a little hard & therefore not so proper for washing as the soap is decomposed in some measure: the same state of the weather continues, the thermr. at 3h. p.m. being at 30°, and at 8h. p.m. 28°. At bed time the weather still continues dark and threatening more snow.[59]

HUNTER

Therm in the morning 30°[.] at 3 p m. 30° & at 8 p.m 28

Thursday 20th

DUNBAR

Thermr. 30°. wind in the valley West, there appears over head driving light clouds from the N. W. the snow still continues lying on the ground, the night was very cold, but has greatly softened towards morning, from appearance we expect a thaw, it becomes a little clearer: The Dor. & myself both a little indisposed, probably from cold and wet feet & the inclemency of the weather. After breakfast, some hopes of the clouds dissipating, the Sun has shewn himself thro' the veil of clouds for a moment, Prepare for observation but disappointed the heavens are again completely veiled in clouds, and a thaw comes on, the thermr. being at 36° at 3h. p.m. Engaged writing great part of the day. Examined some water of one of the hot springs, which stood a little stagnated on one side, its temperature 132°. found no living animal in it, by the aid of an excellent microscope, examined also some of the green matter and the white coagulum lying under it, which I shall further prosecute with day light, being unable yet to determine whether the green matter is

59. The average snowfall in Arkansas is 5.2 inches, and the average rainfall is 45 inches. The January average temperature is 39.5 degrees. www.sosweb.state.ar.us. "Carbonic," or carbonic acid, H_2CO_3, is a substance that contains tetravalent carbon. It is formed when carbon dioxide dissolves in water.

vegetable or merely a feculum. Thermr. at 10h. p.m. 32°. The weather continues cloudy & the snow lies upon the ground the thaw having stopped.[60]

HUNTER

Therm at 8 a.m. 30°[.] at 3. p m. 36°. & at 8 p m 32°.[61]

Friday 21st

DUNBAR

Thermr. 32°. Wind N. no favorable change as yet in the weather; cloudy, dark, damp & cold; the snow still lies upon the ground so that the Dor. is unable to undertake another more considerable excursion as he intended. We were in hopes also of making another set of astronomical observations for the Long. of this place, but as the time is now much advanced we shall be desirous of getting away as soon as the weather permits the transport of our baggage—in the mean time the Doctor is desireous of making another excursion, while we are preparing to move: observed a spot of ground on the same side of the Creek with the hot Springs, covered with herbage which had not lost but partially its verdure; upon this spot no snow lay, it appeared to thaw as soon as it fell, altho' on other places even very near some of the hot Springs the snow remained undissolved; as soon as the weather permits, I shall examine this ground & ascertain the temperature which resists the rigours of winter. What a fine situation for a green or hot house, where at a small expence all the tropical fruits may be propagated; Thermr. at 3h. p.m. 36° it has rained a little we were in hopes of seeing the snow carried away, that it might afterwards become dry under foot. yesterday our pilot & some of the people went out a hunting & fell in with some buffalo; two of them were shot at and grievously wounded, the blood streaming from their sides, as this happened in the evening, they were unable to follow the chase but returned to the pursuit this morning, they discovered the tracks & blood which they followed great part of the day without coming up with the buffalo and were obliged to return without success; it appears that the great strength of this animal enables him to carry off on many occasions several shots without falling, it is necessary to shoot him thro' the heart to make him fall speedily; we are told that a rifle bullet is by no means certain (if ever so well directed) of penetrating thro' the scull into the brain, or if it does, provided the ball only reaches into the front or fore part of the brain, the animal will not

60. "White coagulum" is a clot, a clump, or a congealed, curdled mass.

61. Hunter was apparently still feeling the effects of his attacks of gravel. In the "Dunbar Trip Journal, Vol. II," Dunbar calls Hunter's discomfort "bowel complaints." See also Dunbar's journal entries of December 20 and 23.

fall; some even assert that the thickness & strength of the scull with the immense quantity of hair which covers the head of the buffalo, will resist the penetration of an ordinary rifle bullet. Some venison was brought in, so that we are never without fresh provisions. The Turkeys are not plenty in this neighbourhood, keeping near the river. found a myrtle wax tree covered with its fruit, which must have hung since July or August, the wax is no longer green having changed its colour to a greyish white by being so long exposed to the atmosphere; examined the berries with the microscope; the whole berry is a little oval & less than the smallest garden pea, the nucleus or real seed is as large as a raddish seed covered all over with a number of brownish kidney shaped glands of a brown colour & sweetish taste, those glands secrete the wax, which completely envelopes them & gives the whole the appearance at this season of an imperfectly white berry; this is a valuable plant and merits cultivation; its favorite position is a dry soil rather poor & looking down upon the water, it is excellently adapted to ornament the margins of Canals, lakes or rivulets; the Cassina Yapon is equally beautiful & proper for the same purpose, it grows here along the banks of this stoney creek intermingled with the myrtle, & bears a beautiful little red berry, very much resembling the red currant.[62] Thermr. at 8h. p.m. 31°.

HUNTER

Therm at 8 a m 32°[.] at 3 p m. 36° & 8 p m 31°

Saturday 22nd

DUNBAR

Thermr. 31°. Wind N. dark & cloudy, the Snow continues upon the ground, without any prospect of favourable change; after breakfast it began to rain, the water of the rain froze as it fell upon the branches of the trees; many limbs broke down around us in consequence of the weight of the Ice adhering to them; we are still confined within doors by the inclemency of the weather which greatly retards us, to that we can not even prosecute our intended researches respecting the hot springs. Engaged writing great part of the day; we had 10 quarts of the hot Spring

62. The "myrtle wax tree," or wax myrtle (*Myrica cerifera*), is also known as bayberry or candleberry. This plant is a small tree or shrub with 2–4-inch narrow tapering leaves that can be both toothed and untoothed. It produces a nutlike fruit one-eighth inch in diameter with a white wax coating. The wax of the fruit has been used to make candles. The "red currant" may be any of several varieties in the genus *Ribes*, including buffalo currant (*Ribes odoratum*), prickly gooseberry (*Ribes cynosbati*), and Missouri gooseberry (*Ribes missouriense*). Grimm and Kartesz, *Illustrated Book of Wildflowers and Shrubs*, 397, 434–441, 490–491.

water evaporated which produced about 10 grains of matter, of which the chief part appeared to be carbonated lime with some feculum, the greater part disolved with effervescence in the muriatic acid.[63] The Thermr. at 3h. p.m. 36°. The day continues unfavorable & keeps dropping rain from time to time; yet the snow does not melt: The temperature of the hot springs remains the same as in the former trial & the temperature of boiling water was ascertained to be 212°, hence it appears that this place is not elevated so as sensibly to alter the pressure of the atmosphere, otherwise water would boil at a lower temperature. Caused a number of the grape vines to be dug up, ready to carry along with us. The Doctor goes on with some more experiments upon the spring water, the results will be hereafter given. Thermr. at 8h. p.m. 43° Snow falls again this evening—no prospect of a change.

HUNTER

Therm. at 8. a m. 31°. at 3. p m. 36°. & at 8. p m. 34°

Sunday 23rd

DUNBAR

Thermr. before sunrise 30°. Wind N. W. by the clouds, blows down the valley reflected from the side of the hill N. N. E.—This morning some appearance of a change. The clouds (scudding from the N. W.) begin to dissipate. the blue celestial sky appears in several parts of the heavens, the snow still lies partially on the ground—but we hope it will soon disolve as the Sun appears, prepare for taking equal altitudes, in which I succeeded so far as to take the triple contact in the morning for the regulation of the watch and also one azimuth with time & altitude for finding the variation of the magnetic needle; prepared for a meridian observation in order to complete my set of 4 observations for the Latitude of this place, but was disappointed by the intervention of clouds;[64] Seeing no prospect of taking correspondent altitudes in the afternoon, determined on visiting the hot springs & adjacent places: it requires a length of time to form a good judgement of a new object such as the curious one now before us; on the first view we see a creek with a margin of rock & the hot springs here & there trickling over or passing thro' them; the Creek seems to be undermining the rock, which frequently cracks, divides & falls into the Creek; upon a closer examination it will be found that the water of the creek does not undermine the rock, but on the contrary the rock is con-

63. "Carbonated lime" describes a sediment of magnesium, iron, and/or calcium, usually in the form of dolomite or limestone. Bates and Jackson, *Dictionary of Geological Terms*, 74.

64. For a definition of "azimuth," see Dunbar's journal entry for December 16.

tinually encroaching upon the breadth of the creek, the hot water is perpetually depositing calcareous matter, perhaps some silicious matter also: the new formed rock by those means is continually augmenting & projecting its cliffs & promontories over the running water, which prevents this formation below its own surface: wherever the calcareous crust is seen spreading over the bank & margin of the creek, there most certainly the hot water will be found, either passing over the surface or thro' some channel perhaps below the new rock or dripping from the projecting edges of the overhanging precipice; the progress of nature in the formation of this new rock is curious & worthy the attention of the mineralogist; when the hot water issues from the fountain it frequently spreads over a superficies of some extent; so far as it reaches on either hand, there is a deposition of dark green matter which may either be a plant or only a feculum, I have not yet been able to pronounce which: several laminae of this green matter will be found lying over each other immediately under & in contact with the inferior lamina which is not thicker than paper, is found a whitish matter resembling a coagulum; when viewed with the microscope, this last is also found to consist of several, sometimes a great number of laminae, of which that next the green matter is the thinest & finest being the last formed, those below encreasing in thickness & tenacity, untill the last terminates on a soft earthy matter, and this last reposing on the more solid rock; each lamina of the coagulum is penetrated in all its parts by calcareous grains, which are extremely minute & divided, in the more recent web, but much larger & occupying the whole of the inferior laminae: I think it probable that the coagulum is silex & no doubt the grains are lime the understratum is continually consolidating & adding bulk & height to the rock; when this acquires a certain elevation, the water always seeking the quickest descent will find its way over another part of the rock, hill or margin of the creek & forms accumulations by turns over the whole of the adjacent space; the green matter is also designed by nature for a useful purpose, when the water by seeking new channels has entirely forsaken its former situation, the green matter which acquires sometimes a thickness of half an inch, is speedily converted into a rich vegetable earth & becomes the food of plants the calcareous surface itself decomposes and forms the richest black mold intimately mixed with a considerable proportion of silex (formed as I have supposed from the coagulum) plants & trees of every kind now vegetate luxuriantly upon this soil; many however thrive upon the rock, where very little earth is to be seen particularly the cedar which seems to grow from between the clefts of the hard rocks. The Grape vine also seems to prosper in the unpromising situation.[65]

65. Also see Dunbar's entry for December 10.

I proceeded to examine the piece of ground (above mentioned) upon which the snow would not lie: I found it covered in great measure with herbage, which was in part turned brownish by the season, altho' there was on a part of it a very fine small grass which was green, a calcareous crust appeared in some places at the surface, but in general there was a depth of 5 or 6 inches & in some places a foot of the richest black mold, the surface was manifestly warm to the touch; the thermr. in the air was then at 44°, when placed 4 inches under the surface & covered with earth, it rose rapidly to 68° & when placed at 8 inches or upon the calcareous rock & covered up it rose to 80°, this result was very uniform over the whole surface which was about a quarter of an acre; in searching we found a Spring about 15 inches under the surface, which raised the thermr. to 130°: under the black mold was found a brown mixture of lime & silex very loose & divisible, which appeared to be advancing in its progress of decomposition towards the formation of black mold, under the brownish mass it gradually became whiter & harder & at the depth of six to 12 inches was nearly hard calcareous stone sparkling with silex: it was evident from every thing we saw around that the water had passed over this place & formed a flat superficies of silicious limestone, & that its position nearly level had facilitated the accumulation of earth in proportion as the decomposition advanced: Similar spots of earth were found higher up the hill resembling little savannahs, near which were always found hot springs, which had once flowed over the savannahs; it seems probable that the hot water of the springs at an early period had all issued from its grand reservoir in the hill at a much higher elevation than at present, the Calcareous crust may be traced up in most situations on the west side of the hill looking down upon the creek & valley to a certain height, perhaps 100 feet perpendr.: from that division the hill above rises precipitously & is studded all over with hard silicious stones, below the descent is more gradual, the soil calcareous black earth, the rock itself very often at the surface, & frequently there is a precipice on the margin of the Creek or a very precipitous descent along the calcarious newformed rock.

The Thermr. at 3h. p.m. was at 44° and at 8h. p.m. 38°. Doctor Hunter continues indisposed.

HUNTER

Therm. at 8. A.M. 30°. at 3. p.m 44° & at 8 p m. 38°

Monday 24th

DUNBAR

Thermr. before sun rise 32° . Wind moderate from N. W. Some prospect this morning of a favorable change, the moon is visible and the sun yet behind the hill, announces his approach with a bright blaze: prepare for observation—took the Sun's triple contact, hoping to obtain correspondent observations in the afternoon to regulate the watch, the moon was already eclipsed by the Pine tree tops on the western hill before the sun was risen high enough in the East to enable us to take their distance, we were therefore obliged to wait with patience and ordered all the intervening trees to be cut down to facilitate future observation; at noon obtained a good altitude of the Sun, but soon afterwards it became cloudy, so that we got no corresponding altitudes for the regulation of the watch.

The Doctor found himself a little better; we agreed to walk up the hot spring hill to make new observations on this natural curiosity: we now found it easy to trace out the separation between the primitive hill and that which has been accumulated upon its west side by precipitation from the waters of the hot Springs; this last is entirely confined to the west side of the hill washed at its base by the waters of the Creek, no hot spring being visible in any other part of its circumference; by actual measurement along the base of the hill, the influence of the springs is found to extend 70 perches in a direction a little to the eastward of North; along the whole of this space the springs have deposited stoney matter, which is probably principally Calcareous, but there is also evidence of silex and iron; all the springs deposit red calx of iron in their passage to the creek; the existence of silex does not appear to me to be so fully decided; there is certainly sparkling chrystals mingled with the lime, particularly remarkable in the calcarious matter partially decomposed, but having observed by the aid of the microscope that the whole of the calcareous rock exhibits nothing but a mass of congregated sparry matter,[66] it is not improbable that those shining chrystals may be chrystalised lime; the Doctor is now employed upon an analysis which will probably decide the point; from some specimens I shall carry home with me I shall hope to investigate the matter more at leisure. The accumulation of calcarious matter is much more considerable at the north end of the hill than towards the south, the first may be above one hundred feet perpendicular, but sloping much more gradually than the primitive hill until it approaches the Creek where it frequently terminates in a precipice of from 6 to 20 feet; the difference between the appearance of the primitive & secondary hill is so striking that the most superficial observer cannot avoid taking notice of it; below on the secondary hill, which carries evident marks of recent formation, no si-

66. Silex is a finely ground form of quartz. The term "red calx of iron," or calc, describes a metal oxide produced by the application of heat to a metal or mineral. The "sparry matter" usually refers to mineral containing a majority of calcium carbonate, $CaCO_3$. Bates and Jackson, *Dictionary of Geological Terms,* 69, 70, 479; Parker, *Dictionary of Geology and Mineralogy,* 276, 284. The "Dunbar Trip Journal, Vol. II" included on this day the statement "it therefore does not appear to me decided whether the sparkling particles are really silicious chrystals or only calcareous spar."

licious or flinty stone is to be seen, the calcareous rock has concealed all from view, & is itself frequently covered by much fine rich black earth; it would seem that this compound which is precipitated by the hot water, enclosed in its own bosom the seeds of its destruction, for it is remarkable that when the waters have ceased to flow over any portion of the rock, a superficial decomposition will there speedily take place; tho' I am inclined to suspect that heat communicated from the interior of the hill below contributes much to this operation of nature,[67] because it is observable, that insulated masses of the rock remain without change.

The Cedar, the wax-myrtle and the Cassina Yapon, all beautiful evergreens attach themselves particularly to the calcareous region & seem to grow and thrive in the clefts of the solid rock; at small intervals along the line of separation between the primitive and secondary hill, we discover many sources of hot water; some flowing with some degree of freedom, & others in a manner stagnated and shut in by the accumulations of stoney concretion extracted by their own operation from the bowels of the hill: any spring enjoying a freedom of position proceeds with great regularity in depositing its solid contents, the borders or rim of its bason forms an elevated ridge, from whence proceeds a glacis all around; when the waters have flowed for some time over one part of the brim, this becomes more elevated & the water can no longer escape on that side, but is compelled to seek a passage where the resistance is least, thus it proceeds with the greatest regularity forming in miniature a Crater resembling in shape the conical summit of a volcano; the hill being steep above, the progress of petrifaction is at length stopped on that side, & the waters continue to flow and spread abroad, encrusting the whole face of the hill below. I am persuaded that the accumulations and extent of the calcareous matter would have been vastly greater, perhaps the whole valley might have been filled up with it, did not the continual running of the creek put a stop to its progression on that side:[68] the last formed calcareous border of the circular bason, (covered by the green feculum) is soft & easily divided, a little under it is more compact, and at the depth of six inches, it is generally hard white stone: if the bottom of the bason is stirred up, a quantity of red calx of iron arises and escapes over the summit of the crater.

67. This is one of the current, inconclusive, theories about the source of the heated waters at Hot Springs. Bergfelder, "Origin of the Thermal Water of Hot Springs," 62.

68. A "glacis" is a sloping embankment or gentle slope leading downward. On this day Dunbar recorded in the "Dunbar Trip Journal, Vol. II": "slope of the secondary hill would have been greatly more extended if the continual running of the creek waters had not put a stop to its progress on that side, nay I have no doubt but the whole valley might in time have been up to a level nearly with the upper springs, had not this opposing cause intervened."

It is surprising to see plants, shrubs & trees with their roots absolutely in the hot water; this circumstance being observed by some of the visitants of the hot springs has induced some of them to try experiments by sticking branches of trees into the run of hot water; we found some branches of the wax-myrtle thrust into the bottom of a spring-run, the water being at a temperature 130° of Farheneit thermometer, the foliage & fruit of the branch were not only sound and healthy, but at the very surface of the water fresh roots were actually sprouting from the branch; the whole being pulled up for examination, it was found that the part which had penetrated into the hot mud was decayed: this phenomenon is so new & singular, that few persons will at first be disposed to believe, judging that deception or want of accuracy has led us into error; it is however in the power of every curious person who will give himself the necessary trouble to try the experiments himself; in the mean time Doctor Hunter and his son are evidences of the truth of the above statement.—A luxuriant vegetation clothes the decomposed surface of the Calcareous region, the black rich mold being of a good depth in some few places (6 or more inches) & in others shallower, and the rock in other situations is nearly unchanged, giving nourishment however to a mass of very short moss, which is gradually forming a soil different in appearance from that which is generated from the decomposed lime. The primitive part of the hill is greatly inferior in fertility to the secondary or recent portion, but it is far from being sterile: grape vines abound in both particularly in the calcareous soil.

It may be proper to pause for a moment and enquire and enquire what may be the cause of the perpetual fire which keeps up without change the high temperature of so many springs flowing from this hill at considerable distances from each other. Upon looking around us, no data present themselves sufficient for the situation of the problem; nothing of a volcanic nature is to be seen in this country, neither have we been able to learn that in any part of the hills or mountains connected with this river, there is any evidence in favor of such a supposition.[69] [A]n im-

69. Scientists are still not completely sure about the flow system or the cause of the hot waters of the springs. According to a National Park Service, Water Resources Division, report by James Peterson and David Mott, "almost all of the hot-springs water is of local, meteoric (i.e. atmospheric) origin from recharge of the Bigfork Chert and the Arkansas Novaculite . . . the water slowly percolates to depth, resides in the heated part of the system for a relatively short time (no more than a few hundred years), and then travels rapidly to the surface. Based on carbon-14 dating, estimates of the age of much of the water exceeds 4,000 years." In a master's thesis, Bill Bergfelder stated that "the meteoric water percolates through a fracture zone (location unspecified) associated with the margin of the pluton [a large body of intrusive igneous rock] and then (because the heat source is unknown) either takes in heat from the pluton, percolates below the pluton to depths of about 8,000–12,500 feet and takes in heat,

mense bed of blue or blackish schistus appears to form the basis of the hot-spring hill and all of those in its neighborhood. the bottom or bed of the Creek is composed of scarcely any thing else; I have frequently taken up pieces of this stone, rendered soft by decomposition and possessing a very strong aluminous taste, it seemed to require nothing but lixiviation and chrystalisation to complete the manufacture of alumn;[70] As all bodies which suffer chemical changes generally produce an alteration of temperature, it may be enquired whether the decomposing schistus is capable of generating a degree of Caloric corresponding to the temperature of the hot springs. Another cause we shall notice which perhaps will be thought more satisfactory: it is well known that in several positions within the Circle of the waters of this river, vast beds of martial pyrites exist; they have not yet been discovered in the vicinage of the hot spring, but it is extremely probable that they may be accumulated in immense strata under the bases of those hills, and as we have noticed at one place at least some evidence of the existence of bitumen, we cannot doubt that due proportions of those principles united, will in the progress of decomposition by the admission of air & moisture produce the degrees of heat necessary to support the phenomina of the hot springs. No sulphuric acid is present in this water; the springs may be supplied by the vapor of heated water assending from the Caverns where the heat is generated or the heat may be immediately applied to the bottom of an immense natural Cauldron of rock contained in the bowels of the hill, from which as a reservoir the springs may be supplied.

Thermr. at 8h. p.m. 34°. Extremes 32°–45°.[71]

---

or percolates below the pluton to a lesser depth and takes in heat from underlying magma." Petersen and Mott, *Hot Springs National Park,* 6; Bergfelder, "Origin of the Thermal water of Hot Springs," 62.

70. Lixiviation is a process of leaching. There has never been any mining of alum in the immediate vicinity of the springs. Marcus Phillips, Garland County Historical Society, interview by Berry, February 12, 2004.

71. The following entry was included in the "Dunbar Trip Journal, Vol. II" on this day:

> this same schystus is discovered in all situations near the same level & it most probably pervades a great tract of Country & forms the basis of many of the adjoning hills; it is known to chemists that aluminous schystus being moistened in due degree by water, generates in the progress of decomposition a very great degree of heat; I leave it to Scientific men to decide whether this Cause may be sufficient to account for the heat which keeps up the temperature of the water; it is in vain to search for any other external mark which might lead to the solution of this problem; no volcanic appearance is to be found in any part of the hills or mountains of the Washita; it has already been observed that they are insulated & form no connection with the grand chain of mountains to the westward, being enclosed on that quarter by the lofty plains or prairies which unite the western parts of the red river to those of the ar-

HUNTER

Therm. at 8. A.M. 32°. at 3. p m. 45° & at 8 p m. 34°

This is the first good day we have had since the 17th; The intermediate six days have exhibited a series of rain, sleet, snow & dirty dissagreable weather, which confined us about our camp. as for myself, having had a severe atack of the gravel for about 5 hours on the evening of the 18th & afterwards, for about two or three days indisposed with pain & griping. I could not, even if the weather had permitted, have continued my intended excursions till this time. The ground is still very wet under foot, the creeks & rivulets pretty full; to morrow is Christmass & our soldiers have requested an holiday then, which considering their toils & exposures cannot well be denied them. It is now evening & they have already begun to celibrate by social songs & glee Christmass eve.

After Christmass, if the weather shall permit, I intend to take three or four men & go into the wilderness with our Pilot (who by the by has never been farther himself) for a few days, to see if any thing further is to be seen worthy of notice.

At noon took another observation for the Latitude viz ⊙ doub. Mer. Apt. Alt. 63°.37′42″. Ind. Error.—0.1.12 Lat 34°.30′.48.5″

My hands, or rather the fingers of my right hand are now healed up, altho yet very tender in the frost, by the accident of the pistol shot—This afternoon made a circuit of that part of the Hill from whence the Hot water issues, & find a variety of more springs than we saw at first. The region of hot water is very conspicuous; The coarse porous limestone forms there an immense mass deposited evidently by the hot water on the side & foot of the Mountain, making it in an manner lopsided. Hot springs issue here & there from the upper & lower parts of this region. There is no part of the hill covered with this crust (which is from ten to forty feet thick & extends three or more hundred yards long & about 150 ft. high.) except where the hot water does or manifestly has issued. All the rest of the Mountain above & on each side is composed of hard flinty stones, in their pristine form, & of strata of Pyritous Scistus,[72] some black, blue & grey. These Scisti occupy generally

---

cansa, which are much frequented by the Osages; the difficult approach of the Washita waters being supposed the cause which protects this place from their unfriendly visits.

Thermr. at 3h. p.m. 45°—at 8h. 34°.—Having thrust a stick down into one of the hot springs up the hill, several drops of petroleum or naptha rose 'spread upon the surface, the quantity was small,' upon repeating the experiment, it ceased to rise after 3 or 4 attempts.

"Naptha" is a secondary fuel or a distillate used in the petroleum and chemical industries today. For more information, see Langenkamp, *Handbook of Oil Industry Terms and Phrases.*

72. Schist is a coarse-grained metamorphic rock. It is unlikely that Dunbar saw metamorphic schist and more likely that he was viewing a shale or slate. A distinctive feature of the geologic area

the lower parts of this & the other neighboring Mountains, & some of which taste very strongly of Alum being sweetish & very astringent. Such pyritous & bitumenous Scisti are said to give rise to & form the forms of most Volcanoes.—

On stirring up one of the hot springs high up on the hill observed some bituminous oil come to the top of the water & presently dissapear & sink, as if specifically heavier than water. In this same spring which was so hot that I could not bear my hand in it two seconds, found some branches of different shrubs which some person had stuck there in the summer, some of these twigs were dead apparantly[;] that is they had cast their leaves: but there was a myrtle with berries on it, which was not only alive but had begun to shoot out roots on the surface of the water half an inch long This spring & several others had formed to itself a cup or bason by depositing the lime round the edges, which as it runs slowly over it continually raises. Saw this day 12 or fifteen different Issues to the hot springs, perhaps the whole might be forced thro a pipe of one foot diameter with a pressure. There is only one & that is the largest which seems to leave an iron coloured sediment; This we have not yet tried. The former trials were made by candle light. but since I have found by the Oxalat of ammoniae an evident sediment of Lime. In that spring which we tried we found by evaporation of lbs. 16 water, ten grains of a grey powder, three of extractive gelly like mater. This powder effervessed in marine acid which dissolved the greater part of it. The solution turned a deep blue, by pouring a drop of Prussian Alkali into it.[73]

Tuesday 25th

DUNBAR

Thermr. 34° Wind N. N. Cloudy—The state of the heavens did not admit of any astronomical observations in the morning: it cleared up before noon so that we had a good mer: alt: of the sun, which was scarcely over, when the clouds overspread the heavens & it rained a part of the afternoon; this being Xmass we were obliged to indulge the men with a holy day for which purpose they had hoarded up their ration whiskey, to be expended on this day; a great deal of frolick was the consequence; but perfectly innocent: we amused ourselves with some farther experi-

---

known as the Ouachita orgeny is that there is almost no sign of volcanic or metamorphic activity and/or deposits. James S. Aber, *Ouachita Mountains*, 2; Bates and Jackson, *Dictionary of Geological Terms*, 449; Chesterman and Lowe, *Field Guide to Rocks and Minerals*, 621, 629, 714, 733.

73. The "bituminous oil" was possibly a substance called naptha rose (see note 71). "Oxalat of ammoniae" is oxalic acid, or ethanedioic acid, HO_2CCO_2H. Prussian blue is a complex salt compound created from the oxidation reaction of iron sulfate, $FeSO_4$, with potassium ferrocyanide, $K_4Fe(CN)_6$. It is used in inks and dyes.

mental enquiries into the qualities of the hot waters, the conduct of which being left to Doctor Hunter as a professed chemist, I shall give the results when completed. Thermr. 51° at 3h. p.m. and at 8h. 44°.

HUNTER

Therm. at day light 34°[,] at 3 p.m. 51° & at 8 p.m. 44°. The weather in the forepart of this day was fair & clear; but as soon as it was midday the sun was overcast so that we could not finish our meridian observation to satisfaction. & in the afternoon it began to rain which continued till bed time, & then it increased.—Being Christmass our Soldiers had previously divided themselves into two messes or parties, one of which remained at the Springs & the other half went to the river Ouachita to keep their holiday at the camp by the boat with the sergeant. They had made a reserve of their liquor for the occasion,[74] with which & a Saddle of Venison they made themselves very merry, dancing, Hooping in the Indian Manner & singing alternately, not forgetting to serenade us from time to time with a volly from their riffles, wishing us an happy Christmass with all the compliments of the season &cc The night came at length with the heavy rain which put a period to their mirth, & sleep closed their joys for the day.—

In the mean time made some examination of the water which rus[h]es hot from these Springs, beginning with that one which discharges the most water & which by the Thermometer shews 150° degrees of Farenheit.

1st. It deposites a rusty Iron coloured thin crust on the bottom of its course to the creek—this crust dissolved in Muriatic Acid with effervescence became a deep blue by a drop of Prussian alkali being poured into the solution.—
2nd The hot water at the spring, deposited a copious white cloud, on the addition of a drop of Oxalat of Ammoniae.
3rd. Nitrat of Silver produced the same effect on the waters
4th Sulphuric Acid, threw down a few detached particles.
5th The hot water shewed a very slight & scarcely perceptible tinge of green by the addition of the Prussiats of Lime or potash.

Hot Spring N 2. Temperature 140°. also deposits a rusty coloured sediment which on being dissolved in any acid becomes blue by the addition of Prussiat of lime

74. The supplies listed for the trip included "38 Galls. Whiskey" and "17 Gall. Brandy," as well as "1 Case Gin." On November 10, the expedition sold Lieutenant Bowmar at Fort Miró 30 gallons of whiskey. McDermott, "Western Journals of Hunter," 68, 71; Hunter journal entry, November 10.

This spring throu out its course to the creek down the hill gives rise to a sort of green moss, very tender whilst in the water which acquires a little consistance when in the air some days.

This plant grows at the 130° temperature & shews a network appearance in the microscop. I have described this moss before. It gives support to many microscopic shell animals resembling clams of the size of a pin point ar about the 50th part of a inch in length. It shows four legs & a double tail.[75]

1st. The hot water of this spring gave a copious white cloud with the Oxalat of Ammon.

2nd. The blue paper tinged with litmus became red on being dipped into the hot water.

3rd Nitrat of Silver produced no change

Sulphuric Acid after three day[s] shewed a few distinct particles.

5th. Nitrat of Barytes after the same time, a scarcely perceptable white powder

6th. Prussiats of Lime & potash shewed no perceptible change in the color of the water

7th Lbs 16 of the hot water evaporated to dryness left ten grains of a grey powder of which the greatest part dissolved in murriatic acid with effervessence leaving a small quantity of insoluable residuum This solution turned blue by a few drops of Prussian Alkali.[76]

And in order to have further investigation Preserved samples of the deposited limestone by the hot water, both white & coloured & penétrated by Iron, of the Scisti at the base of the hill & in the Bayu, of the scilicious stone of which the upper part of mountain consists, of the white, black, cream coloured & reddish ones found in the Bayu, & on the hill.

Wednesday 26th

DUNBAR

Thermr. 34°. Wind N. W. clear. Prepare for observation. Took the sun's contacts in the morning hoping to get equal altitudes in the afternoon; but as that is not al-

75. Dunbar also records the presence of these organisms in his entry of December 28, explaining, "I found this evening upon the green matter a minute shell animal shaped like a muscle or kidney, it is about the size of the smallest grain of sand." These may have been ostracods, minute crustaceans found throughout the world. The discovery of these organisms in such a hostile environment may be one of the most significant scientific finds of the expedition.

76. "Prussian Alkali" is Prussian blue.

ways certain, I make it a rule to note down the sun's attitude, so that the apparent time may be calculated; & if the corresponding altitudes are taken after noon, the calculation of the correction from change of declination during the interval is greatly facilitated by noting the altitudes. Before instruments were brought to their present state of perfection, the method hitherto in use was to be preferred; but no reason can be assigned. Why we should now adopt a mode equally correct, which saves half the labor, and more especially that by using the altitudes, we do not require that the Latitude should be previously known.[77]

This afternoon took the altitude of the hill west of the camp by measurement of a base and two correct angles of elevation with the circle of reflection, and found it to be 300 feet, which is less than we had supposed: very steep hills are extremely imposing; the ascent of this hill was not much more than double its perpendicular height, i.e. about 700 feet of inclined plane and the angle at its base made by the summit with the horizon above 26°. We had no favorable position to ascertain by the same means the height of the hill of the hot springs, but having been on the top of both distinctly seen from each other, we judge them to be of equal elevation.

In the morning between 10 & 11h. made a set of lunar observations, by taking twelve distances of the Sun and moon's limbs:[78] the moon being advanced within less than 60° of the Sun, appeared with a very faint light in presence of the Sun's image altho' darkened considerably, and it required very particular attention to obtain fine contacts' the eye remained greatly fatigued.—, . The afternoon being cloudy prevented taking the correspondent equal altitudes for the regulation of the watch. Thermr. at 8h. 44°. Extremes 34°–50°.

HUNTER

Therm. at day light 34°[.] at 3 p.m. 50° & at 8p.m. 34. Clear blowing weather. This day was occupied taking Observations for the Latitude & Longitude of this

77. In the "Dunbar Trip Journal, Vol. II," this passage was added:

> in fact the calculation is not half so laborious; I have therefore wondered that astronomers have neglected this mode as the instrument generally used gives the altitude correctly; the other mode was no doubt thought better when coarse instruments only were in use, but the case is now altered, and the observer is at present certain of his observation to less than ¼ of a minute; it not necessary to know the Lat: when the altitude is noted for corresponding altitudes for in fact those corresponding altitudes do give the Latitude & supply the want of a meridian altitude to a very great certainty as I have ascertained by observation for amusement.

78. In the "Dunbar Trip Journal, Vol. II," he said that he made "14 distances of the Sun" and "the eye remained greatly fatigued—, which are supposed to be very correct, altho & no doubt created an uncertainty of some seconds in time, which I believe has never arisen to 12 seconds & can therefore influence the result only 3 minutes in Longitude." He also observed the "Thermr. at 3h. p.m. 50°."

place—at Midday the Suns Apparent double Alt by my Sextant was 63°.44′.55″ Ind. err.—1′.22.5″ Latitude found 34°.30′.48.8″.—

This corresponds with the observation of the 24th Inst. exactly, & appears to be as near the truth as my Instrument can shew.—

Afterwards the Altitude of the Hill[79] opposite that which yeild the hot springs was taken by the Circle of reflection & by the compass: The results which were nearly alike, gave three hundred feet for the perpendicular height.

The hot Spring hill may be computed to be about the same height.—That is that part of these two hills which can be seen from the base; as for the true tops they are considerable higher as may be seen from the opposite hills—

Thursday 27th

DUNBAR

This morning being fine Doctor Hunter prepared to make his long meditated excursion of 3 or 4 days into the mountains, which the unfavorable state of the weather has hitherto prevented: the thermr stood at 26° before the sunrose, and the face of the hill and creek were shrouded in condensed vapor. After breakfast the Doctor set out with our Pilot and three of the people;[80] the rest were dispatched with loads of baggage to the river. Took a set of observations for equal altitudes but we were again disappointed in obtaining the correspondent afternoon observations by the intervention of clouds; the morning's altitudes of yesterday and this day will nevertheless be sufficient for the regulation of time by the watch and obtaining her rate of going. At noon had a very fine altitude: of the Sun, which is the seventh observation for the latitude of this place, and concludes our astronomical observations here, from which will be deduced (it is hoped) with sufficient precision the Latitude and Longitude of this point of Louisiana, rendered remarkable by the presence of so great a natural curiosity as the Hot-Springs. The mean of the seven observations whose respective results were all very near to each other makes the Latitude of the Hot-Spring No. 3 to be 34° 30′ 59″. 82. This may be farther corrected by introducing the deviation in north polar distance, occasioned by the mutation of the Earth's axis; this being common to the Sun and to all the Stars ought not to be neglected when great precision is required. The series of observations above mentioned being reduced to the 21st December as the mean or middle time of the series; it will be found that the Sun's Right ascension was

79. The "hill" was West Mountain.

80. Dunbar recorded a morning temperature of "26°" in the "Dunbar Trip Journal, Vol. II." "Our Pilot" means Samuel Blazier.

then 9 Signs and the place of the Moon's ascending mode 9 Signs 27 degrees; from whence results a correction in the Sun's declination of -4″, 34 which quantity being additive to the Latitude deduced, gives for the true Latitude 34° 31′ 4″, 16. The Longitude will be calulated at leisure & will be hereafter noticed.

After the Doctor set out I amused myself with pursuing experiments on the Analysis of the hot waters, etc.

Thermometer at 8h. 38°. Extremes 26°–45°.[81]

HUNTER

Therm. at 7. a m. 26°[.] at 3. p.m. 45° & at 8 p.m. 38 Clear & cold weather.

I set out from our Camp at the hot Springs at half past 9. a m. on an excursion of three or four days according as it might prove interesting to explore the circumjacent country, taking with me besides the Guide, three of our Soldiers, who carried a tent, 2 riffles, a spade, a matlock an ax & two days provisions, depending upon what the woods should afford for the rest. I put a small compass in my pocket, to serve in cloudy weather.

As in the two former short tours of a day each which I had taken before we could only proceed for about half a day from camp at a time, the other part of the day being necessarily occupied to return back[;] of course could see only a small distance round; the days being now at their shortest, I therefore now determined to proceed in a straight line for two days or so, except circumstances should point out otherwise, & then return by a circuitous way. We directed our course to the N.W. & continued in that direction all this day, thro, stoney thin Land bearing pines & dwarf timber of the usual kinds, now & then assending steep craggy hills & mountains so close to each other as to leave but very small valleys, & these generally or the greatest part under water during the inundations between them. Thro these valleys commonly run small or large branc[h]es of the creeks which take their rise here & empty themselves into the Ouachita. In these valleys altho the soil is but thin & stoney being washed away by the torrents the timber is more various & larger than on the higher ground, & when one assends the hills he can percieve as far as the eye can reach at the height of from 60 to 100 feet perpendicular up the hills the visible commencement of the piney region allmost, as streight as a line from the above level to the tops of the mountains, This is the more visible now as the other trees ar[e] deprived of their leaves. Not that the hill tops are destitute of other trees, or the valleys intirely without pines. but that the pines almost exclusively occupy the upper regions & leave the valleys for the other timber

81. Dunbar measured the temperature "at 3h. p.m. 45°." "Dunbar Trip Journal, Vol. II."

cheifly. This day went only about twelve miles, being without a path & often turned aside & interrupted by waters, briars, fallen trees & rocky precipices. Encamped a little past 4 p.m, pitched our tents as the weather was now raw & likely for a storm, which took place in the night; yet having a good fire we passed the night comfortably, near the head or source of the main branch of the Califat, along which we had come the latter part of this day.

We raised a large Sclate out of the bottom of the rivulet[82] which by driving three pegs into the ground, served as a table & smaller ones as plates Our flour we kneaded on our table into a flat cake, & baked in the ashes like a potatoe & by some sharp pointed wands run thro our venison & wild Turkey & then stuck in the ground before the fire, we were at no loss for spits[.] Saw many signs of Deer, Bears Buffaloes & wild Turkies this day, but could not get a shot at any of them. The first part of this days journey the stones were as u[su]al silicious, whitish grey, on the hill tops, now & then white flint, redish, cream coloured &cc. The beds of the runs & sometimes for a good way up the hills shewed immense masses of scistus both blue & grey, some of the former were effloressing & tasted strongly aluminous.—& the latter falling to peices seemed to form the bulk of clay of the soil at the surface. The latter part of the day we passed over & between hills of a black opake hard compact flint in shapeless masses,[83] & immense layers of scistus now & then below.

Friday 28th

DUNBAR

Thermr. 34° Wind S. W.—Cloudy—appearance of rain or snow—Dispatched six of our people with loads to the river Camp: after breakfast set out upon a geographical tour round the Hill of the hot spring; young Mr. Hunter, with one of the people and my negro servant attended;[84] in the course of this survey there was no indication of any hot spring but those of which we have already spoken, all of them on the same side of the hill within a space of 70 perches as has been already noted: Every new inspection of those Curious springs brings forth some addition to the limited knowledge we have acquired of them; we find it now pretty evident that most of the springs if not all have flowed from a more elevated part of the hill than

82. By "Sclate" Hunter possibly means slate or shale.

83. This may be a description of Big Fork chert, which is one of the strata that make up the Ouachita ridge system. For more on the various minerals that comprise the Ouachita Mountain Range, see Dunbar's entry for December 13.

84. This reference is the first that Dunbar made concerning the presence of one of his slaves. Early in his journal (October 18), Hunter revealed that Dunbar brought his servant and two slaves on the expedition. It is unclear whether the servant was a freeman or also a slave.

at present; and the perpetual accumulations of calcareous matter confining the sources have probably elevated them to nearly the level of the grand recervoir within the bowels of the hill, during this process the calcareous rock has been formed which we now see attached to the side of the hill; at length however the issues of the waters have become so obstructed and probably the level of the water in the grand reservoir so elevated, that by the superincumbent pressure of the waters, new passages have been forced in lower situations: it is evident that the springs which now break forth along the margin of the Creek, cannot be supposed to have flowed for a long time (comparatively) in their present situation ; the formation of Calcareous rock created by the springs in their actual position, resembling only small excresences growing from the base of considerable precipices, is a proof of what we have advanced:[85] Some of these new springs have formed small flats of 20 to 30 feet extent; in general they have formed little elevations of 5 to 6 feet perpendicular, with a glacis of 10 or 15 feet terminated by a precipitate fall into the Creek. Those small accumulations when compared with the great mass of rock spreading along the face of the hill to the perpendicular height of one hundred feet, are certainly a demonstrative proof of the recent existence of the inferior springs: an ingenious observer of Nature, by some years attention might determine the quantity of Calcarious matter precipitated in a given time from some one spring, which would furnish us with a datum, from whence to form a proximate Calculation of the antiquity of the Springs. We have already noticed that some springs still exist even at the very limit which separates the calcareous region from the primitive hill; their temperature is similar to those below, they are all feeble and are soon lost upon the face of the hill, & perhaps contribute to augment the inferior springs.

We found the Circuit of this hill to be about 3 1/5 miles, measuring round its base as correctly as the uneven surface would permit: altho' this hill when seen from the hill to the west of the valley appears to represent a handsome conical monticule in an insulated situation, yet our geographical survey discovered to us that it is connected in the rear by a very narrow ridge, with a chain of inferior hills dividing the Creek of the hot-springs from a branch of the Calfat. We find invariably the upper half of the hills to be filled up with the hardest flinty rocks, with an admixture of the hardest freestone; much of both particularly the first have rolled

85. An excrescence is an abnormal outgrowth or projection. Dunbar also added in the "Dunbar Trip Journal, Vol. II," "if we may be permitted to judge from the small accumulations of Calcareous matter which appear evidently to have owed their existence to them, when compared with the precipices immediately behind them & the vast bulk of the same kind of deposition, which in some places ascend gradually up the mountain to the height of 100 feet perpendicular, altho' it is generally less than this sloping gradually to the South end of the hill."

down and are found all the way to the base: At the foot of those hilss & at some elevations are also found immense strata of schistus, some of a yellowish color, which forms by decomposition an earth of the same color, presenting at first view the appearance of clay, but it is greatly deficient in tenacity. The base of the hills and the vallies contiguous to the hot-spring hill seem chiefly occupied by a bluish black schistus, altho' there be veins of the silicious genus crossing this last in several places: there is no doubt that a manufacture of Alumn might be established here upon an immense scale; the Schistus under foot is frequently found in a state ready to yield alumn, as appears from the astringent and sweet taste it possesses.[86]

After our return to Camp, I determined to have another microscopic examination of the green matter and hot water before leaving finally this place. I procured some of the green matter of a very beautiful kind, resembling a moss whose fibres were more than half an inch in length; a film of the same green matter was spread upon a calcareous base, & from the film srping the fibres representing a beautiful vegetation completely immersed in water of 130° temperature; this moss (if it shall be found to be vegetable) was brought to this state of perfection by growing in a small natural bason containing some depth of water in a state of comparative repose, communicating freely with one of the springs, but no current passed thro' it.[87] This moss sparkled before the microscope with innumerable globules of chrystalized lime/as I suppose/ the texture of the green matter seemed to be some what fibrous but did not present a decided organized form: the inclemency of the weather while we have been here & other employments, have not permitted me to give so much attention to this curious enquiry as it merits, some future opportunity may prove more favorable to some naturalist competent to decide; I would however wish to recommend to other inquirers, not to be too hastily carried away in their opinions from a superficial view; for there is no doubt that an inexperienced person would at once decide, that this green matter is a real vegetable; the fine green colour & the mossy appearance appear to be a strong evidence; but when one considers the wonderful productions of nature, from chemical attractions, such as the beautiful configurations of some chrystalised salts perfectly resembling

86. Again, by "the Calfat," Dunbar is referring to Gulpha Creek. He also describes the ridge they explored this day as "rather a double headed hill, a second being in the rear connected by a ridge, nay there is a continuation of this ridge, which runs up between a branch of the Calfat and the hot spring branch, but this ridge is extremely narrow & low near to the hill where the line of the Survey crossed it." "Dunbar Trip Journal, Vol. II." The "bluish black schistus" is probably shale. "Veins of the silicious genus" refers to stones that contain an abundance of silica. In fact alum, $KAl(SO_4)_2 \cdot 12\ H_2O$, does have a sweet-sour astringent taste. Bates and Jackson, *Dictionary of Geological Terms*, 276, 16.

87. This could have been any of a wide variety of mosses or algae.

vegitation; the arbor Veneris and the arbor Dianae, which can be produced by art in a few minutes, & which may be seen under the microscope shooting forth a trunk branches, foliage & even an appearance of fruit; I say from the consideration of those & similar productions, we are taught to hesitate & to doubt. If I have not been able to decide upon the vegetable properties of the green matter, I have been fortunate in another research; I have always thought it probable that minute animalcules might be found in this water & have always looked attentively to that object; at length I found this evening upon the green matter a minute shell animal, shaped like a muscle or kidney, it is about the size of the smallest grain of sand, the colour of the shell is purplish brown, it opens the shell & thrusts out two articulated & very slender sharp clawed legs before; two more behind, but differently formed, the extremities of the hind legs having the appearance of some breadth as if composed of several hairs, from each shell behind there is a kind of tail of forked hairs 3 or 4 in number, which the animal has the power of moving; it probably feeds upon the green matter, or at least lodges in it; the sharpness of the claws before indicates that the animal finds its food by piercing & dividing longitudinally some tender matter.[88]

A considerable quantity of snow fell while we engaged on the Survey & after our return. Thermometer at 8h p.m. 30°. Extremes 30°–34°—3h p.m. 32°.

HUNTER

Therm. at 7 a m. 34°[.] at 3 p.m. 32° & at 8.p m.30°

Left the encampment at 8h.30m. a.m. The weather raw cold & like for snow[.] continued our course N.E. In half a mile passed the source of the califat, & the hill afterwards which gave it birth; we had scarcely proceeded two miles when a storm set in from the N.E. in our faces accompanied with rain & sleet, returned to camp & remained till the storm abated, when we set out again a 11h.15m. a.m. & proceeded N.E. till one p.m. Then E. half an hour over a ridge of mountain; Then S.

88. "Arbor Veneris" means tree of copper; copper deposits sometime appear as treelike branches. "Arbor Dianae" is a branching structure that forms in a mixture of nitric acid, mercury, and silver. "Minute animalcules" are microscopic organisms. The microscopic organism he describes, perhaps the most important scientific find of the expedition, is an ostracod, a minute crustacean (from the subclass *ostracoda*) found throughout the world. These creatures may be grouped into the broad category of thermophiles, which are heat-loving organisms that thrive in temperatures of 50°C and above. They are also included in a category of organisms called extremophiles, organisms that survive and thrive in extreme environments (hot, cold, acidic, high-pressure, etc.). This term usually, but not exclusively, refers to unicellular organisms. Many of the unicellular extremophiles are members of the *Archaea* family. Russell-Hunter, *Life of Invertebrates*, 243; Madigan and Marrs, "Extremophiles," 82–87; "Extremophiles," 157–158; Adams and Kelly, "Microorganisms That Grow in Extreme Environments."

for ¾ of an hour till ¼ past 2. p.m. Then S.W. over the hills till half past two; when our Guide shot a Doe. the skinning & dressing of which took up half an hour; when after each man had got his proportion to carry, set out again at 3 p m. SW for half an hour, when we killed another Doe & after dressing & giving each his proportion to carry, looked out for a place to pitch our tent, which we found shortly on the banks of a small brook which our Pilot said was a branch of the Bayu de saline which stretches towards the river Arkansa & empties into the Ouachita many leagues below.[89]

This day went about 12 miles by estimation, without a path or the sight of the Sun, guided by the compass. The land passed over is if any thing rather poorer more stoney & mountainous than yesterday The soil stones &cc, of which I collected samples being much alike

Saturday 29th

DUNBAR

Thermr. 25°. Wind at N. W. Strong all night, some flying clouds appear in the morning.—Got the people ready with their loads between 9 & 10h a.m. and I set out with them my self for the river camp; it began to snow at 10 o'clock, but did not continue; the weather continued cloudy, but the exercise of walking rendered the temperature (tho' cold) very agreeable; the low grounds thro' which we passed were a little watery in consequence of the rains which had fallen, but not more so than when we first walked out to the hot springs; the soil of the flat lands under the stratum of vegetable mould was chiefly yellowish and was evidently decomposed schistus, of which there were immense beds in every stage of its progress from the hard stone recently uncovered, partially decomposed and down to the yellowish earth apparently homegenious. The covering of vegetable mould between the hills and the river/ to constitute a good soil, being from four to six inches and it is the opinion of the people upon the Washita that wheat would grow here to great perfection.[90] Altho' the highest hills (300 to 600 feet) are very rocky, yet the inferior hills & sloping bases of the first are generally clothed with a Soil of a middling quality, the natural productions are sufficiently luxuriant, consisting chiefly of black and red oak intermixed with a variety of other woods and a con-

89. This may have been the South Fork of the Saline River. The South Fork rises in central Garland County and flows through that county. It meets the Alum Fork and the Middle Fork to form the Saline River in western Saline County. Arkansas Quadrangle Maps.

90. Wheat has never been grown in any significant amount along the Ouachita River. Some farmers plant winter wheat (genus *Triticum*) today to provide added resources for their farming operations. It is a significant cash crop elsewhere in the state.

siderable under growth; and even on those rocky hills, Nature has bestowed a soil which will reward the future labors of the industrious Vigneron: Nature herself indeed unaided by man has already planted on them three or four species of vines, which are said to produce annually an exuberance of excellent grapes. A great variety of plants, some of which in their season, I am informed, produce flowers highly ornamental, would probably reward the researches of the Botanist.

[W]hen the Doctor comes with the last party, I have appointed two good hands to chain the same distances to be noted down by young Mr. Hunter—At 8h. p.m. the thermr. was down at 24°—the wind blew strong all the afternoon, but fell calm by night.[91]

I omitted to observe in its proper place that having observed from the bottom of one of the hot springs a frequent ebulition of gas,[92] we should have collected some for examination, no apparatus was provided for the purpose, it was so unfortunate that we had not even a funnel at the Springs, which with a bottle might have sufficed. It was not hydrogen, because I failed in several attempts to inflame it by a lighted torch, there can be no doubt of its being Carbonic acid,[93] by which the lime and iron were disolved in the water. With respect to the quantity of hot water delivered by the spring I made the following rough estimate.—There are four principal springs, two of inferior note, one rising out of the gravel and a number of drippings and drainings all issuing from the margin or under the rock which overhangs immediately over the creek. of the four first mentioned, three deliver nearly equal quantities, but one (N°. 1) the most considerable of all and the hottest delivers about five times as much as one of the other three, the 2 of inferior note may be equal to one, & all the drippings and small springs are probably under-rated

91. Invented in seventeenth-century England by the mathematician Edmund Gunter, the surveyor's chain (also called a Gunter's chain) was made of thin metal rods with circular eyes on each end; the rods are connected by metal links. These chains were 66 feet long and contained 100 metal rods. Andro Linklater, *Measuring America: How the United States was Shaped by the Greatest Land Sale in History* (New York: Penguin Group, 2002), 5, 13–18.

Dunbar stated in the "Dunbar Trip Journal, Vol. II," "On the way into the river I took the courses by Compass and the distances by time; tomorrow or next day."

92. "Ebulition" means seething, overflowing, or boiling up, sometimes referring to an outburst of gas.

93. Carbonic acid, H_2CO_3, is formed when carbon dioxide dissolves in water. It forms carbonate and bicarbonate salts. Dunbar used litmus paper to determine the acidic nature of the water and explained this in the "Dunbar Trip Journal, Vol. II," saying that "it being pretty evident that by as excess of this acid the lime and iron ore both dissolved in the hot water of the springs; to litmus paper the water always communicated a pale red and no other acid could be detected by repeated trials of Doctor Hunter & myself." Litmus paper that turns a pale red usually indicates that a substance is slightly acidic.

at double the quantity of one of the three, that is, taking all together, the whole will amount to a quantity equal to eleven times the water delivered by the standard spring, which was the only one commodiously situated for measurement; I neglect the springs up the hill, because it is probable that what is not evaporated unites with the Springs below. we found a Kettle containing eleven quarts was filled by the standard spring in eleven seconds; Hence the whole quantity of hot water delivered by the springs issuing visibly from the base of the hill may amount in one minute to 165 gallons and in 24 hours 3771½ Hhds of 63 gallons each which is equal to a handsome brook and might work an overshot mill. In cool weather condensed vapor is seen arising out of the gravel bed of the Creek from springs which cannot be taken into the account; during summer and fall I am informed the creek receives little or no water but what is supplied by the hot-springs: at those seasons probably many small springs may be seen rising out of the bed of the creek , which are no invisible; during that time the Creek itself is a hot bath, too hot indeed near the springs, so that a person may chuse the temperature most agreeable to himself by selecting a natural bason nearer to or farther from some of the principal springs; at 3 or 4 miles below the spring, the water is tepid and unpleasant to drink.:[94]

HUNTER

Therm. at 7 a m. 25°[.] at 3 p m [blank in MS.] & at 8 p.m. 24° Weather Overcast & raw.

As we saw no appearances of minerals or mettals in this part of the country worth further search, & as all that I could find hitherto amounted only to a little iron ore, not rich enough to pay the expence of being worked, as the face of the country was like that we had seen, & the season far advanced, the time for which provisions the Soldiers had drawn, would expire in three days, & we were still 300 miles beyond the Post of the Ouachita where alone we could expect a supply, I thot it most prudent to return to Our Camp at the hot springs, which we effected by an ot[h]er rout that same evening; where I found Mr Dunbar had removed with all the soldiers & baggage to our old encampment at the boat on the banks of the Ouachita, leaving My Son & one Soldier to wait our Arrival. I now packed up the sam-

94. Between 1970 and 1973, M. S. Bedinger and others measured the mean monthly combined flow of the water from the springs and found a volume of 750,000 to 950,000 gallons per day. The Bedinger group also reported that the flow was heaviest during the winter and spring months. In a U.S. Geological Survey study between October 1998 and September 1999, a mean monthly flow of 617,000 to 695,000 gallons per day was recorded. Bedinger et al., *Waters of Hot Springs National Park*, 33; Petersen and Mott, *Hot Springs National Park*, 13. On this day, Dunbar reported the temperature at 8 p.m. to be "24°." "Dunbar Trip Journal, Vol. II."

ples, of stones ore &cc, the venison we had killed & got every thing ready to follow him in the morning.

Sunday 30th

DUNBAR

Thermr. in air 9°—in river water 36°—wind very light at N. W. This morning & the night past are the coldest we have experienced this winter. The People set off very early to bring in Doctor Hunter's baggage from the springs. Employed myself in bringing up my journals, etc. The Doctor arrived with the people about 3h. p.m.[95]—The sky was most serenely clear this day, its color over head was that of the darkest prussian blue and during last night the stars shone with uncommon lusture. People have conceived an idea that they see more stars here and at the hot springs than any where else; which idea arises from the extreme transparency of the atmosphere, which causes the stars to strike the eye with greater brightness, & no doubt stars of inferior magnitude will be seen in a pure sky which are invisible in an ordinary one. This evening some light clouds appeared about the Sunsetting which is an indication of change of weather; we now anxiously expect rain, as we wait only for the first rise of the river to go down with safety over the falls and rapids; 5 or 6 feet perpendicular will be sufficient. At night the atmosphere became again extremely bright—at 8h. p.m. the themr. was at 21°. Extremes 9°–38°—It became very cold at 10h.p.m.—

HUNTER

Therm. 10° at 7 a.m. at 3. p.m. 38°[.] at 8 p.m. 21° Clear & cold.

Set out from the hot springs about 10 A.M. carrying baggage & everything left the day before, which we were enabled to do as Mr Dunbar had sent all the men that could be spared for that purpose, where we arrived about 3 p.m. All in good health (except Tuttle who had been long troubled with a dysentery & griping.)[96] yet he was now much better. Being now arrived at the banks of the Ouachita we found the river about 1½ foot higher than when we came up, yet Mr Dunbar judged it safer to wait for a further rise, which might be expected to take place at this time as it begain to cloud over & appeared likely for rain or snow[.]

95. This entry shows that there was a second volume to Dunbar's trip journal that continued past December 10, 1804. "Thermr. at that hour [3 p.m.] 38°." "Dunbar Trip Journal, Vol. II."

96. Hunter earlier identified this soldier as William Tutle. McDermott, "Western Journals of Hunter," 65.

Monday 31st

DUNBAR

Thermr. in air 29° in river water 36°.—Wind S. E. During the night the Weather altered greatly, the temperature was much molified and the stars disappeared; in the morning one general cloud enclosed the horizon, and from the damp penetrating chilliness of the morning we look for snow: ordered setting poles to be made and every thing to be prepared for the first favorable moment to depart. The day continued cloudy & in the afternoon the thermr. having risen to 32° it began to snow and continued all day and part of the night:—Examined some of the green moss from the hot-springs with a view to shew Doctor Hunter one of the Bi-valved testaceous animals,[97] found a large one which under the microscope measured 1/50 of an inch in length by the micrometer.

HUNTER

Therm at 7. A M. 29°[.] at 3. p.m. 32° & at 8 p.m. 32° Snowed all this day & night, the snow was about one foot deep next morning which by measurement yeilded about 1½ Inches water. We are now waiting for a thaw to raise the water in the river & enable us to decend it with safety.

Whilst I was out in the last excursion of three days; Mr Dunbar in company with my Son run the chain round the hot spring hill & found it to be about 3 1/5 miles in circumference, this includes another hill behind it, which at the base appear as one & the same tho at the tows they separate, appearing double headed.—Tuttle continues better.

Yesterday as we returned to our boat at Ellis's camp on the banks of the Ouachita, my son with the aid of two of the soldiers measured the distance from the hot Springs here. & found it to be by the Chain 8 1/5 miles nearly by the following courses Which were taken the day before by Mr Dunbar viz S 15 E 788 perches—to 1st. knoll 122 p. to 1st branch 162 to 2nd. branch 282 p, to 3rd.do 322 p., to 4th. do 502 p, to crossing hot spring creek 614 p, to branch at the station.—N 80 E, to the top of the ridge 70 p. S 69 E 184 to 2nd branch. S 25 E 160 p. S 68 E, 80 p. to big lick, N 55 E 200, at 160 p. 3rd branch, 2nd lick at Station—N 82 E 534, at 168, 4th branch 5th branch at station. S 84 E 122, at 56 cross last branch, Califat at station, (course of Califat S 38 E) S 74 E 178 to 3rd lick, S 54 E 304. at 94 a branch & to camp at Ouachita Total 2620 perches equal to 8 miles 60 perches—Distance from the post to the hot springs 311 miles 61 perches

97. "Thermometer at 3h. p.m. at 32° and at 8h. p.m. at the same freezing point." "Dunbar Trip Journal, Vol. II." The "Bi-valved testaceous animals" are, again, probably ostracods. See Dunbar entry for December 28, 1804.

January 1805

Tuesday 1st

DUNBAR

This morning the thermometer was at 26°—it had ceased snowing in the night, but recommenced after day light; the snow was sounded and found in most places to be from 11 to 13 inches; we are in hopes that the melting of this snow united to the rain which will probably accompany the thaw, will be sufficient to take us down in safety; being desireous however of ascertaining what aid we had to expect from the Snow, I made the following experiment—I took a Cylindric Kettle 10 inches deep & having by sounding found a flat piece of snow of the same depth, I pressed down the Kettle perpendicularly to the ground, I was thus enabled to return the Kettle completely filled with its column of snow, and having thawed it gradually to the temperature of 33°, I found the water to measure exactly 1.07 inches, that is 9.346 inches of snow will yield one inch of water in the circumstances above mentioned: it is observable that the snow fell lightly without wind, it is therefore probable that the proportion of ten to one may be adopted as a general standard to the varied according to circumstances. The snow continued frozen all day & the thermr. at 3h. p.m.did not fall below the freezing point and in the evening at 8h. p.m. it was fallen to 18°.[1]

HUNTER

Therm at 7. a.m. 26°[.] at 3 p.m 32° & at 8 p m. 18°. Thick Snowy weather. no observation.

The snow which had been falling yesterday & last night began again this day but did not last long. We are here still waiting for rain to raise the water in the river to facilitate our progress down.

This day finishes the period for which the soldiers had drawn provisions. & as

1. This would be a very large and rare snowfall measurement for Arkansas today, but in the nineteenth century such snowfalls were more common. The average annual snowfall today in south Arkansas is 2.8 inches. www.sosweb.state.ar.us.

their flour was already expended we Issued to them about half a bble of our own. Their Bacon, having been saved since we came to this place by the game killed, will last some time longer.

The river continues to fall about 1 foot every twenty four hours, or more,

Wednesday 2nd

DUNBAR

Thermometer in air 6°. in river water 32°. Calm—The night proved extremely cold; large fires with all the covering that could be conveniently used were necessary to render our situation comfortable in a bad tent, negligently chosen at New Orleans. The sun arose bright and shone with splendor upon the surface of snow which covered every object upon the ground; the river alone presented a bleak appearance with a condensed vapor floating upon its surface, the temperature of its water was at the freezing point; a Kettle of the water being brought up to the Camp and placed on the ground four feet from a large fire, its surface began immediately to shoot into icy chrystalizations.—

Our hunters are tolerably successful, bringing in every day abundance of venison and turkies.—The day became pleasant and agreeable, the temperature at 3h. p.m. being 45°, and at 8h. p.m. the thermometer fell to 32°.

HUNTER

Therm. at 7. a.m. 6°[.] at three p.m. [blank in MS.] & at 8 p m. 32° The river water 32° in the morning. Weather clear & cold.—

Being still detained by the lowness of the river, at noon I made the following observation for the Latitude Doub. Appt. Mer. Alt. 64°. 46′.36″ Index error—1′.9.6″ Lat found. 34°.27′.8″.—N.B. I used this day Mr Dunbars Artificial horizon with plate glass cover, whereas I generally use a cover of thin Talc or isinclass fixed in a tin frame.—[2]

Our guide brought in a fat Doe from the woods which he shot—Our boat is now all in order to go down, having altered the situation of the cabbin, so as to make it more commodious & more suitable for using a steering oar, & having also cut a number of setting poles to replace those broken & lost coming up. The river still continues to fall.

2. "Thin talc" is hydrous magnesium silicate, $Mg_3(Si_4O_{10})(OH)$. Similar to mica, talc is a soft green-gray mineral that may be sliced into thin semitransparent pieces. By "isinclass" Hunter means *isinglass*, a common term that referred to a semitransparent covering of mica.

Thursday 3rd

DUNBAR

Thermometer in air 22°. in river water 34°—wind moderate at N. W. The atmosphere became cloudy in the night and we looked confidently for a change of weather, but this morning it has become serene & fine. The vicissitudes of the weather have of late been frequent, a change is now extremely desirable but the season obstinately bent against all change. The day became pleasant and of an agreeable temperature, the thermometer at 3h. p.m. being at 48° and at 8h. in the evening 30°.

HUNTER

Therm. in air at 7. a.m. 22°[.] in the river 34°[.] at 3. p.m.45° & at 8 p.m. 30°—Clear. Wind N.W.

The River now falls but slowly. The snow melts in the heat of the day by such insensible degrees as not to be felt in the river. At noon made the following observations for the Latitude viz ⊙ double Appt. Mer. Alt. 64°.57′.45″ Ind. Error, 1′.9.6″ Lat found 34°.27′.17.5″.

3 of our people went out to hunt with the Guide & brought to the encampment a fine fat Doe & a young cub.[3] Bear.—

Friday 4th

DUNBAR

Thermometer in air 22°.[4] in river water 36°.—Calm—During the night it became cloudy, not a star was to be seen, but before day it cleared away & became perfectly serene and cloudless, The day proved fine, the sky over head of a bright but deep prussian blue. [T]he temperature mild, the thermometer at 3h. p.m. being up to 50°. In the afternoon the Doctor made an excursion upon the river to examine some of the neighbouring hills: I continued to bring up & arrange my Journals. The evening was fine, the thermometer at 8h. p.m. was at 32°—no favorable appearance as yet of rain to raise the river; the snow is disappearing without producing any beneficial effect: we continue here as prisoners, waiting for what is usually called bad weather, to bear us away from this place.

3. These cubs were probably born in the early months of 1804, since January and February are the usual months for bear births. Sow bears usually give birth to two cubs (though occasionally one or three), and cubs normally stay with their mother for one year. Sutton, *Arkansas Wildlife*, 226–227. Dunbar also wrote on this day, "our hunters last night brought in some venison and a Cub-bear of the last spring, the mother and two other Cubs having escaped from the hunters." "Dunbar Trip Journal, Vol. II."

4. In the "Dunbar Trip Journal, Vol. II," he recorded a temperature of 21°.

HUNTER

Therm. at 7 a.m. in air 21°[.] in River 36°—at 3 p m. 50° & at 8 p.m. 32° Clear & pleasant.

The river still continues to fall a little; altho the snow g[r]adually melts a little in the day, yet it is not felt in the river. I made the following observation for the Latitude viz ⊙ d. Appt Mer. Alt. 65°.10″.6′ Ind. error 1′.23.4″ Lat. 34°.27′.25.3″

I took a short turn up & down the river in the boat to search for Whetestones as samples & brought in a few of the yellowish & flesh coloured flints like the turkey oil stones. They did not well bear dressing into form, as we had not tools suitable. I ground a face upon one of them with some difficulty & it seemed to answer tollerably well with oil to give an edge to my tomahawk—Another of the slates, [a] kind almost black but soft, gave a fine edge to a razor, when dressed smo[o]th. The flesh coloured flint above seemed too hard rather, this was picked up on the beach where it had been long exposed to the weather. I believe they ought to be dressed as soon as they are dug up or quarried.—The hunting parties brought in two deer in good order & my son shot a young swan flying which proved very good eating being fat.

Saturday 5th

DUNBAR

Thermometer in air 22°. in river water 36°. Wind N. W. The atmosphere became cloudy in the night but was perfectly serene and clear at day-break, so that we have yet no near prospect of our departure. The day became fine and seemed to invite us to recommence astronomical observations, and altho' a sufficient series had been made both for Latitude and Longitude at the hot-springs connected by survey with this place, yet we commenced a new series. Equal altitudes of the sun were taken before and after noon; three distances of the moon and Sun's limbs were taken near 2h p.m.[5] and in the evening three distances of the moon's west limb from Aldebaran were taken between 6 and 7h p.m.—a greater number would have been taken, but in the first case the Sun got behind some trees and in the second case, the moon was in a similar situation, if tomorrow proves fine we shall prosecute the same operations to more advantage, having ordered several trees to be cut down which stood in the way—Wind S. E.

The day continued fine and of a mild temperature; some few clouds keep up our hopes of change—Thermometer at 8h. p.m. 28°. Extremes 22°–55°[6]

5. These are the outer edges of the Sun's corona. It was impossible to look directly at the Sun to make observations, so readings were taken by covering the Sun and calculating from the "Sun's limbs."

6. "At 3 p.m. the thermr. was at 55°." "Dunbar Trip Journal, Vol. II."

[TABLE 5]

Equal Altitudes

⊙ dble Appt. Alt. 43°.18′.30″ (In er. +13″.15″)

A M		PM	
upper limb	9h.43m.10″	Lower limb	2h.59m.22″
Center	45 .12	Center	3 . 1 .27
Lower limb.	47 .19	upper limb	3 . 3 .33

at 10h.3m.42″ (In er. +13″.15″)

⊙ Mag.Az S46.[1] E

⊙ appt. dble alt L L [Sun's Lower Limb] 47°.21′.10″

Note (added by editors):

1. For more on magnetic azimuth see Hunter's entry for December 17.

HUNTER

Therm. at 7. a.m. in Air 22[.] in the river 36 & at 3 p.m. 55[.] at 8 p.m 28 Clear & cold.

We are still detained by the want of water of the river, at Ellis's Camp Our Hunters brought in two Dear a Turkey & rabbit,[7] this last appears to be of a size between an European Hare & rabbit, it is of a brownish grey on the back, belly inclining to white, thick fur.

during the day & evening the following double Altitudes of the sun were taken, by Mr Dunbar, also the distances between the sun & moon & in the evening the distance between the moon & the Star Aldebaran whilst I took the Moons Altitude.—[tables 5 and 6]

Sunday 6th

DUNBAR

Thermometer before sun-rise in air 28°. in river water 38°. This morning proved cloudy contrary to expectation and revived our hopes of a change of the weather favorable to our descent. This State of the atmosphere continued all day; from time to time, there was a little very fine rain or mist.[8] The rain increased a little after dark, but still very light. The snow seems now melted away to about one fifth or sixth of the original quantity, we began to apprehend that the whole would

7. The rabbit could have been the eastern cottontail (*Sylvilagus floridanus*), length 13–20 inches, or the swamp rabbit (*Sylvilagus aquaticus*), length 17–22 inches. Sealander and Heidt, *Arkansas Mammals,* 110–114.

8. "At 3h. p.m. the thermr. was at 50°." "Dunbar Trip Journal, Vol. II."

[TABLE 6]

Lunar Observations
distance between the ☉ & ☽ limbs [Sun & moon's right limb]
(Ind. E. + 13′.15″)

at	Distance
2h.22m.45″	54°.1′.0″
" 25 .50	" 2 .0
" 28 .45	" 3 .0

Distance between the ☽'s west limb [Moon's west limb] from Alldebaran
(Ind. E. + 13′.15″)

at	Dist.
7h.1m.56″	84°.52.0″
" 4 .0	" 51.0
" 6 .6	" 50.0

dbl.alt ☽ 1 limb. 64°.17′.3″ (Ind Er. —1′.20″)

disappear without any influence upon the river, but now it has risen about 12 inches. Thermometer at 8h. p.m. 44°. Extremes 28°–50°.

HUNTER

Therm. at 7 A M. in Air 28°, in river 38°[.] at 3 p m. 50° & at 8 p.m.—44° Thick drizly weather with small rain.

We are still waiting at Ellis's camp for the river to rise which the present weather bids fair to do.—no observation

Monday 7th

DUNBAR

Thermometer in air 64°. in river water 44°. Last night it rained by intervals, but so slightly that a cylindric vessel placed to receive it, did not contain enough to be measured. During the night the temperature was extremely warm, and the weather continues to be cloudy but not very dark, so that our prospect of rain is not very flattering; the river has nevertheless risen 18 inches since last night, which has no doubt been caused by the melting of the snows. The sun shews himself at intervals between the clouds: it became so warm, that we dined abroad under the shade of some pine and oak trees, upon the wild game of the forest and the river, such as Venison, wild Turkey, bear, cygnet, &c., the thermr. at the hour if dinner was 75°, which at this season produces the sensation of a summer's sun of 90°;

the river continues to rise, and we have taken our resolution to wait the issue of the present state of the weather; and set out at all events; if there be not water enough to go over the falls with safety by the oars, we shall pass along by letting ourselves down by the help of a rope, step by step, until the danger is passed. Thermometer at 8h. p.m. 38°. Extremes 38°–78°. In the evening the river continued to rise.[9]

HUNTER

Therm. at 7. a.m. in Air 64°[.] in the river 44° At 3 p m 78° & at 8 p.m. 38° Cloudy no observation

The river has risen about 3 or 4 feet perpendicular at 3 p.m. & continues to rise gradually.—

Tuesday 8th

DUNBAR

Thermometer in air 28°. in river water 46°. Last night was cloudy, moist and cold, the river rose considerably in the night; as we suppose it to be about 6 feet perpendicular, higher than the level of the river when we came up, we now think ourselves secure of going down with speed and safety; orders were therefore given to embark our baggage and prepare for departing. We had the satisfaction of taking with us an abundance of fresh provision chiefly venison, to supply us to the Post of the Washita. We accordingly set off between 9 & 10 o'clock, and landed a little below upon the opposite side, and went to examine the first rapids, which we found to be very safe; we re-embarked, and by directing our course between the breakers, passed along with the rapidity of an arrow in perfect security; we continued with great rapidity on the face of the current but thought it prudent to land and view a second rapid, and after exploring the best passage we passed down in perfect safety.

We got over the great 'Chutes' about 1 o'clock, two of our oars having been violently dashed overboard by the willows, the Pilot thinking it safest to keep the eastern shore on board; we halted below and regained our oars by sending up the Canoe. Here we dined and went on & stopped a little below to examine the flinty promontory already noticed on the 3d December. [W]e took some specimens of the rock resembling the Turkey oil stone; it appears to me to be too hard; I re-

9. The "cygnet" was a young swan, probably a trumpeter swan (*Cynus buccinator*). James and Neal, *Arkansas Birds*, 100–101. "Thermometer at 3h. p.m. 78°." "Dunbar Trip Journal, Vol. II."

marked that the strata of this chain ran perpendicularly nearly East & West, crossed by fissures at right angles, 5, 6, to 8 feet apart, the laminae were from ¼ to 4 or 5 inches thick: About a league below on the same side landed at Whetstone hill, and took several specimens; this projecting hill consists of a mass of greyish blue schistus of considerable hardness and about 20 feet perpendicular, near the top, it was in a state of progression towards decomposition, being there extremely crumbly and part of it changing into a dirty yellowish color: the laminae were perpendicular in general perpendicular, but not regularly so, and from ¼ of an inch to 2 inches in thickness, but did not split with an even surface. Went on and encamped about ten leagues below Ellis' Camp.[10] Thermometer at 8h. p.m. 37°. Extremes 28°–37°. It rained lightly after we encamped, which rendered the flat ground of our encampment very wet and the wood difficult to burn.

HUNTER

Therm. at 7 a m in air 28°[.] in the river 46°[.] at 3 p m. 37° & at 8 p.m. 37°. Cloudy Wind N.E.

This day, having put every thing on board the boat, we struck tents & took our departure from Ellis[']s Camp on our way back to the Mississipi to the no small joy of all hands, as their provisions, especially of flour & whiskey were for some time past expended, at half past 9. a.m. We soon ran over the rapids, falls & rocky cascades which had so much interrupted our progress up; yet as our pilot appeared to be very much afraid of touching the rocks with the boat which with the present velocity would have peirced her bottom, it was thought most prudent to stop always above such places & go along the banks to examine them to find the best channel, These stops delayed us so much that with them & stopping two or three times to examine & take samples of the stones we made only ten leagues to day; for it came on to rain this afternoon (yet we made the same distance this part of a day that took us four days in coming up) which obliged us to encamp sooner than we otherwise would have done, a little above the prairie de champignole.[11]

Wednesday 9th

DUNBAR

Thermometer in air 42°, in river water 44°—The river fallen about 6 inches—During the night it rained by intervalls, but very lightly, the air was moist and cold,

10. "Whetstone hill" is probably below the town of Malvern in Hot Spring County. In the "Dunbar Trip Journal, Vol. II," he refers to their camp site as "little Bayu or Ellis' Camp" and records a temperature at 3 p.m. of 37°.

11. The "prairie de champignole" is near the community of Social Hill in Hot Spring County. Arkansas Quadrangle Maps.

the soil here immediately under the vegetable stratum is yellowish and of little consistency, resembling greatly the understratum observed near the hot springs, produced probably by the same cause, the decomposition of shistus. Last evening ordered provisions to be dressed for the day to save the time of landing during the day for that purpose; about two miles below our Camp landed to examine some freestone and blue slate in sight of Bayu de la Prairie de Champignole mentioned the 2d Decr. The free stone of which we took specimens, seems proper for grind stones, scythe-stones &c., but the blue slate as it is called is only bluish shistus, hard and brittle and not proper for the roofing of houses; we have not seen slate good for that purpose, except some discovered on one of the Doctor's excursion on the Bayou Calfat. Much game on the river, such as Geese, Ducks, Swans &c., they continue equally wild and difficult of approach as before, so that we derive but little benefit from that source.

The Day continued dark, cloudy & cold with the wind at North at 11h a. m. it began to snow and hail rain by intervals. We observed nothing this day meriting remark, different from what we saw on the way up. Towards evening it began to clear away, and soon after we encamped the sky became serene. By the Pilot's estimation we made 19 leagues, which probably do not exceed forty miles: we propose five of our night encampments on the way up. Encamped a league above 'Cache à Maçon'[12]—Slept a little higher on the 27th Novr. Thermometer at 8h. p.m. 24°. Extremes 24°–42°. At 3h p.m. 36°. The moon & stars shone with uncommon lusture.

HUNTER

Therm. at 7 a m 42° in air, & in the river 44°[.] at 3 p.m 36° & at 8 p m. 24° Wind N E. Rain hail sleet

Set out & [at] 8 A M. having taken an early breakfast, after having proceeded down a few miles, stopped & gathered a few samples of a sort of slatey whetstone & at another place of a sandy freestone. The river has fallen about a foot during the night, which the present rain will soon replace. We are now past all difficult rocky places. At noon passed the saline de Bayu des Roches, this was the last saline I visited where the water was not quite so strong as the former.—

at ¼ past noon passed the Bayu or Fourche de Cadeaux on the right going down.—About 3 p m. passed the Grand Glaise (great lick) & at 4 p m. the Bayu de leau Froid. (Cold Water River)—passed also the first place where I went & found the strong salt water. We passed this day about five days journey going up, & en-

12. This location is about 4 miles southwest of Dalark in Dallas County. Ibid.

camped about a league above the Cache de Macon near the Island of Ouachita below the great chutte.—

Thursday 10th

DUNBAR

Thermometer in the air 23°, in river water 42°—river fallen 7 inches. The face of the heavens changed much in the night; it became extremely dark and cloudy, and this morning with the wind at North, it is cold, damp and penetrating; the river fallen seven inches during the night. After setting out, the clouds began to dissipate & the Sun to shew himself, a very agreeable sight to travellers in cold and unpleasant weather; it continued never-the-less cold all day, the sun not possessing power to soften the rigorous cold which prevailed, the thermometer not rising above the freezing point from morning until night. We made this day by the Pilots account fourteen leagues and encamped at 'Auges D'Arclon' (Arclon's troughs) three leagues below the little misouri; slept near this place on the 23d November:[13] it appears by reference to the Journal that we were thirteen days in going up from this place to Ellis' Camp, which has required but three broken days to come down, having made several stops to examine certain objects on our way down, and to day we made a considerable delay at the Camp of a M. Le Fevre.[14] This was an intelligent man, a native of the Illenois, now residing at the Arcansas, he is come here with some Delaware & other Indians whom he has fitted out with goods & receives the peltry, fur &c at a stipulated price, as it is brought in by the hunters. This gentleman informs us that a considerable party of the Osages from the Arcansa river have made an excursion round by the prairie towards the red river, & down the little misouri as low as the fourche d'Antoine,[15] and there meeting with a small party

13. In his entry for this date, Hunter explains that Arclon's Troughs was so named because "one Arclon made troughs to carry down his bears Oil to market."

14. Pierre LeFevre settled near the mouth of the St. Francis River and along the Mississippi around 1802. Prior to his bankruptcy in 1810, LeFevre was a prosperous landowner near Arkansas Post, holding five tracts of land that totaled more than 1,000 acres and a substantial colonial style home with elaborate furnishings. In 1812 he received a land grant along the Cache River in eastern Arkansas. He traded heavily with Indian hunters and extended credit to them, including the Quapaw, the Cherokee, and the Chickasaw. LeFevre apparently accompanied the Dunbar-Hunter expedition to Fort Miró. From the journals of Dunbar and Hunter it seems obvious that LeFevre identified numerous new sites (some labeled by the names of hunters and traders) along the river. McDermott identified this man as possibly John or Jean LeFevre. "Certificate #1161," in Lowrie, *Early Settlers of Missouri*, 2:598; Shinn, *Pioneers and Makers of Arkansas*, 43, 120; Arnold, *Rumble of a Distant Drum*, 52, 56–58; McDermott, "Western Journals of Hunter," 110.

15. The "Arcansas" probably refers to Arkansas Post on the Arkansas River. Established in 1686 by

of Cherokees, are supposed to have killed four of their number and others are missing; Three Americans and ten chicasaws went a hunting into that quarter, who may also have been in danger; those Osages being no respecters of persons. M. Le Fevre possesses considerable knowledge of the interior of the Country; he confirms the accounts we have already obtained that the hills or mountains which give birth to the various sources of this little river, are in a manner insulated, that is they are entirely shut in and enclosed by the immense plains or prairies which extend beyond the red river or at least some of its branches river to the South & beyond the Missouri (or at least some of its branches) to the north and range along the eastern base of the great chain or dividing ridges commonly known by the name of the Sand hills, which separate the waters of the Missisippi from those which fall into the western pacific ocean. The breadth of this great plain is not well ascertained, it is said by some to be at certain parts or in certain directions not less than two hundred leagues, but I believe it is agreed by all that have a knowledge of the Western Country, that the mean breadth is at least two thirds of this quantity;[16] A branch of the Misouri called the river platte or shallow river takes its rise so far South, as to derive its first waters from the neighbourhood of the sources of the Red and arcansa river.[17] By the expression planes or prairies in this place is not

---

Henri de Tonty as a trading post for the French, Arkansas Post continued in varying states of prosperity and became the territorial capital in 1819. In 1820 Little Rock became the territorial capital, and Arkansas Post quickly declined. Whayne et al., *Arkansas*, 48–50, 54–55, 96–97. The Antoine River begins in northern Pike County and runs southeasterly, forming the western boundary between Clark and Pike Counties. It joins the Little Missouri near the present community of Okolona in Clark County. Arkansas Quadrangle Maps.

16. Dunbar recorded in the "Dunbar Trip Journal, Vol. II" on this day a different passage concerning the Great Plains: "in breadth due west, & extending probably from the heads of the rivers to the South of the red river, which fall into the bay of Mexico, as far northward as the Missouri & perhaps farther." The North Platte River runs from northeast Colorado through southeast Wyoming and through Nebraska, where it meets the Mississippi. The South Platte begins in northeast Colorado and runs through Nebraska. It joins the North Platte to form the Platte River near North Platte, Nebraska. *National Geographic Atlas of the World*, 39–40.

17. The two forks that converge to make the Red River are Prairie Dog Town Fork, which originates in western New Mexico, and North Fork, which rises in the Texas Panhandle within Carson County. Dunbar, like many other Americans of the early 1800s, was mistaken in thinking that the Red River had its source in the Rocky Mountains. The Red River's source had been theorized to be close to that of the Arkansas River. Many people believed that the Red River rose just east of Santa Fe or just north of Taos, New Mexico. Even Zebulon Pike used this information when he began his western voyages in 1806. The famed Russian scientist and explorer Alexander von Humboldt, in his "General Map of the Kingdom of New Spain," also placed the river sources close together. Pike soon learned that the true source of the Red River was, unfortunately, hundreds of miles to the east (Pike never discovered the sources). William Clark also place the sources fairly near each other in his 1805 map. The Arkansas River originates in central Colorado near the town of Leadville, approximately 450–500 miles from

to be understood a dead flat resembling certain savannahs, whose soil is stiff and impenetrable, often under water & bearing only a Coarse grass resembling reeds, very far different are the western Prairies, which expression signifys only a country without timber: These Prairies are neither flat nor hilly, but undulating into gently swelling lawns and expanding into spacious vallies in the center of which is always found a little timber growing upon the banks of brooks and rivulets of the finest water. The whole of those prairies is represented to be composed of the richest and most fertile soil, the most luxuriant and succulent herbage covers the surface of the Earth interspersed with millions of flowers and flowering Shrubs of the most ornamental and adorning kinds: Those who have viewed only a Skirt of those prairies, speak of them with a degree of enthusiasm as if it was only there that Nature was to be found in a state truely perfect; they declare that the fertility and beauty of the rising grounds, the extreme richness of the vallies, the coolness and excellent quality of the water found in every valley, the Salubrity of the atmosphere and above all the grandeur and Majesty of the enchanting landscape which this country presents, inspires the Soul with sensations not to be felt in any other region of the Globe. This Paradise is now very thinly inhabited by a few tribes of Savages and by immense herds of Wild Cattle (Bison) which people those countries; the Cattle perform regular migrations according to the seasons from south to north, and from the planes to the mountains, and in one time taught by their instinct take a retrograde direction on: those tribes move in the rear of necessity. Herds and pick up stragglers & such as lay behind, which they kill with the bow and arrow for their subsistence; should it be found that of this rich and desirable Country there is 500 miles square, and from report, there is probably much more, the whole of it being cultivated, it will admit of the fullest population, and will at a future day vie with the best cultivated & most populous countries on the Globe; in this particular the province of Holland exceeds perhaps all others; there, one million of acres support two millions of Inhabitants, but as Maritime Countries enjoy superior advantages respecting population, by the interchange of their manufactures for the necessities of life, which last in an inland country must be totally drawn from the product of the proper soil, we shall suppose this new Country to be populated in the proportion of one tenth only of that of Holland, in which case it will be capable of subsisting a nation composed of twenty six millions of souls. This County is not exposed to be ravaged by those sudden and impetuous deluges of rain which in most hot countries and even in the Missisippi Territory, do some-

---

the Red River sources. Ibid.; New Mexico Quadrangle Maps, Colorado Quadrangle Maps, and Texas Quadrangle Maps, U.S. Geological Survey, Denver, Colorado, 1983; Clark, "Map of Part of the Continent of North America"; Allen, *North American Exploration*, 3:33, 36–43.

times tear up & sweep away with irresistible fury the crop and the soil together; on the contrary, rain is said to become more rare in proportion as the great chain of mountains is approached and it would seem that within the sphere of attraction of those elevated chains little or no rain falls upon the adjoining planes; this relation is the more credible, as in the respect our new Country may resemble other flat or comparatively low countries similarly situated, such as the Country lying between the Andes and the Western Pacific: the planes are dappled with nightly dews so extremely abundant as to have the effect of refreshing showers of rain, and the spacious vallies which are extremely level may with facility be watered by the rills & brooks which are never absent from those situations: such is the description of the better known country lying to the south of the red river, from Nacokdoches towards St. Antonio in the province of Texas: the richest of crops are said to be produced there without rain, but agriculture in that quarter is at low ebb, the small quantities of maize furnished by the Country, is said to be produced without cultivation, a rude opening is made in the earth just sufficient to deposit the grain at the distance of four or five feet in irregular squares, and the rest is left to nature; the soil is naturally tender, spongy and rich, & seems always to retain humidity sufficient with the bounteous dews of heaven to bring the crops to maturity.

The red and Arcansas rivers whose Countries are very long, pass thro' portions of this fine Country, they are both navigable to an unknown distance by boats of proper construction; the Arcansas river is however understood to have greatly the advantage over its neighbour with respect to facility of Navigation: some difficult places are met in the red river below the Nakitosh, after which it is good for 150 leagues (probably the computed leagues of the Country of nearly two miles each) there the Voyager meets with a very serious obstance, viz the commencement of the Raft as it is called, that is, a natural covering which conceals the whole river for an extent of 17 leagues continually augmenting by the drift wood, supports at this time a vegetation of every thing abounding in the neighbouring Forest, not excepting trees of considerable size, & the river may be frequently passed without any knowledge of its existence; it is said that the annual inundation is opening for itself a new passage thro' the low grounds near the hills, but it must be a long time before Nature unaided will dig out a passage sufficient for the reception of the waters of the red river; about 50 leagues above the natural bridge is the residence of the Cadeaux or Cadadoquis Nation, of whose good qualities we have already spoken; the Inhabitants estimate the Post of Nakitosh to be half way between New Orleans and the Cadeaux Nation:[18] above this point the red river is said to be em-

18. "Nakitosh" is Natchitoches, Louisiana. Louis Juchereau de Saint Denis founded Natchitoches in 1714 as a French trading post—fort. Denis had selected a site along the Cane River and near the vil-

barrassed by many rapids, falls and shallows, none of which are said to be met with in the Arkansa river as high as it is known, except in the very lowest state of its waters; the navigation is reported to be safe and agreeable, the lands on either side are of the best quality & well watered with Springs, brooks & rivulets, & many situations proper for mill-seats; from the description it would seem, there is along this river a regular graduation of hill and Dale presenting their extremities to the river; the hills are gently swelling eminencies and the Dales are Spacious Vales with living water meandering thro' them: the forests consist of handsome lofty trees, & chiefly what is called open woods, without cane-brake or much underwood; the quality of its land is supposed much superior to that of the red river until it ascends to the Prairie Country, where the lands are probably very similar.[19] About 200 leagues up the arcansa is an interesting place called the Salt Prairie; there is a considerable fork of the river there, and a kind of Savannah where the salt water is continually oozing out & spreading over the surface of a plane; during the hot dry summer season, the salt may be raked up into large heaps; a natural crust a hand-breadth in thickness is formed when the dry season prevails; this place is not often approached on account of the danger from the Ozage Indians;

---

lage of the Natchitoches tribe of the Caddo Indians. Natchitoches became one of the final vestiges of European civilization for those people ascending the Red River or traveling overland to the Texas frontier. In 1804, just a few months before the Dunbar-Hunter expedition began, the United States had begun to station troops at the small fort. By 1806 the population at Natchitoches was estimated at 500–600. The town is now the oldest continually occupied European settlement in the Louisiana Purchase. Sibley, "Report from Natchitoches"; Robertson, *Louisiana,* 1:125; Arnold, *Rumble of a Distant Drum,* 4; Flores, *Jefferson and Southwestern Exploration,* 11, 119.

The "Cadeaux nation" may mean the Kadohadacho village near the old French fort of St. Louis des Caddodoches along and above the great bend of the Red River, or some of the other Caddo Indian villages. By 1805 most Caddo were living in an area near Caddo Lake (called Tso'to Lake by the Caddos) south of the great bend of the Red River and northwest of present-day Shreveport, Louisiana. Additional scattered groups lived upstream from the great bend. Carter, *Caddo Indians,* 218.

19. In the "Dunbar Trip Journal, Vol. II" Dunbar described the western lands adjacent to the Arkansas and Red Rivers this way:

> both of [the rivers] run a very long course thro' immense regions of the richest and most fertile lands, and are likewise both navigable for flat built & well constructed boats to an unknown distance; the great obstacle upon the Red river, has already been mentioned, viz the covering which has been laid over it by drift logs, which is continually increasing from above, and is said to extend 17 leagues, it exists between the Nakitosh and the Cadauz nation, about 150 leagues above the first: but no obstacle of any kind is found on the arcansa river; the navigation of which is safe and agreeable without shallows or rapids; the lands on either side are of the best quality well watered with springs, rivulets & creeks sufficient for the turning of Mills.

much less do the white hunters venture to ascend higher where it is generally believed that silver is to be found. We have also been informed that high up the arkansa river, salt is to be found in form of a Solid rock & may be dug out with the Crow-bar.[20] The waters of the arcansa like those of the red river, are not potable during the low state; they are both charged highly with a reddish earth or marl, and are also extremely brackish, this inconvenience is not greatly felt upon the Arkansa, where springs, rills & brooks of the finest fresh water are so frequent; the red river I believe is not so favorably situated. Every account seems to demonstrate that immense natural magazines of salt must exist in the great chain of mountains to the westward; all flowing from those mountains during the dry season, all rivers flowing from those mountains during the dry season retain a strong impregnation of salt, until that property becomes imperceptible by the accession of the fresh waters of many other rivers.—The great western prairies, besides the herds of wild Cattle (Bison commonly called Buffalo), are also stocked with vast numbers of a species of wild goat, (not resembling the domestic goat) extremely swift of foot; as the description given of this goat, has not been very perfect, I have supposed from its swiftness it might be the antelope, or it may possibly be a goat which has escaped from the Spanish Settlements of new Mexico: I have conversed with a Canadian who has been much with the Indians to the westward; this man told me that he hade seen great flocks of an wood-bearing animal larger than common sheep; the wool is much mixed with hair. This is probably the same animal which has been described & of which a plate has been given in the medical repository of New York. The Canadian pretends also to have seen an unicorn; the single horn he says rises out of the forehead & curls back according to his description so as to convey the idea of the fossil Cornu Ammonis; this man says he has travelled beyond the great dividing ridge as far as to have seen a large river flowing to the west-

20. In the "Dunbar Trip Journal, Vol. II," the explorer presented the following description of the Great Plains:

> The plains to the westward are not to be understood as a dead flat, but as a Country almost withot wood, with gentle eminences & valleys in which last is found abundance of water, with rich soil; very little rain falls upon those plains, but the dews are extremely abundant, and the rivulets may be led all over the valleys; such also is the situation of the Spanish settlements at Nacodoches & others westerly, & their crops are generally made without the aid of rain; depending on the dews and the industry of the people in conducting the water of their brooks over their rich level lands. From all the accounts we have been able to procure the two rivers adjacent viz the Red & arcansa rivers are both very interesting, with respect to the immense bodies of the richest lands, thro' which they pass, and which united to the great western plains, will hereafter support a prodigious population.

ward; the great dividing mountain is so lofty that it requires two days to ascend from it base to its top, other ranges of inferior mountains lie before and behind it; they are all very rocky & sandy, large lakes and vallies lie between the mountains; some of the lakes are so long as to contain considerable islands, and rivers flow from some of them: great numbers of fossil bones of very large dimentions are seen among the mountains, which the Canadian supposed to be of the Elephant;[21] he does not pretend to have seen any of the previous metals, but has seen a mineral which he supposed might yield Copper: from the top of the high mountain, the view is bounded by a curve as upon the ocean and extends over the most beautiful prairies which seem to be unbounded particularly to the East; the finest of the land he has seen are on the Misouri, no other can compare in point of richness and fertility with those of that river. This Canadian as well as M. Le Fevre say that the Osages of the tribe of White hairs in the mount of December (early in the month), plundered all the white hunters and traders upon the Arkansa river. All the old french hunters agree in accusing the Osages of being extremely faithless, particularly those on the Arcansa, the others pretend to make peace & enter into terms of amity, but on the first favorable occasion, they rob, plunder and even kill without any hesitation; the other indian tribes speak of them with great abhorrence, and say they are a barbarous uncivilized race. The different nations who hunt in their neighbourhood, have been concerting plans for their destruction.

M. Le Fevre informs me that the Nation of the aransas [Quapaws] always waging a deffensive war with the Osages, propose sending in the spring of the year a deputation of three Chiefs to the Government of the United States. They say that the Country from the Washita river on the South to the river St. Francis on the North, is their property;[22] that they propose to say to the Government of the U. S.

21. "Cornu Ammonis" is the obsolete name for an ammonite, a fossil shell that is curved like a ram's horn. O'Connell, "Phylogeny of the Ammonite Genus"; Kennedy, "Earliest Tissotiid Ammonite." By "Elephant" Dunbar is possibly referring to the wooly mammoth (*Mammuthus primigenius*), the Columbian mammoth (*Mammuthus columbi*), or the mastodon (*Mastodon americanus*). The mastodon was a western-hemisphere species. Its range included most of the contiguous forty-eight states of the United States, the possible exception being the states bordering Canada. The mammoth's North American range mostly coincided with the range of the mastodon. Gingerich, *Pleistocene Extinctions*, 200–222. Silverberg, *Mammoths, Mastodons, and Man*, 1–25.

22. Indeed Chief White Hair, who was also known as Cheveux Blancs to the French, had visited Jefferson in the summer of 1804. In his speech to Jefferson, White Hair said he hoped for harmony with the United States and healing of the split in his own nation among what was called the "Big Osage." In late 1806, six chiefs of the Osages, accompanied by Indian Agent Pierre Chouteau, visited President Jefferson, and Jefferson expressed to the delegation his own hope for reconciliation among the factions

"We will relinquish to your people all our lands to the North of the arkansa river, on the white river & on the river St. Francis; we will also relinquish our lands upon the Mississippi lying between the rivers arcansa and Washita to an extent westerly far beyond any settlements which have been attempted by the White people, the limits of which we will ascertain; but we request that the powerful arm of the U. S. will defend as their children in the possession of the remainder of our hunting ground lying between the Arcansa & Washita rivers."[23]—Thermometer at 8h. p.m. 19°. Extremes 19°–32°. The moon & stars shine with uncommon splendor.

HUNTER

Therm at 7 a m 23°[.] in the river water 42°[.] at 3 p m.32° & at 8 p.m. 19° Weather Clear & Cold Wind N.W.

Set out at 8 A.M. We rowed down the river as usual, passing the Bayu de Cypre, Isle de Charbon, where in going up I found the Coal half formed & also the coalified wood. shortly after [we passed] the little hills near which [we] came to a camp of about ten Indians who had come from the river Arkansa to hunt for a Mr Le Fevre who had accompanied them & had Ammassed a large quantity skins particularly Deer, & Bear. This Le Fevre informed us that the little Ozages from the Arkansa had killed a party of ten Cherokees as was supposed on the waters of the little Missouri not far from us, He also said that intelligence was received that some of the grand Ozages who with White hairs their cheif had been to visit the President of the U.S.,[24] had on their return met a number of white hunters whom they had robbed & plundered.—We passed in the afternoon the little Missouri & encamped about 3 leagues below it, at a place called auges d'Arclon, (arclons troughs) where one Arclon made troughs to carry down his bears Oil to market.—We have now made in these 3 days going down what took us twelve days to go up.—

within the tribe. "Speech of White Hairs," July 12, 1804; Jefferson to Chiefs of the Osage, December 31, 1806, both in Jefferson Papers; Din and Nasatir, *Imperial Osages*, 150–151, 160, 190–191.

The St. Francis River begins in southeast Missouri and today forms the western border of the Missouri boot-heel. The river continues southward through eastern Arkansas and enters the Mississippi River just north of Helena, Arkansas, in Phillips County. Arkansas Quadrangle Maps.

23. The "Arcansa Nation" (rendered "Nation of the aransas" by Dunbar in this paragraph) is a synonym for the Quapaw Indians who lived along the lower Arkansas River. For more than one hundred years, the Quapaw sent numerous war parties into Osage territory and the Osage sent war parties into Quapaw territory. The Quapaw Treaty of November 15, 1824, ceded lands to the United States, and the Quapaw moved to Caddo territory in Texas and northern Louisiana. However, the Caddo never fully accepted their new immigrants, and eventually the Quapaw settled in northeastern Oklahoma. Arnold, *Rumble of a Distant Drum*, 4, 20, 60, 73, 92, 100, 108, 114–116; also see Baird, "Reduction of a People."

24. For more on LeFevre, see Dunbar's entry for this date.

Friday 11th

DUNBAR

Thermometer in air 11°, in river water 39°. river fallen 4½ inches. Wind moderate at North, The morning is fine and, but the air very cold and penetrating; passed the petit ecor à Fabri,' the osier which grows abundantly upon the beaches above is not seen any lower upon this river,[25] and at this place we begin to see the small tree called 'Charnier' which grows only at the waterside, and is to be seen all the way down the Washita below this place, the Latitude here is about 33°. 40′ which is the limit Nature seems to have placed to those two vegetables, one on the north, the other to the south.

I have already remarked in my Journal of the 17th November that we saw no long moss Tillandsia) above Latitude 33°, & conjectured that Nature had limited its vegetation to that parallel; having this circumstance in my recollection, I asked M. Lefevre for information respecting its existence at the Arcansa Settlement, which is known to be not far beyond 33° of Latitude; he informed me that about ten miles to the south of their settlement the growth of the Tillandsia is limited and that so curiously as if a line had been drawn East and West for the purpose, as it ceases all at once & not by degrees; hence it would appear, that Nature herself has marked with a distinguishing feature the line which Congress has thought proper to draw between the territories of Orleans and of Louisiana. It is a question of curiosity at what Latitude the limit of the Tilansia is found in the atlantic States, and also the Cypress, which last upon this small river is not found higher than 34° of latitude, it is believed to be much higher on the Missisippi: our maps represent a Cypress swamp on the confines of the States of Maryland & Delaware, in Latitude 38½° at the sources of Pocomock River. Q. Is it the same species of Cypress which is found in the Carolinas, Missisippi Territory, etc?[26]

The weather continued clear & very cold all day, we landed at the Cadaux path to make a fire and dine,[27] the thermometer at 3h. p.m. 32° and at 8h. p.m. it fell to 26°—Encamped 1½ league below 'petite pointe coupéé, being nearly the same place where we found the latitude on the 21st November to be 33° 29′ 29″; having

25. Dunbar's "osier" is a French term for a "water willow." It usually refers to the basket or silky willow (*Salix viminalis*); however, this is an introduced species to Arkansas. Little, *Field Guide to Trees*, 337–339.

26. "Arcansa Settlement" refers to Arkansas Post. For more on the Orleans and Louisiana Territories, see Dunbar's entry for November 15. The "Pocomock River" is the Potomac River. The bald cypress (*Taxodium distichum*) does grow throughout the southeastern United States. Petrides, *Field Guide to Trees and Shrubs*, 4, 20, 150.

27. For more on the Caddo Trace, see Dunbar's entry for November 22.

made by the pilot's reckoning about 15 leagues, we stopped twice today, which has retarded us nearly 2 hours; our rate of going has been about 2¼ of those leagues pr. hour.

HUNTER

Therm at 7 A m. 11°[.] in river 39°[.] at 3 p m. 32° & at 8 p.m. 26°—Clear & Cold, Wind N.W.

The river fell about 5 Inches during the last night, yet by noon this day we overtook the height of the flood—In about an hour passed the great cut point, shortly after, or about another hour we came to the drunken Islands & in about another hour the great Beach, & in 1½ hours more the little ecor a Fabri (little hills of Fabri) about another hour to the great ecor a Fabri (big hills of Fabri)[28] where the Lead is said in old times (tho not beleived) to have been put as a line mark with the name & arms of Louis of France between the French & Spaniards. & in about another hour to the two creeks, also about 1 hour more came to little cut point & in about two hours more to Ross's Camp & in another hour encamped on the right bank going down, having passed about the same distance this day which took us 2½ days to go up. about ten o clock A M. passed Campbel & a party of 4 Canadiens who had been a hunting for Bears of which they had killed about 40, & a couple of Panthers. up the little Missouri which they had left about 15 days ago having seen no signs of hostile Indians or indeed any body at all. Our Pilot informs us that about 26 leagues up the little Missouri at the Fork of Antoine, on the lower and [word illegible] side upon a bayu that runs up to the hills and that there is a saline on the other side or upper side between the Missouri & the fork on a branch which runs into the fork & also at the three forks of the Ouachita & at the cote blue on the little Missouri 5 leagues above the fork Antoine there are to be seen many trees called Bois jaune or Bois d'arc. (yellow wood) or Bow wood, which grow about 15 or 20 feet high[.][29] the wood is of a reddish orange colour, & gives a fine yellow dye; This tree resembles the chinkapin tree in external appearance, bears in the fall a fruit resembling an Orange but twice as large filled with seeds, this tree has a very knotty scrubby appearance.

28. Hunter's journal entries for November 22 and 23 provide more information concerning these locations and the name origins. Also see Dunbar's entry for November 22.

29. Two Bayou ("the two creeks") begins slightly northeast of Bearden, Arkansas, and forms the border between Calhoun and Ouachita Counties. It joins the Ouachita about 4 miles below Camden. Arkansas Quadrangle Maps. The "Panthers" were mountain lions or cougars (*Felis concolor*). Sealander and Heidt, *Arkansas Mammals*, 234–237. For more on the "Bois d'Arc" tree (*Maclura pomifera*), see Dunbar's entry for November 20; and Moore, *Trees of Arkansas*, 67.

Perhaps it is the famous tree which yeilds the yellow dye in so much esteem in Europe & reckoned so valuable & rare, capable of dying the finest scarlet.—Mr Le Fevre gave the same account of this tree.—

Saturday 12th

DUNBAR

Thermometer in air 20°, in river water 40°—river risen an inch. Much vapour ascending from the river. Part of the night was cloudy and this morning the heavens are not entirely cloudless, we therefore expect an approaching change of weather. The air is damp & penetrating, so that it continues yet very cold on board the boat, as the day advanced, it proved more cloudy and disagreeable and altho' at 3h. p.m. the thermometer was found at 43°, the sensation of cold to the human body was greater than in a dry air at 22°—the face of the heavens was overspread with clouds & the atmosphere extremely moist. We made a good encampment in the evening called 'Campement des bignets' (fritter camp) being about 18 of the pilots leagues, tho' not much exceeding two days of our voyage up, about 37 or 38 miles by our own reckoning; we passed this place between breakfast and dinner on the 19th November.[30] The Thermometer at 8h. p.m. 30°.

HUNTER

Therm at 7 a m. 26°[.] in the river 40°[.] at 3 p.m. 43° & at 8 p.m. Wind N.E Cloudy & cold

Set out at 8 a m, & in two & an half leagues by our Guide's estimation passed a place called Ross's camp on the right side going down, came one league further to the Bayu D'Acacia (Locust Bayu) on the left, & in one league from thence passed a Cote a faine (beach mast hill) on the right or west side; thence three leagues to the petite Bayu on the left or west side, thence 2 leagues to the Vieille abbatis (the old clearing of timber by an hurricane) on the right[,]thence two leagues further to La Piniere (the Pine forrest[)], thence 2 leagues to the Chemin couvert (the Covered way[)],on the West, thence one league to the bayu de cabane a Champignole on the east side, thence one league to John Skinners camp on the west[,] thence one league to the hill of the cabbin of Champignole on the west, thence 2 leagues to the camp of flitters [fritters][31] where we are now encamped for the night. hav-

30. The location of "fritter camp" was probably near the present town of Calion in Union County or near the community of Champagnolle in the same county. In his January 11–13 entries, Hunter gives a more detailed description of the various hunter-trader sites along the river.

31. "Ross's camp" would have been in Ouachita County, or on the west bank. In Hunter's reference to "our Guide," it is not certain whether he is referring to Samuel Blazier or Le Fevre. Locust Bayou

ing come down better than two of our days journey up.—One of our people killed a Swan which proved poor. The river rose one inch last night, yet the current here is Moderate & has been so all this day, & the river is considerably increased in size. The banks are becoming gradually lower, the Soil sandy on a bottom of clay, no stones appear on the banks of the river[,] no evergreen trees except Holleys which are very numerous & pines. The lower branches of cypress & others where they are subject to be overflowed produce an abundance of green shaggy moss for these two days journey down; but as yet we are not come to the region for the production of the long moss called spanish beard or Carolina moss.—

Sunday 13th

DUNBAR

Thermometer in air 27° , in river water 40°—river risen 1½ inches—Calm. The morning is very fine and the atmosphere dry, consequently the temperature not cold to the human body. These two mornings the river has risen a little, notwithstanding that we have been without rain for several days passt, & it will be remembered that the three first days of this voyage the river was found each morning to be fallen; this is to be accounted for by the boat gaining upon the velocity of the stream more in the day than it loses in the night. Since we have got below the rapids, the Current is more gentle and we make only two of the Pilot's leagues pr. hour, which does not exceed perhaps 4 english miles; it appears that in nine hours (one day's) rowing down we have made the same distance which we made in 13 hours coming up, the current at the time of our ascent being nothing and the space passed over being 36 miles, it will be found from those data that in each 24 hours we gain upon the current about 6½ miles, we have therefore reason to conclude that we have got beyond the apex of the tide or wave occasioned by the fresh, and are descending along an inclinded plane but as we always encamp at night, it is not surprising that in the morning we find ourselves in deeper water because the Apex of the tide is constantly endeavouring to overtake us, and in the morning we find ourselves in a more elevated part of the inclined plane, which we had left behind in the evening before.

This morning no condensed vapour was visible on the surface of the river, yesterday it was considerable; hence it appears that 13° difference of temperature:

---

enters the Ouachita on the east bank in Calhoun County. "Vieille abbatis" is located in present-day Calhoun County. "Chemin couvert" is Smackover Creek. It forms one of the borders between Union and Ouachita Counties and joins the Ouachita River north of the town of Calion. Champagnolle Creek is in Calhoun County. Both "John Skinners camp" and the "cabbin of Champignole" would be located in Union County. Arkansas Quadrangle Maps.

(the river being highest) does not condense vapor with sufficient rapidity to render it visible, altho' 20° are more than are necessary: it must not be omitted to be mentioned that this morning the atmosphere was extremely dry, & therefore greedy of moisture, and yesterday it was very moist and consequently not disposed to disolve water rapidly. The day proved cool, tho' not disagreeably so; the wind in the afternoon N. E. and air moist. Made this day by the computed distances about 15½ leagues and encamped about one league below where we found our Latitude to be 33° 13′16.5″ on the 17th November, so that we have again completed two days voyage ascending in one descending. Thermometer at 8h. p.m. 30°. Extremes 27°–53°.

HUNTER

Therm. at 8 a m 27°[.] in the river 40°[.] at 3 p.m. 53 & at 8 p.m. Cloudy, moderate weather.

Set out at 8 a.m. find the river rose last night 1½ Inches by which, we have got ahead of the fresh or last flood, & go somewhat faster, or farther in the day time, that the current does in a day & a night. about 4 Leagues from our place of our encampment come to cote de Hachis on the west, thence 3 leagues to bay Moreau on the east, thence 2 leagues to Cache la Tulipe on the west side. (where a person of the name of Tulipe concealed or hid the skins acquired by hunting till he returned again to the same place), passing 3 leagues further down to sort of cleared place on the east called L'aigle (the Eagle). Thence 2 leagues to la pirague d'Auguste on the east. (Augustes boat) About 1½ league below this last place we encamped for this night. We are now come to the low lands,[32] where the high freshes overflow all the banks yearly—We have come this day the same distance nearly as took us two days to come up—

Monday 14th

DUNBAR

Thermometer in air 23°, in river water 40°—river risen 1½ inch. Wind very light at N. W. The atmosphere is dry, and the temperature to the human body

32. Hunter identified "cote de Hachis" on November 19, during the ascent, as "Bayu de Hache." It was probably located in Union County. Moro Creek forms the border of Calhoun and Bradley Counties. It joins the Ouachita through Moro Bay at Moro Bay State Park. Hunter labeled it "Bayu Moreau" in his November 18 entry. Regarding "Cache la Tulipe," see Hunter's entry for November 18. L'aigle (the eagle) Creek in Bradley County meets the Saline River approximately 3 miles above the Saline-Ouachita confluence. There is evidence to suggest that L'aigle Creek could have moved parallel with the Saline at an earlier time and may have entered only one-half mile north of the Saline-Ouachita confluence. The "low lands" would be today within Felsenthal National Wildlife Refuge in Bradley, Union, and Ashley Counties. Arkansas Quadrangle Maps.

seems not very cold; there is a thin condensed vapor upon the surface of the river, the difference of temperature between the river water and air being 17°; yesterday the atmosphere being nearly in the same state 13° were insufficient to render the vapour visible. If our hygrometers[33] were instruments of a less dubious nature and capable of indicating by a scale the absorbing, disolving or attracting power of the atmosphere for water, without being influenced by heat and cold we should then be able to determine à priori at what difference of temperature between water and air corresponding to a given degree of the hygrometer, ascending vapour will be visibly condensed.

A green moss is found upon the branches of trees which are immersed in the waters of the inundation, none of same species appears in a more elevated situation; when the waters subside vegetation does not seem entirely in thoses passages which are but a foot or two above the surface, they continue to be of a lively green & hang to the length of 5 or 6 inches: the vegetation of this moss must commence under water; it may be of the same nature with the green matter deposited in fresh found in fresh water conduits, which has been examined by Priestly[34] and others & which here has arrived to a higher state of perfection from its free & open situation; it is evident this moss must vegetate under the impulse of a considerable current.

In the afternoon passed the Latitude 33° and the Island of Mallet noticed in the Journal of the 15th of November:[35] made about 19 leagues this day, being about 2½ days voyage ascending; since we have got into the low alluvial country, the channel is narrower and the velocity of the current greater; we are now encamped where we passed in the afternoon of the 14th of November.

The day continues fine and of an agreeable temperature; at 3h. p.m. the thermometer was at 53°, at 8h. p.m. 32°, an eclipse of the moon will take place this night after nine at night, we prepare to observe it; regulated the watch as near as possible to the apparent time at the setting of the Sun; tomorrow we shall give an account of our observations, the sky is perfectly serene.

33. A hygrometer measures the water vapor or humidity content of the atmosphere of a certain gas. Leonardo da Vinci created a crude hygrometer in the 1400s; Francesco Folli invented a more practical apparatus in 1664.

34. For more on Joseph Priestly, see Dunbar's entry for December 10.

35. See the November 15 entry for more information on the Island of Mallet. Hunter wrote that the team also passed the "Marais de Cannes," which was possibly Caney Marais Bend in Bradley County; "Bayu de la Saline," or the Saline River, at the conjuncture of Bradley, Ashley, and Union Counties; or "Marias de la Saline," or the Marais Saline, in Bradley County. Arkansas Quadrangle Maps.

HUNTER

Therm at 7 A M. 23°[.] in the river 40°[.] at 3 p m 53 & 8 p m. 32° Clear, little wind. Moderate. Set out at 8. a.m. & shortly after passed the Marrais de Cannes on the east side of the river, about ½ league from our encampment. Thence to the petite Marrais (little Swamp[)] on the west 2 leagues[,] thence 1½ leagues to the Bayu Poiles (Hair Creek) on the west Thence 1 ½ leagues to the Bayu de la Saline (saline creek[)]on the east. Thence 1 league to the Marrais de la Saline (saline swamp) on the east. Thence 3 do to the Bayu de la pelle (mortar Creek) Thence 3 do to the 3 battures (3 beaches) thence 1½ do to the Isle de Mallet. When the line strikes between the Territories of Orleans & Louisiana viz Lat 33°.0′.0″. Thence 1½ to the bayu de long vue (long reach creek) on the east, Thence 1½ do to the bayu Franqueure (after a person of that name who lost his life there in chase of the Buffaloes[)][36]—All these leagues are to be understood as the nominal measure of the country, which we have found to contain about 2 english miles to the league.—Here we encamped on a flat piece of ground about 4 feet above the present water, but much overflowed in the freshes like all the rest of the lowlands; The river rose last night 1 ½ Inches, The current tho more rapid than when we came up, yet is still but moderate perhaps about 1 mile pr hour.—We dayly see before us at the distance of about 200 yards large numbers of Wild Gees & ducks who fly at our approach, so that it is difficult to get a shot at them.—

we have passed this day also about the same distance as we did in two coming up.—At our camp observed an eclipse of the Moon when the total darkness commenced at 13h.36m.29″. by the silver watch[37]

Tuesday 15th

DUNBAR

Thermometer in air 30°, in river water 40°—no vapour visible on the surface of the river: River risen 1½ inch—wind light at S. E. cloudy. Prepared last evening to observe the Eclipse of the Moon with a very indifferent Spy-glass magnifying about 8 times. The commencement of the Eclipse was not correctly noted, occasioned by the very strong effects of the penumbra in our perfectly serene & clear sky, the moon being not far removed from the Zenith, which induced a belief that

36. The "Marrais de Cannes" was possibly near Caney Marais Bend in Bradley County. The "petite Marrais" and "Bayu Poiles (Hair Creek)" both would have been in present-day Bradley County. Ibid. "Bayu de la Pelle" is possibly Lapile Creek in Union County. Both "bayu de long vue (long reach creek)" and "bayu Franquere" were probably in Morehouse Parish, Louisiana. Louisiana Quadrangle Maps.

37. Dunbar waited until his January 15 entry to record the eclipse and the events on the evening of the 14th.

the Eclipse had actually commenced at 12h 32′, this circumstance produced some inattention at the instant of the true commencement, which was supposed to have happened at 12h 40′; but the commencement of total darkness was observed with due attention, and is believed to be as correct as circumstances with our instruments would admit, and took place at 13h. 37′:[38] It is believed that the uncertainty of the moment of observation did not exceed half a minute, I am rather disposed to say a quarter of a minute, for the transparency of the atmosphere was as perfect as can ever be expected in situations not more elevated than ours. We shall ascertain the error of the watch below at some known point whose latitude or position can be deduced by referrence to our Geographical Journal; & and we shall again perform on our arrival at the post of Washita, from which we shall gain the rate of the water's going & the whole may be referred to the meridian of the Post & will serve to compare with the results of our lunar observations made there on our way up.

[T]his morning the heavens are veiled by clouds; during the night the thermometer was down to 28° with a pure serene sky & the atmosphere was so dry that the cold was not very sensible; this morning with a higher temperature & moist atmosphere, it is cold and penetrating. We saw this morning the first long moss (Tilandsia) called generally by the french 'barbe espagnole (Spanish beard) on trees growing on the margin of the river about 2½ leagues (5 miles) above the 'Bayou des Butes.'[39] At this time also we emerge from the alluvial country noticed in the former part of this Journal, the banks are now of a good elevation, about 15 to 18 feet above the present level of the river & probably not liable to be inundated, whereas the alluvial lands we have just quitted, are subject to be overflowed from 8 to 12 feet; we saw none of the green moss along the alluvial tract, which I much regret, having intended to take some specimens for examination, I am in doubt whether any of the same species grows below, as yet we do not see it at the 'bayou des butes.' The Sun at last broke forth and we landed to take this altitude for the correction of the watch, the position was recognized by the mouth of a Creek so that by a reference to the Geographical Journal, we found that the latitude of this point is 32°. 49′. 24″ being the same which will correspond with N10° W 8h 8½ on the 14th. Novth. ascending, the Sun dble alt. Lower limb was 66° 36′ 45″ Ind: err. +12′ 20″ taken at 10h 56′ 24″ a.m.—

38. A penumbra is a shadowy or marginal area where the light from the source of illumination is only partially broken. Dunbar wrote the following sentence about the eclipse in the "Dunbar Trip Journal, Vol. II": "By calculation it ought to have happened about 13h. 32′; the difference of 5′ is probably the error of the watch as regulated by the setting of the sun, indistinctly seen thro' the woods."

39. In the "Dunbar Trip Journal, Vol. II," he names the moss "Spanish beard." "Bayou des Butes" is Bayou de Buttes in Morehouse Parish. Louisiana Quadrangle Maps.

The day became cloudy in the afternoon, & the thermometer rose to 63°, which awe consider as a indication of rain.

We made this day nearly 15 computed leagues, being now the eighth day from Ellis Camp and are now encamped within 5 of those leagues from the Post of the Washita, being about a mile above the place where we dined on the 12th November, Latitude then found was 32° 34′ 47″. The moon and stars shine with a mild lusture, no appearance of change in the weather notwithstanding the increased temperature of the atmosphere. Thermometer at 8h p.m. 43°.[40]

HUNTER

Therm at 7 a m 30°[.] in the river 40°[.] at 3 p m. 63° & 8 p m. 43° Cloudy moderate weather

at 11h.4m.0″ A m Took an Altitude of the sun, to correct the silver watch viz ⊙ d. Alt. apt. 69.59.10 Ind. error—1′.37″

Supposed Lat. 32′.48′.24″

True time found 11h.0m.20″
Error of watch too fast −3.40
Time pr watch h.11,4m.—

Time of Eclipse pr watch	13.36.29
Error watch too fast	−3.40
	13.32.49 true time
	19.40—time at Greenwich
hours	6. 7.11 equal to Longitude West 91°, 47 , 45″

Set out at 8 a.m. as usual after breakfast viz less than an hour, at the distance of 1 ½ leagues passed the Bayu Batture a pierres (bayu of the stoney beach.) on the west. from thence at the distance of two leagues passed Bayu de buttes (Bayu of Indian mounts, or cementery) on the east side of the river, thence 2 to Bayu Mercier on the west[,]thence 1 do to bayu Asemine on the east, thence 1 do to Bayu Bartheleme on the east, thence Bayu de la L'Outre three leagues on the west, thence 1 do to grand Roquerau on the west, thence 1 do to the Bayu de l'eau noir (black water Bayu) on the east, thence passed the point aux oufs (egg Point) 2 leagues to the east. In all 14½ leagues about the same distance that took us two days to go up—About the middle of this day's journey came again to the region of

40. "I found a little of the Green moss today which forms under water, but not so long nor so beautiful as it appeared higher up." "Dunbar Trip Journal, Vol. II."

Spanish beard (Carolina moss) for above this latitude there is none to be found on the trees.[41]—The river rose 1½ Inches during the night.

The Soldiers have already expended all their pork & venison. We gave them a temporary supply out of our own rations having before given them all the flour we had left, which is also now consumed. Yet we expect to receive a supply at the garrison which is now only about 5 leagues off.

Wednesday 16th

DUNBAR

Thermr. 36° in river water 41°—river risen 1¼ inch.: a thick fog proceeding from atmospheric moisture, being very different from what we see arising out of the river under considerable differences of temperature—Arrived at the post of Washita about noon—The day proved very fine and warm, the thermometer at 3h p.m. being 65° and at 8h. p.m. it remained at 60°—Found all well at the post—No news of any importance—Our people all in good health except one Soldiere who has been a good deal incommoded by a dysentery; but he is not in danger. [R]eturned the hired boat.[42]

HUNTER

Set out after breakfast at 8 a m. as usual, & arrived at the Garrison about 11 a.m.[43] here we delivered up the boat Mr Dunbar hired & took possession of our own again which looked more weather beaten than if we had used her all the time. The men were set to clean her out, cut & make six more oars to replace as many broken in our journey, to cut & form a new mast in the woods, to bend the sail & put all the rigging in order, & to take back again from the Garrison all our spare articles which on account of their weight we had found it necessary to leave behind, to make the boat draw as little water as possible—In the mean time we were very civilly & politely entertained by Leiut. Jos. Bomar commandant of the Garrison who studied to give us every facility in his power. Here we drew one months ra-

41. "Bayu Mercier" could be Possum Bayou in Morehouse Parish. Persimmon Bayou meets the Ouachita River 1 mile north of the Bayou Bartholomew confluence. In his November 13 entry, during the ascent, Hunter had labeled the stream "Bayu Assmine." Bayou de Loutre joins the Ouachita in Union Parish approximately 5 miles below Sterlington, Louisiana. The "Bayu de l'eau noir" may have been Lonewa Bayou, which flows between the Ouachita River and Bayou Siard. It meets the Ouachita north of Monroe. Louisiana Quadrangle Maps.

42. The group also parted with Samuel Blazier, their pilot, and with LeFevre.

43. "The Garrison" was Fort Miró / Ouachita Post / Poste du Ouachita. Hunter made no entries for January 17 or January 18.

tions for the soldiers commencing the first & ending the last of Jany. & as it was not convenient for him to spare of his small store all the Bacon necessary he procured some fresh beeff & some fresh pork & we delivered them lbs 100 Bacon for which I got a duebill on the Contractor at Natchez or Orleans for the payment, he also gave a due bill which I delivered the Sergeant for 1 bble flour & some whiskey salt & vinegar which he could not spare—I drew nothing for ourselves as I had money to get what we might stand in want of—These operations took up two days besides the part of the day we arrived here.

Thursday 17th

DUNBAR

Thermometer in air 60°, in river water 44°—river risen one inch. Wind at S. W.—very clear during the night, but cloudy this morning—made the following observation to correct the watch & ascertain her rate of going . At 8h. 53′ 7″ sun's apparent double altitude of the lower limb 36° 44′ 45″ Ind: err: + 12′ 30″.

Friday 18th and Saturday 19th

DUNBAR

Employed in getting mast and oars for our large boat. Judging it of importance to get to Natchez as soon as possible, I determined after being disappointed in procuring horses, to take the Canoe with one soldier and my own Domestic and push down to Catahoola, from whence there is a road to Concord, about 30 miles to Concord opposite to the town of Natchez.[44]

Saturday 19th

HUNTER

This morning Mr Dunbar set out in the canoe with one hand & his servant at 7. A.M. with a veiw of getting to Cadets at Catahoula as soon as possible & from thence to hire horses & take a short cut home to his family whom he had received letters from at his his Arrival here; He desired me to follow as soon as the boat

44. Dunbar's reference to "my own domestic" probably means the servant he took with him on the expedition. It probably does not refer to either of the two slaves he had with him. "Catahoola" is today Jonesville in Catahoula Parish, and Concord is the current town of Vidalia, in Concordia Parish. Around 1798 a Spanish citizen of Natchez named Don José Vidal received a considerable grant of land to build a fort on the opposite side of the river. It became known as the Post of Concord or the Post of New Concordia. It is possible that the name came from the title of the home of Spanish governor Gayoso de Lemos in Natchez ("Concord"). The Post of Concord was transferred to U.S. control on January 12, 1805. For more see Calhoun, "A History of Concordia Parish"; Louisiana Quadrangle Maps.

would be ready to the Mouth of the Red river & then assend the Mississippi to St. Catherines Creek (his landing) near Natchez where I should deposit all the remaining stores &cc to serve for the next expedition, & then to go down the Missisipi & deliver the boat & detachment to Colo. Freeman.—from whence I expect to go by water to Philada. to visit my family.—At half past twelve on the same day the boat being got ready & everything on board, I set off on my way down the Ouachita. The wind Southwardly & blowing fresh is right ahead & retards our progress.

Therm. at 3 p.m. 58°. in Air, in the river 43° & at 8 p.m. 50°. This day came about 4 french leagues when the darkness of the night prevented further progress. I slept with my son on board the boat, the soldiers encamped ashore.

Sunday 20th

DUNBAR

Set off about day-break, and arrived after night at the lower Settlement, about 20 computed leagues from the Post. Called at the house of an old hunter, with whom I had conversed on the way up: This man informs me that at the place called the mine on the little Misouri, there is a smoke, which ascends perpetually from a particular place, and that the vapor is sometimes insupportable; the river or a branch of it passes over a bed of mineral, which from the description given is no doubt martial pyrites. In a creek or branch of the Washita called 'fourche à Luke,' (three leagues above Ellis Camp) there is found on the beaches and in the cliffs a great number of globular bodies, some as large or larger than the head of a man; which when broken, exhibit the appearance of Gold, Silver & precious Stones, this most probably is pyrites with chrystalized spar: also at the 'fourche des glaises à Paul,' (higher up the river than "Fourche a Luke) there is near to the river a cliff full of hexagonal prisms terminated by pyramids, which appear to grow out of the rock, some an inch in diameter and six to eight inches long:[45] there is also beds of pyrites found in several small creeks communicating with the river Washita: but it appears that the mineral indications on the [Little] Misouri were most considered, some of the hunters actually worked upon it and sent a parcel of the ore to New Orleans as observed above: it is the belief of the people here that the mineral contained precious metal but that the Spanish Government did not chuse that any

45. "The Post" is probably "Olivots first settlement" as identified in Hunter's journal entry of October 23. Dunbar's "fourche à Luke" is Fourche a Loup Creek, which enters the Ouachita from the south and above Gulpha Creek, just east of Highway 7. The "fourche des glaise à Paul" is Glazypeau Creek, which enters the Ouachita on the east bank just below the community of Mountain Pine in Garland County. The confluence of Glazypeau Creek is also about 1 mile below Blakely Dam, which forms Lake Ouachita. Arkansas Quadrangle Maps. The "hexagonal prisms" are quartz crystals.

mine should be opened so near to the British Settlements, for which reason an express prohibition was issued against any farther work being done upon the mine; since which time it has not been no more spoken of.[46] This man procured me some small roots & a few seeds of the patate à chevreuil; he also took me to the next house where I saw a solitary tree of the 'bois d'Arc,' (bow-wood) or yellow wood, which was raised from a seed brought from the little Misouri;[47] I requested some large branches, but could only obtain from the Old Lady Mistress of the place, two very small ones; the fruit fallen before maturity lay upon the ground, some were of the size of a small orange, with a rind full of tubereles; the color tho' in appearance faded, still retained a swemblance to pale gold: the tree in its native soil, when loaded with its golden fruit (nearly as large as the Egg of as Ostrige), presents I am told the most spendid appearance, the foliage of the finest deep green greatly resembling the varnished foliage of the orange tree, and upon te whole no forest tree can compare with it in respect [illegible word in manuscript] grandeur. The bark of the young tree which I saw resembled in its texture externally the Dog-wood bark, but its colour is a reddish or brownish yellow; the appearance of the wood recommends it for trial as an article which may yield a yellow die:[48] I hope to suc-

46. On this day Dunbar recorded in the "Dunbar Trip Journal, Vol. II" another passage concerning the speculation of mineral deposits along the Ouachita: "whether it was, that the precious metals were found in the ore, or out of regard to the poor hunters who might be tempted to misapply their time in a research attended with no advantage; an express prohibition was sent up by the Genl. Govt. to the Commandant, forbidding all persons to work upon the Mine . . . this circumstance happened 30 years ago. The genl. belief is that silver exists in those places." There are no silver deposits in the Little Missouri valley; however, cinnabar (HgS), or quicksilver, a rhombohedral mineral and a principal ore of mercury, was discovered in the southern border areas of the Ouachita Mountains in 1930. According to geologists, most of the cinnabar deposits were located in the Jackfoot, Stanley, and Aleka formations along the Athens Plateau and ran east to west for more than 25 miles. The deposits were primarily in Pike County, but additional ones were located in Clark and Howard Counties. Among the several subsequent mines in the area were some located in the vicinity of Amity and Alpine, Arkansas, in northwestern Clark County. Some of the companies that eventually mined the region included the Arkansas Quicksilver Company, the Arkansas Cinnabar Mining Company, the Southwest Quicksilver Company, and the C. Mining Corporation. Other deposits were found at Gap Mountain in southwestern Montgomery County and at Wall Mountain in northeastern Pike County. Branner, *Cinnabar in Southwestern Arkansas*, 1–9, 30–35; Reed and Wells, *Geology and Ore Deposits*, 15–87.

47. The "patate à chevreuil," or chevrenil, could have referred to a number of possible flora: man of the earth, or wild potato vine (*Ipomoea pandurata*); deer's potato (*Liatris acidota*); and even the sweet potato (*Ipomoea batatas*). Moerman, *Native American Ethnobotany*, 275, 304–305. For more information on the bois d'arc, or Osage orange (*Maclura pomifera*), see Dunbar's entry for October 20.

48. Indians of the region did use the roots of the bois d'arc for making yellow dyes. Moerman, *Native American Ethnobotany*, 327.

ceed in raising trees from the cuttings and a small Cion, which I have procured; the people suppose this tree too young to mature its fruit; as it has always hitherto fallen when of the size of an orange; I am inclined rather to suspect that the failure may be occasioned by its open and exposed situation as it naturally grows under the shade of the forest, this tree is about six inches in diameter, it is deciduous and appears to be in a sound and healthy state; the branches are numerous and full of short thorns or prickles, it seems to recommend itself as highly proper for hedges or live fences, which are greatly wanted in many parts of the United States, this tree is known to exist near the Nakitosh (perhaps Lat: 32°) and upon the river arcansa high up (perhaps in Lat. 36°) it is therefore probable it may thrive from Lat: 28° to 40° and will be a great acquisition to a great part of the U. S. should it possess no other merit than that of being ornamental.[49]

On my way down I endeavoured to discover a place said to produce Gypsum, but being without a proper guide I failed in the research; I have no doubt of the existence, and having noted the places where it has been found, one of which is the first hill or high land which touches the river on the west above the large Creek called Bayou Calumet and the other is the second high land on the same side;[50] as those are two points of the same continued ridge, it is probable that an immense body of Gypsum will be found in the bowels of the hill connecting those two points and perhaps extending far beyond them; it has been said that fossil coal is found on the east side of the river opposite to the second hill; it is probably Carbonated wood only: a person who pretends to have been up among the sources of the

49. For more information on Natchitoches, Louisiana, see Dunbar's entry for January 10. Dunbar also expounded on other uses and importance of the bois d'arc tree, explaining, "from the information of a man from that place I met at Catahoolo, he says that attempts have been made to dye Cotton with it, but without any mordant, the dye disappears by washing, but is handsome. It appears therefore that this tree is produced naturally much lower than the petit missouri, as it is found at the Nakitosh, tho' rarely & grows there both on the high and low land; the Latitued of Nakitosh is unknown to me, but as from information the Post of Nakitosh lies from that of Washita, about West S.W. 40 or 45 leagues in direct line and from the Catahoula W. by N. about 35 leagues, the Latitude of that post may be about 36° 44′.45″." "Dunbar Trip Journal, Vol. II."

50. Gypsum contains hydrous calcium sulfate, $CaSo_4 \cdot 2H_2O$. It is a very common sulfate mineral that was and is used in making plaster. Bates and Jackson, *Dictionary of Geological Terms*, 227–228. This mine was probably in Caldwell Parish. During their ascent, both Dunbar and Hunter identified a coal mine on the opposite bank from the purported gypsum deposit. On October 23, Hunter recorded the settlement as being called "Olivots first settlement." Traveling separately from the expedition, Dunbar also outlined a long conversation he had at the home of "an old hunter." McDermott speculated in his footnotes that this hunter may have been "Olivo" or Olivot. McDermott, "Western Journals of Hunter," 113; see Dunbar and Hunter entries for October 23. "Bayou Calumet" could be Bayou Calamus in southern Caldwell Parish. Louisiana Quadrangle Maps.

Washita 100 leagues higher than the hot springs declares having found true mineral coal, which burns with a strong heat and bright flame without the aid of other fuel, a property which carbonated wood does not possess. I do not give entire credit to this last report, the person who informed me, being a fond of the marvellous.

HUNTER

Therm. at 7 a m. 56°, in the river 43°. at 3 p m. 51° & at 8 p m. 40°. cloudy with small rain drizly.

Set out before sunrise & proceeded down the river with a moderate current, tho twice as much as when we went up here & the river about 6 feet higher, yet it is about 20 feet below the high floods in the Spring. Came this day to the lowest habitations or Settlements about 16 estimated leagues, which makes what they call 20 leagues from fort Miro.—& 32 from the solitary house of Cadits at the mouth of Catahoula.[51] Here we staid for the night. Going ashore to enquire, learned that Mr Dunbar left this place after breakfast on his way down in the canoe. The sun was set, all was still & silent as death I saw a small encampment with two fires & apparently two families of Choctaw Indians, I heard some melancholly mourning in a female voice, it seemed to come from the heart & was very expressive. I turned to where the sound proceed[ed] from & saw a person on the ground wrapped entirely in a blanket, & leaning on a small heap of dead branches rudely piled together, to protect from the wild beasts of the wilderness, the remains of her first & only child, which I was informed died six months ago. Joy & Greif are the same in all languages—This night the wind came round to the N E & the weather grew raw & cold. in the forenoon passed a large covered boat going up rowed by 6 oars, under the direction of an elderly french Gentleman called Capt. Tousard (or Boussard I could not distinguish which[)] from Baton Rouge, on his way to the hot springs for the recovery of his health. He appeared to be much emaciated & complained much of pains in his lower limbs. I thought it best to give him hopes of recovery, he was very eager to know if the waters were salutary which [I] assured him they were.—I also passed the camp of an Hunter called Louis Francisque the brother in law of our later Pilot of whom I bought three bearskins for three dollars.—

Monday 21st and Tuesday 22nd

DUNBAR

Continued my voyage with contrary winds and arrived the evening of the 22nd at the Catahoola, which by computation is fifty leagues from the post of Washita:

51. This is the place Dunbar refers to as the "lower settlement" in his entry of this date. Hunter had previously named "Cadits" "Monsr Cades."

At this place a french man named Hebrard is settled, who keeps a ferry across the black river:[52] [H]ere the road from Natchez forks, one branch of it leading to settlements on the red river and the other up to the Post of the Washita: The proprietor of this place has been a hunter and great traveller up the Washita & into the western countries, he confirmed generally the accounts we have received ; it appears from what he and others say, that in the neighbourhood of the hot-springs, higher up among the mountains, and upon the little Misouri, during the summer season, Explosions are very frequently heard proceeding from under ground, and not rarely a curious phenomenon is seen which is termed the blowing of the mountains, that is, confined elastic gar forced a passage thro' the side on top of a hill driving before it a great quantity of earth and mineral matter: it appears that during the winter season the explosions and blowing of the mountains entirely cease, from whence we may conclude that the cause of those phenomena is comparatively superficial, being brought into action by the increased heat of the more direct rays of the summer-sun.

Upon my arrival at the house of M. Hebrard,[53] I enquired for horses to carry me across the low country to Concord opposite to Natchez, the distance by the road is computed 30 miles, but it is probable the direct distance falls short of 25, and it is remarkable that the river Washita preserves a kind of parallelism to the Missisippi until it comes within the influence of the highlands of the arkansa, & thence it is deflected to the North west & probably holds a middle ground between the red river and the arcansa; the inclination of the missisippi is such that the

52. McDermott referred to this man as "Don Juan (Caddy) Hebrard." As stated before, Hebrard had been given a Spanish land grant of 2,000 acres and permission to operate a ferry at that location. Hunter called the man "Cadet" or "Cadi" or "Cadits," and he said that the man had built his home on one of the Indian mounds in the vicinity. McDermott, "Western Journals of Hunter," 82. Dunbar added what appeared to be an afterthought to the "Dunbar Trip Journal, Vol. II" on this date:

> I forgot to mention in its proper place that during the summer season, explosions are heard frequently, which are supposed to be under ground in the upper parts of the Washita in the neighbourhood of the hot springs & upon the 'petit Missouri,' and it happens not rarely as I am informed that the mountains blow as it is termed that is the Earth opens & a quantity of earthy & mineral matter is thrown up with considerable force, occasioned no doubt by confined gas which escapes by the place of least resistance; it appears that the explosions & blowing of the mountains have not been observed during the winter season, from which we may conclude, that the cause of those phenomena is comparatively superficial, being influenced by the increased heat of the more direct rays of the summer-sun.

The phenomenon that Dunbar reported has not subsequently been seen or reported.

53. In the "Dunbar Trip Journal, Vol. II," he called this man "M. Hebrard dit Cade."

walnut-hills are 30 miles to the east of the Natchez, the Post of the Washita will be found there for nearly under the same meridian with that of Natachez very contrary to the general idea.—M. Hebrard very obligingly promised to furnish me with horses which it was necessary to hunt up in the woods, In the meantime I went to view the Indian mounts spoken of in the beginning of this Journal:[54]

54. This site was probably what archaeologists call the Troyville Site in Catahoula Parish. The site was first studied by Dr. Cyrus Thomas in 1883 and excavated by Winslow Walker of the Smithsonian Institution in the twentieth century. Archaeologists attribute most of the mound complex and its remaining artifacts to the Troyville—Coles Creek Culture (A.D. 395–1250). The last of the mound complex was destroyed by a highway project in 1931. Neuman, *Introduction to Louisiana Archeology,* 169–177; Walker, *Troyville Mounds,* 14–40. For more on this site, see Hunter's entry for January 24. Dunbar described the mounds as follows in the "Dunbar Trip Journal, Vol. II":

> At this place are several indian mounts being mostly covered by a thick cane brake, it was difficult to examine them with due attention: there are 5 of the usual form placed within the angle formed by the black river & the Catahoula, another lies beyond the Catahoula; those are oblong, about 50 yards long by 25 wide on the top, with a rapid descent about 12 feet perpendicular; there exists a sixth mount of a very particular construction, the base is nearly square, & consists of three stories; M. Heberd the proprietor thinks the whole is 80 feet high, but I cannot persuade myself that it exceeds 40 or 45 feet, the ascent of the first story is not very rapid, & may be estimated at 15 feet perpendicular; a flat of 5 or 6 feet wide reigns all around the mount, from which arises the 2d story, the ascent of which is not more rapid than the 1st & may be about 8 feet perpendicular; a 2d flat of the same breadth is found above the 2d story passing in like manner around the mount from whence arises the third story, whose ascent is extremely steep, it is necessary to support one'self by the Canes, which cover this mount to be able to get to the top; the form of this 3d story is that of a very regular cone, terminated at the top by a circular flat of about 8 feet diameter, which has probably been less, the perpendicular height of the cone may be about 20 feet having brought no instruments with me from the boat & moreover the mount being entirely covered by thick canes I had it not in my power to make an exact survey, which I hope to do upon some future occasion: The proprietor says that the base covers a square of about 180 feet to each side, & at each angle, there is a kind of abutment or projection, from which an imperfect idea may be formed of the curious form of this singular mount; which may have been a temple for the adoration of the Supreme being; or it may have been a monument erected to the honor of some great Chief; or it may have been barely a watch tower. The country all around being alluvial, or at least subject to inundation, it is extremely probable that the five oblong mounts were places of residence, composing a considerable village, there is also the appearance of an embankment, which composes two sides of an imperfect square, the black river & the Catahoula forming the other two: this embankment has probably been nearly perpendicular without & in form of a glacis within; but as this formation of soil is found upon the margin of every alluvial Creek or brook in this low Country, I reserve my opinion respecting this apparent embankment, untill I can examine the whole more at leisure; in the meantime I have taken from the proprietor a sketch which he has formed of this remarkable place.

I find this to be a very interesting place, it is the point of confluence of three navigable waters viz the washita river, The tenza and the Catahoola,[55] the second communicates with the missisippi low lands by the intervention of other creeks and lakes & by one in particular called Bayou d'argent which enters into the missisippi about 14 miles above Natchez, during high water this is navigation for batteaux of any burthen along those bayoux, a large lake called St. John's lake occupied a considerable part of this passage between the Missisippi and the Tenza; it is a horse-shoe form, & has been at some former period the bed of the Missisippi, the nearest part of it is about one mile removed from the river of the present time: this lake possessing elevated banks similar to those of the river has lately occupied & improved; many similar possessions and improvements have been made since the first news of the cession of Louisiana by the french to the American Government; I omitted to mention in its proper place that it is understood, that even the hot-springs included within a tract of some hundreds of acres were granted by the late Spanish Commandant of the Washita to some one of his friends, but it is not believed that a regular patent was ever issued for that place, & it cannot be asserted that residence with improvement can be set up as a plea to claim the land upon.[56]

The Catahoola bayou is the third navigable stream; during the time of the inundation there is an excellent communication by the Lake of that name[57] & from thence by large Creeks to the red river; The Country around the point of union of those three rivers is altogether alleuvial; but the place of M. Hebrard's residence is no longer subject of inundation for reasons which have been already assigned; there is no doubt that as the country augments in population and riches, this place

55. Dunbar's "Catahoola" river is the Little River. The Little River, the Black River, the Ouachita River, and the Tensas River all converge at Jonesville in Catahoula Parish. Louisiana Quadrangle Maps.

56. Land claims in the Hot Springs area began within fifteen years of Dunbar and Hunter's visit. Under the New Madrid Land Act of 1815, Francis Langlois (who reported that he had been injured in the famous New Madrid earthquakes) first claimed 200 arpents, or approximately 168 acres, of land, which included the hot springs area, on November 26, 1818. On February 19, 1819, Elias Rector purchased Langlois's 200 arpents and later willed the land to his son Henry M. Rector. Almost immediately the Rector claims came into dispute with the U.S. government. Over the next fifty years Rector was involved with various legal maneuvers to legitimate his claims. Finally on April 24, 1876, the U.S. Supreme Court ruled the Rector claims invalid. *Hot Springs Cases* (1874) 10 U.S. Ct. Cl. 310–358; Albert Pike to Henry M. Rector, Lake Village, Arkansas, October 1870, in A. Howard Stebbins Collection, Arkansas History Commission, Little Rock, AR; Albert Pike, Robert W. Johnson, and J. B. Sanborn, *Claim of Title to the Hot Springs of the Ouachita: Brief and Arguement, General Jurisdiction*, Case No. 6245, Records of the Court of Claims, National Archives, Washington, DC, 150–151; Albert Pike, Robert W. Johnson, and others, *In the Supreme Court of the United States, No. 646, Henry M. Rector vs. The United States, 1876*, National Archives, Washington, DC, 176.

57. Catahoula Lake in La Salle Parish. Louisiana Quadrangle Maps.

will become the site of a commercial inland town, which will hold pace with the progress and prosperity of the country. On this place are to be found a number of indian mounts, one of which is of considerable elevation, with a species of rampart surrounding a very large space which is no doubt the position of a fortified town; having taken some notes respecting this place, the whole will be digested and introduced into an Apendix which will be added to this Journal.[58]

Monday 21st

HUNTER

Therm. at 7 a.m. 21°, in the river 40°, at 3. p.m. 36° & at 8 p.m. 26°. Clear & cold & raw. Wind easterly & variable.

Set out at day light, current as yesterday, I had previoussly to leaving the post, & also at the last settlement got directions to find a quarry of plaster of Paris said to be ten leagues from where we left this morning on the right side of the river going down about ten acres inland 1½ leagues below the prairie de Cote (Hill prairie) & a coal mine opposite to it on the left; yet altho I landed now & then to search, I was not more fortunate than we were going up, for I could not find the place of either the one or the other.—saw near the supposed place some encampments of Pascagula Indians, who did not or would not understand, english french or spanish & of course could get no information from them.—bought of them two Swans skins for two bitts.—[59]

Came this day by estimation about 14 Leagues french.

The sun set by my watch at 18 minutes past five—

The wind being ahead the greatest part of this day our progress was thereby considerably impeded—

58. The appendix that Dunbar refers to was printed in *Discoveries Made in Exploring,* 160–169. Dunbar's appendix also included navigational statistics entitled "List of Stages and Distances, on the Red and Washita rivers, in French computed Leagues," as well as a section called "Of medical properties of the Hot Springs."

59. In his reference to "a quarry of plaster of Paris," Hunter is writing of the possible gypsum mine. Gypsum is a mineral that contains calcium sulfate, $CaSO_4 \cdot 2H_2$, which was and is used for making plaster. Bates and Jackson, *Dictionary of Geological Terms,* 227–228. The Pascagoula Indians lived alongside the Biloxi Indians just north (approximately 30–40 miles) of present-day Biloxi, Mississippi, and along the Pascagoula River. According to anthropologist John R. Swanton, the Pascagoula may have been either Sioan or Muskhogean people. During the mid-1700s the Pascagoula moved to Louisiana and lived along the Red and Boeuf Rivers. By the year 1817, much of the tribe had moved to Texas in or near Angelina County. For more see Swanton, *Indian Tribes of the Lower Mississippi Valley,* 31–32; and Swanton, *Indian Tribes of North America,* 190–191, 208. Two bits equals twenty-five cents.

Tuesday 22nd

HUNTER

Therm at daybreak 21°. in air, in river 39°. at 3 p.m. 48° & at 8 p m. 40. Cloudy, Wind N.E. & E, raw weather.—& cold.—

Set out at ¼ before 7. a.m. Little current all this day & the wind being cheifly a head, went but slowly.

The rapids & shoals which we passed in coming up, were not to be seen at the same places now, the water being over them, they were all as smooth and still as a mill pond, & it was just visible which way the current run. Came this day to the rapids where we cut the channel for our boat when going up, but all is smooth now, having made about 14 leagues—here we encamped for the night. sunset 15′ past 5. p m.

Wednesday 23rd

DUNBAR

This morning is cloudy and threatens rain, the horses are not found, therefore no prospect of setting out today; A little rain fell about 9h a. m.—in the afternoon, one of the horses only is found.

HUNTER

Set out at half past six A.M. Therm. at daybreak 49°, in air, in the river 42°[.] at 3 p.m. 64° & at 8 p m 54 Wind S. Easterly[.] drizling rains, cloudy. The water still rises, so much so, as to take away all the current & it is now difficult to tell which way is up or down the river. I attribute this to the back water from the Mississipi. The want of current & the head winds together with the rains contrubuted to shorten our progress this day so that we made but about 7 french leagues & encamped about 1 league above the Bayu ha, ha,[60]

Thursday 24th

DUNBAR

Last night there was much thunder and lightening, and this morning the rain falls very fast: Having no other employment I endeavoured to collect information, Here I met with an American who pretends to have been up the Arcansa river 300 leagues; the navigation of that river he says is good to that distance, for boats draw-

60. Ha Ha Bayou joins the Ouachita 3 miles above the confluence of the Ouachita and Tensas Rivers in Catahoula Parish. Without Blazier and LeFevre, Hunter only mentioned the more significant features during the remainder of the voyage. Louisiana Quadrangle Maps.

ing 3 or 4 feet water: I do not give implicit faith to his man, when he speaks largely of the silver which he pretends to have himself collected upon that river, and even says that on the Washita 30 leagues above the hot springs he has found silver ore so rich that 3 tit of it yielded one of silver & that this was found in a cave: he asserts also that the ore of the mine upon the little Misouri was carried to Kentucky by a certain Boon,[61] where was found to yield largely in silver: This American says he has also been up the red river, that there is a great rapid just below the raft or natural bridge & several others above it:; The Cadaux Nation is 50 leagues above the raft, and near to their village commenced the Country of the great Prairies, of the Misouri., and extend 4 or 500 miles to the Sand mountains as they are termed,[62] those great planes extend to the south far beyond the red river; north over the arcansa river and among the numerous branches, This man confirms the accounts of the beauty and fertility of the Western Country, etc.—[63]

This evening the other horse has been found so that I hope to set out tomorrow morning.

61. The "certain Boon" was possibly from the family of Daniel or Squire Boone. By 1799 Daniel Boone and his family had crossed the Mississippi River and settled in Missouri. Hammon, *My Father, Daniel Boone*, 199–133; Thwaites, *Daniel Boone*, 43, 220–241.

62. Dunbar's "Sand mountains" undoubtedly are the Rocky Mountains. The more popular assumptions of the period placed the Rocky Mountains much closer than their actual distance of approximately 800–850 miles. In his three-volume work *North American Exploration*, John Logan Allen described the general ignorance of the region beyond the Purchase. This ignorance fueled numerous false assumptions about the distance, range, and height of the Rockies. Allen stated, "Jefferson and his contemporaries simply imagined the Rockies as a distant version of the familiar Blue Ridge." Allen, *North American Exploration*, 3:18–19, 135, 138, 180. Dunbar also stated in the "Dunbar Trip Journal, Vol. II" that "On the red river about 150 leagues above the post of Nakitosh the raft or natural bridge is said to commence . . . the great Chain separating the waters of the Mississippi from those which fall to the westward, those great planes extend to the arcansa river & from thence to the Missouri & beyond."

The raft that Dunbar mentions in this entry is the enormous logjam on the Red River known as the Great Raft. The raft extended for approximately 100–150 miles (although the size and area covered changed regularly), with its southern end near Campti, Natchitoches Parish. Some studies have showed the raft to have been formed as early as A.D. 1100. Between 1832 and 1839, Captain Henry Miller Shreve began clearing the Great Raft with his "snagboats." The logjam was not completely eliminated until after the Civil War, and then it was done at a cost of nearly $250,000 to the federal government. Shreve, "Rough Sketch," 14. Flores, *Southern Counterparts to Lewis & Clark*, 127–128, 133–137.

63. In a reference again to the bois d'arc tree, Dunbar mentioned on this day that "the Bois d'Arc' or yellow wood is also found upon the arcansa river about 150 leagues above the post. Rain falls rarely upon those planes but the dews are extremely abundant, & numerous brooks & rills water the surface of this fertile region. the breadth of those mountains is not yet known but supposed to be very considerable." "Dunbar Trip Journal, Vol. II."

HUNTER

Therm at daybreak 55° at air[.] in the river 43°. at 3 p.m. 50°, & at 8 p m 46°. Rain.—

Last night was remarkable for a long & heavy storm of thunder lightening & rain which drenched the men to the skin under their tents. Set out at day light, the rain having abated, but soon after it came on again & continued with very little intermission all this day, altho there was but little or no wind, & that mostly ahead. Went ashore at Monsr. Cadets at the mouth of the Catahoula Lake & opposite to the mouth of the Bayu Tensa[64] one league below the Bayu ha ha.

Here found Mr Dunbar being detained by the storm & waiting to have horses caught from the woods (where they are suffered to range) to carry him & his servant to Natchez which from this place is only 30 miles by Land altho about 150 by water & half of which is against the current of the Mississipi. Horses are now provided & he is set out to morrow morning weather permitting. We remained here the rest of this day, on acct. of the rain, having made only two leagues this day.— Mr Dunbar has received a few cuttings, suckers & seed balls or the fruit of the famous yellow dying tree from the little Missouri which had been transplanted on the Ouachita at the last settlement. but like myself could not find the hill containing the plaster of Paris, tho he landed to look for it.—He has also received a sample of the Briar root which is sometimes used here as food instead of flour, it resembles in shape the yams of the West Indies or rather some thing between that & the Irish potatoe; the flour is extracted from it in a manner simular to what is practised with the Potatoe for that purpose, with this difference, that it gives three different sorts of feculae, first, a reddish, next a grey & then a white, which fall down distinct according to their gravity.—He got also a sample of the Patate de Chevruil (Deer's Potato) said to taste like chesnuts,[65] Those samples as he is near his habitation he proposes to take with him & plant.—The river has risen here last night two feet.

Memorandum[:] the Indian mounts & fortifications to be described which are here omitted[66]—

64. The Tensas River forms the border of Catahoula and Concordia Parishes. The confluence is today across from the town of Jonesville. Louisiana Quadrangle Maps.

65. Hunter is referring to the Osage orange, or bois d'Arc, tree (*Maclura pomifera*). Little, *Field Guide to Trees*, 429. For more see Dunbar's entry for November 20. The "Briar root" was possibly a Chinabrier (*Similax bona-nox*). Hunter previously called it "the China Briar." Grimm and Kartesz, *Illustrated Book of Wildflowers and Shrubs*, 390; "Hunter Official Report," 14–15. For more information on the possible identity of the "Patate de Chevruil," see Dunbar's entry for this date.

66. Hunter described these "Indian mounts" in the "Hunter Official Report" as follows:

> We landed here at a solitary settlement at the confluence of the Catahoula and black rivers, inhabited by a Monsr. Cadet, who has built his house on an Indian Mound . . . If one may judge

Friday 25th

DUNBAR

The horses being late of fetching up we set out only at 9 o'clock; the weather was cloudy but not cold; the meeting of three rivers here which form the black river, has given it a considerable width at this place, little short I think of 400 yards. There is no apparent current here and the river is rising very fast, which is attributed to the Missisippi flowing up into the red river. The rain which has fallen these two days past, has rendered the roads extremely wet & muddy; we made only one league in the hour; arrived at the bayou Crockodile at 2h p.m.[67] This place is considered halfway from the black river to the missisippi, & is one of those creeks which are extremely numerous in the low grounds & serves to assist in venting the waters of the inundations: the whole of the country thru which we have passed today appears to be subject to the annual inundation; there are some places higher than others upon which canes are found growing, the margins of water courses are always found more elevated than the lands at some distance, degenerate into Cypress swamps and lakes.

At this place we found the waters of the Missisippi had already flowed in so abundantly that there was the necessity to prepare the raft for crossing and having in company three white men who understood the business, the raft was prepared of logs of the driest wood we could procure, lashed together with our horse ropes and halters, and after two hours delay got to the other side of the bayou which was about 60 yards wide including the overflowed low margin of the Creek; we had yet 5 leagues to make and it was already 4 o'clock. We pushed on but the roads proved extremely bad being under water for leagues together, it became dark, & we ex-

---

> from the emmense labor necessary to erect those Indian monuments to be seen here, this place must have once been very populus. There is an intrenchment, or embankment running from Catahoula to the black river, enclosing about 200 acres of rich Land, at present about 10 feet high & fifty broad. This surrounds four large mounds of earth, at the distance of a bow shot apart from each other: each of which may now be about 20 feet perpendicular, 199 feet broad & 300 feet long at the top. Besides a stupendous turret, situated at the back part, of the whole furthest off from the waters, whose base covers about an acre of ground, rising by two flats or stories, tapering as you assend, the whole surmounted by a great cone with the top cut off. This Tower of earth on measurement proved to be about 80 feet perpendicular.

This site became known to archaeologists as the Troyville Site. For more see Dunbar's entries for October 23 and January 21–22; "Hunter Official Report," 16.

67. This is possibly a reference to Bayou Cocodrie in Concordia Parish. The bayou runs almost the entire length of Concordia Parish, between and roughly parallel to the Mississippi and the Black Rivers. Louisiana Quadrangle Maps.

pected to be obliged to spend the night in the woods without fire, perhaps without a spot of dry land to rest upon; it was difficult to keep the path, by the sagacity of our horses we had the good fortune about 9 hours in the evening to get to a house four miles short of the river, where we were hospitably entertained with good homely fare.[68]

HUNTER

Therm. in air at daybreak 36°, in the river 40°[.] at 3 p.m. 40° & at 8 p.m. 40°. Cloudy Wind N. Easterly

Set out at 7. A.M. The weather raw & cold & blustering. A considerable part of this day the wind proved favorable & we set our Sail which carried us forward at a brisk rate so that when we encamped at half past five in the evening about sun set we had made about 30 miles this day altho in many of the reaches of the river the wind blew fresh in our teeth. This night proved cold & stormy & very uncomfortable.—

Saturday 26th

DUNBAR

Crossed the ferry and breakfasted at Natchez and arrived at my own house at ten o'clock where I had the satisfaction to find my family all well.[69]

HUNTER

Therm at 7 a m. 32°[.] at the same time in the river 42°. at 3 p m. 36°. & at 8 p m. 33. Wind N Easterly[,] snow, stormy,

Set out about 7 a.m. the wind ahead, but after an hours rowing by the turns of the river, it came favorable. Hoisted sail & made good progress, we rowed & sailed alternately according to circumstances & encamped about ½ past 5 pm. about sun set, about 6 miles below the mouth of black river having made about 30 miles this day.—Yesterday & to day there has been no current & indeed rather a running back of the water which as we came within about 2 leagues of the Red river we plainly perceived by the black river being quite red by the water of the red River which had overpowered its waters thus far.—Saw only one half torpid Alligator in our course down from the Fort.—[70]

68. This house was undoubtedly near Concord (now Vidalia, Louisiana).

69. Dunbar's journal ends on this date.

70. "Torpid" means dormant. The fort Hunter is referring to is Fort Adams. For more information on this outpost, see Dunbar's entry for October 17.

Sunday 27th

HUNTER

Therm. at daylight in air 24°. in the river 44°. & the river at noon 32°. At 3 p.m. 50°. & at 8 p.m. 32°—Clear[.] a light air from the Eastward. cold.—

Set out ¼ before 7. & rowed down, as usual hoisting our sail when the wind served. About 11 a.m. three leagues above the mouth of the Red River found the ice formed quite across half an Inch thick, thro which we forced our way by sailing & rowing four men at the Bow breaking the ice as we went.

At 2h.43m. p.m. Entered the Mississippi & observed that the high land at Fort Adams bears S 85. E distance estimated 15 mi les. Proceeded than S. 85. E till 5h.5m. rate pr Log 5½ perches pr half minute lost. 35′. thenS. 60 E—5.25 rate 3 perches
S. 40 E—5.37
when we encamped on the right shore having made about 17½ miles down the Red River & 3 miles 192 perches up the Mississippi in all about 21 miles this day. rose about 7 a m. & set 5h.32m p m. by the watches The Ice in this Latitude is rather uncommon & shews that this winter has been unusually severe.

Monday 28th

HUNTER

S 40 E. 7h.15m	Set out at 6h.45m. Therm then in air 26°. in the river 34°. at 3 p.m. 56. at 8 p m. 40°. Wind. N.W. Clear moderate weather. ⊙ rose at 7h. am.
S. 15 E. 8. 30	rate 4 perches pr half minute crossed to a point on the left
S. 65 E. 10. 12	lost 1 hour 10. m. aground &cc 2 ½ perches, rapid.
N. 60 E. 10. 50	Breakfast[.] lost. 30m.
N 30 E. 2. 35	to Fort Adams. Wind Fair Set sail[.] rate 5 perches[.] passed the line of demarkation at 11h.21m a.m. at the distance of 7 miles & 5 perches from Red River[.] Lost at the Garrison 45m.—dinner
N 75 W. 5. 40	encamped on the right. passed Buffalo creek[71] on the right at 4h.30. rate 3 perches[.]⊙ set at 5h.32m. Adams fort is 13 miles 35 perches from the mouth of Red River. Came this day 12 miles 43 perches

71. This is a reference to the Buffalo River in Adams County, Mississippi. Louisiana Quadrangle Maps.

Tuesday 29th

HUNTER

Therm. at daylight in Air 34°. in the river 33°[.] at 3 p.m. 56° & at 8 p.m. Wind N. fine weather

Set out at 6h.45m	rowing
N 75 W. 7. 0	rate 3 ½ perches
N 60 W. 9. 0	Breakfast[.] lost 35m
N 20 E. 10. 0	
N 50 E. 10. 45	6 perches Eddy
N 20 E. 11. 50	lost. 15m Wind N W
N 40 W. 1. 30	passes Old River,[72] Homochetto on the right at 11h.45m Dinner[.] lost 34m. rate 4 perches
N 5 E. 5. 15	lost 15m. rates 3 ½ perches
N 20 E. 5. 50	lost 8m crossing to a point on the right[.] here we encamped on a Sand bar for the night, having made 13 miles 287 perches this day.—The water of the Missisipi has fallen yesternight 2 Inches.

Wednesday 30th

HUNTER

Therm. at daylight 36° in Air. in the river 34°[.] at 3 p.m. 55 & at 8 p m. 53.—Light airs from the N E. raw cloudy weather.

Set out at 6h.43m.	A.M.
N 35 E. 7 .38	lost 8m in crossing to a point on the left. Log 3 K
East 9 .30	Breakfast[.] lost 30m.
N. 60 E 10. 10	
N 20 E 12. 45	
N 30 W. 3. 0	Dinner[.] lost 35m.
N 40 W 3. 35	
North 4. 15	
N 30 E 4. 35	passed between the island & the left shore
N 70 E 5. 50	lost in crossing 8 minutes Encamped 1 ½ mile above the big Island on the left shore at Hudsons new improvement Having made this day 9 miles 276 perches

72. Possibly Old River Lake in Adams County. Ibid.

Thursday 31st

HUNTER

Therm in Air at daylight 56°. in the river 38°. at 3 p.m. & at 8 p.m. Wind S.E. Clouds, moderate weather

Set out at	a.m.
6h.45m	
N 70 E 9. 5	3 perches pr Log[.] lost 5m.
North 9 48	lost 8m. crossing to St Catherines Warehouse[73] having come 3 miles & 60 perches this day

Distances from the mouth of the Red River to St Catherines landing

	miles	perches
Jany 27th. Sunday came this day	3,	192
"28th. Monday	12	43
" 29th. Tuesday	13	287
" 30th Wednesday	9	276
" 31 Thursday	3	60
	miles	218
	42	perches

Having arrived at St. Catherines landing, I left the property in charge of my Son & proceeded to Mr Dunbars, 6 miles inland, found that he had come home 5 days before & was occupied making, as he informed [me] the calculations to compleat his journal.—Here it was concluded to send down his waggon to the Landing to bring up the public property to be placed with the bulk of the Indian presents I had left there before to be ready for the expeditions up the Red & Arkansa rivers[74]—& as the mens provision were expended, I was to proceed up the river to Natchez 15 miles with the boat & crew to draw there a few days more provisions to enable them to reach Orleans where I was then to go with them & place the boat &c under the Charge of the Commanding officer there.

73. This warehouse was obviously at St. Catherine's Landing at the confluence of Old St. Catherine's Creek and the Mississippi River in Adams County.

74. This indicates that Hunter still expected the "Grand Expedition" to occur.

February—March 1805

Friday 1st

HUNTER

The waggon being sent down this morning for the goods I followed & delivered everything that was not essential for the use of the boat during the passage to Orleans, which being done we set out after dinner for Natchez, but a strong gale blowing from the N.W. almost right ahead, we made but little progress. & encamped at sun set on the right shore, about 6 miles from where we set out. The wind blew a fresh gale all the night & the weather grew very cold.—

Saturday 2nd

HUNTER

Set out again at sun rise & rowed for three or 4 miles when the course of the river changing were enabled to set our Sail & went against the current at a smart rate & arrived at Natchez about midday.

1805 Saturday Feby 2nd.

Having drawn 10 days provisions for the men commencing the first & ending the 10th. Inst & delivered Mr Dunbars 4 packages of goods to the Collector Mr Bayley,[1] it being then nearly dark concluded to wait here till next morning.

Here I saw Mr John Bringhurst who was very glad to see me & very attentive to serve me in everything I might need.[2] He has promised to make repeated enquiry at the Post office & if he should find any letters for me to forward them to the Care of Wm Donaldson Mercht. at Orleans. I have not received any letters from Mrs Hunter here as I expected The mails are very irregular. perhaps they may have been miscarried—

1. McDermott identified this man as a "D. Ballie." McDermott, "Western Journals of Hunter," 116.
2. Hunter was obviously acquainted with Bringhurst, a Philadelphia merchant. Ibid.

Sunday 3rd

HUNTER

The boat set out in the morning under the charge of my son George & proceeded st. Catherines Landing, whilst I rode to Mr Dunbars Seat accompanied by Mr Bringhurst in Mr. Wilkin's Gig. Mr & Mrs Wilkins were very attentive & polite to me George & I dined [&] Breakfasted there by invitation, we drank Tea at Mrs. Murrays, were kindly received by Mr & Mrs Murray who sent her profiles for Mrs Byrant. Mrs Hitch & Col. Timothy Mattack, with a letter for him & a snuffbox for Mrs. Colo Mattack which I promised to forward. Doctor Seib & Dr Pendergrass were also very attentive & civil. I also visited Mr Postlethwait who was very attentive & polite.—Mr Bringhurst had been on John Hares land,[3] says it is very rich but very broken—We arrived at Mr Dunbars about dinner time but as usual here there was so much company that I could not enter upon business till next day—

Monday 4th

HUNTER

Having lodged at Mr Dunbarrs all night. in the morning he gave me a receipt for the Indian presents & the remains of the stores, & equipments for the expedition, the Medicine chest, Mathematical Instruments watch sextant &cc to be taken care of by him until the next expedition.[4] I paid him also twenty six dollars in cash being the remains of the money unexpended for necessaries which I had received at Orleans in leiu of some parts of our 4 months rations, which were not issued in kind—& he returned back to me the due bill received at the post of Ouachita of

3. "Mr. Wilkin" was possibly James Wilkins, a wealthy Natchez merchant. Seip and Pendergast were Natchez physicians who both served as surgeons in military units in Mississippi and in the Louisiana Territory. Pendergast had published an 1803 work concerning the geography of the Lower Mississippi Valley, and he no doubt shared Dunbar's fascination with that area. James, *Antebellum Natchez,* 150, 86, 235; McDermott, "Western Journals of Hunter," 116.

Samuel Postlethwaite I was a prosperous Natchez merchant and the future son-in-law of William Dunbar. He later partnered in the establishment of the Natchez Steamboat Company. During the 1820s, Postlethwaite was president of the Bank of Mississippi in Natchez. He also served as the executor of Dunbar's last will and testament in 1810 and married William Dunbar's eldest daughter, Anne. Postlethwait to Archibald Dunbar, Forest near Natchez, November 17, 1810; Postlethwaite to Green & Wainwright, Forest near Natchez, November 20, 1810, Dunbar Papers, Jackson; James, *Antebellum Natchez,* 150, 193, 198.

John Hare is possibly the son of Andrew Hare. See Hunter's entry for February 13; McDermott, "Western Journals of Hunter," 24, 43.

4. The next expedition to use these instruments was the Red River expedition of Thomas Freeman and Peter Custis in 1806. Flores, *Southern Counterparts to Lewis & Clark,* 64–65.

Leiutt. Bomar, for lbs 100 Hams we furnished to the Soldiers out of our rations there, which due bill I am to receive at Orleans in money of the contractors.—

After this adjustment, I rode to St.Catherines landing & after delivering a few more of the stores &cc to Mr Dunbars Servant with his Team & a quantity of whetstones & samples of coal &c which we had brought down from the Ouachita, we took some wood for fewel on board & set out with the boat for Orleans, having received on board four recruits sent to Fort Adams by Capt. Cooper at Natchez.[5]—

We set out about 3 p.m. & proceeded down the Mississipi rowing in the day & floating in the night, having divided our crew into three watches for the night.—We run all this night & when we came within sight of Fort Adams it was about 4 in the morning, whe [we] then hauled in & waited for daylight.

Tuesday Feby 5th. 1805.

HUNTER

Set out at day light, for the Garrison where we soon arrived & having delivered the recruits & drawn the rest of the flour & whiskey which remained of the mens due bill at Fort Miro, we prepared to set out again, but a heavy storm of wind & rain came on which induced me to stay a couple of hours more & when it began to abate a little we proceeded in our way down the river again. We had not gone but about an hour when the storm began again with a cold raw rain & head wind, so that even with the current in our favor & ten oars we scarcely made any head way; The people were all wet & drenched with rain, the storm still appeared likely to encrease I thought it best to make for the weather shore, where having got into a safe birth for the boat, the men pitched their tents ashore & soon had good fires & made themselves as comfortable as such circumstances would admit. here we remained till next morning having made but about 8 miles this day.—

1805 Feby 6th. Wednesday

HUNTER

Wednesday, set out at daybreak, the wind still ahead blowing pretty fresh, cold & raw, we continued rowing until by a change in the course of the Mississipi, the wind came favorable when we hoisted our sail & made good progress until evening, when the weather became thick, it rained & blew cold & raw so that we could not see any distance, therefore I put ashore in a proper place for the night under the weather shore, the men pitched their tents & made good fires to dry their cloaths &cc having made about 60 miles this day.—

5. This may have been Captain William Cooper, who was an artillery officer originally from New York. Heitman, *Historical Register and Dictionary of the U.S. Army*, 1:326.

Feby 7th. Thursday

HUNTER

Set out at daybreak & after two hours the wind served to set our sail with a ruff in it, which carried us on finely[.] the weather is still thick, raw & cold & likely for rain—

1805 Feby 8th. Friday

HUNTER

Continued our course down the Mississipi all this day & on the 9th . Saturday about noon arrived at New Orleans, where I delivered the boat to the Orders of the Commanding officer Col. Freeman;[6] The same day paraded all the men with the Sergeant before him in good health, he gave them three days Holiday to rest themselves—Here I delivered also, the sail mast spars, 8 oars poles, rudder & Tiller & the musket with the remains of the equipments not left with Mr Dunbar.—

10th. Febr Sunday

HUNTER

Dined with my Son, (by invitation) with Governor Claiborn.[7]

11th. Monday

HUNTER

Visited Daniel Clark who being unwell, promised in two or three days to settle Mr Hares Acct. with me.—Saw also Mr Donaldson, who had not sold or exchanged the 927 $ of Spanish livrances—They demand 40 prcts. exchange deduction. Dr Zerban has remitted cotton in amount about 200 $ the balance of the former Acct.[8] Therefor the present acct. of 1741.65 Dollars remain outstanding.

6. Colonel Constant Freeman served as the commander of the fairly new American garrison at New Orleans. When organizing the Red River Campaign in 1805–1806, Dunbar had confused this man and Thomas Freeman. Thomas Freeman eventually led the Red River expedition. Flores, *Southern Counterparts to Lewis & Clark*, 52.

7. William C. C. Claiborne was Jefferson's third choice (following the refusals of the Marquis de Lafayette and James Monroe) to serve as governor. Claiborne was indeed in need of company, for only months before he had lost his wife, his daughter, some important staff members, and several acquaintances in the New Orleans yellow fever epidemic of 1804. Garvey and Widmer, *Beautiful Crescent*, 61–67.

Hunter made no entry in his journal for February 9.

8. For more information on Daniel Clark, see Dunbar's entry for October 31. "Dr. Zerban" was probably Dr. Phillip Zerban. McDermott had identified him and placed his residence at 27 rue de Conti in New Orleans. Charles L. Thompson, *New Orleans in 1805: A Directory and Census* (New Orleans: Pelican, 1936); McDermott, "Western Journals of Hunter," 63.

12th. Tuesday

HUNTER

I have recd a Duplicate Order of Survey from Baron de Bastrop Directed to Jas. McClaughlan or any other surveyor conformed by Colo. Lynch for 1940 Acres of land in his Grant from the Spanish Government on the waters of the river Ouachita in leiu of my Account against the late firm of Bastrop & Nancarrow, amounting with interest to $970.6—for which I have given him a receipt in full, the title to the same land being compleated. The original order I have delivered to Colo. Lynch, inclosed in a letter to Capt. Joseph Bommar the Commandant with a 5 dollar bank note, directing Capt. Bommar to present the Above Order to Survey to Colo. Morehouse in order that he may sign the following note at the foot of it,[9] viz "I hereby consent & agree, to confirm the title to the above 1940 acres of land to George Hunter, his heirs or assigns.—" And then to record the same in his office, & transmit it to me at Philada.—as the title to the whole original Grant of 12 leagues square is in these three persons, viz Baron de Bastrop, Colo. Chas. Lynch, & Abram Morehouse. Therefor, having all their signatures to my conveyance, my title will be compleat however they may terminate the dispute between themselves. Morehouse had promised to me, whilst I was at the Garrison in presence of Capt. Bommar the Commandant, to confirm my title to any tract of Land Baron Bastrop should convey to me in satisfaction of my debt due by the Baron.—

13th. Feby.

HUNTER

I have this day enclosed to Wm Dunbar Esqr. the Draft of Baron de Bastrop's grant of 12 leagues square on the Ouachita which Abm. Morehouse gave me for that purpose, & which I had by an inadvertance ommitted to do whilst last at Natchez.—This draft I have sent by Colo. Chas. Lynch. at the same time writing to Mr Dunbar as follows viz. Sir. We arrived at Orleans on the 9th Inst. at noon when I delivered the boat with the remainder of the equipments &cc to the Order

9. McDermott identified James McLaughlin as a surveyor and treasurer at Ouachita Post, and he also clarified that Colonel Charles Lynch was involved in the Bastrop land grant. Carter, *Territorial Papers of the United States*, 9:601; McDermott, "Western Journals of Hunter," 117. By "Capt. Joseph Bommar" Hunter is referring to Joseph Bowmar, who had received his promotion to captain on October 12, 1804. Although Bowmar and the two explorers seemed to be unaware of the promotion during the ascent, Bowmar may have received word of his change in rank by the time the expedition returned to Fort Miró on January 16, 1805. See Hunter's journal entries for November 7 and January 16 and Dunbar's entries for November 6 and January 16, 17, and 18. For more on Bowmar's military career, see the introduction. For more on Colonel Abraham Morehouse, see Dunbar's entry for November 11.

of the Commandant, (Colo Freeman,) before whom I presented the men all in good health, conformably to your directions; I have since learned that the same boat & men, under the command of Leiut. Murray, are sent to Natchitoches with Indian goods, to set out to morrow. As by some innadvertance, I omitted to leave with you the draft of Baron de Bastrop's grant of 12 Leagues square, on the Ouachita, given to me by Colo. Morehouse for that purpose, I now embrace the present safe conveyance by Colo. Lynch, who it appears will eventually become the principal proprietor of that extensive grant; The nature of whose claims on Mr Morehouse, may point out to you the necessity of attending to the amount of your late transactions with him. I purpose to set out with my son to pay a short visit to Attacapa before my return home[.] therefore should any letters for me come to your hands in the interim you will be pleased to transmit them to Orleans to be left at the Post Office. which will be an addition to the many civilities & obligations already conferred on your &c G. Hunter.—respects to Mrs Dunbar & family[10]

16th. Feby

HUNTER

I have been waiting these several days, dancing attendance on Daniel Clark, who had promised to pay me the balance due Mr Roberts of the Illinois country & which Mr Roberts owed to Mr Hare, by which I lost an opportunity of going to Philada. & an other to Baltimore, & now he tells me, that this debt is attached in his hands, by a Patrick Morgan of this place on a bond due him by the late Andw. Hare, & that he has given in his Answer on jany. last to said Attatchment. Also that a Mr Abner L Duncan was employed by the court as Attorney for Hare's estate in my absence,[11] I saw Mr Duncan who appointed me to see him at his house to morrow morning.

In the meantime, I paid a Visit to Mr Wikoff of Appalousi a member of Council here of whom I got a general account of the Attacapa & Appelousi Countries,

10. "Leiut. Murray" may have been Lieutenant W. A. Murray. Originally from New York, he served as an artillery officer from 1802 to 1809. Heitman, *Historical Register and Dictionary of the U.S. Army*, 738; Carter, *Territorial Papers of the United States*, 9:433. Attacapa was an area in southwestern to south-central Louisiana named for an Indian tribe whose territory had formerly been in what is today southeastern Texas and southwestern to south-central Louisiana. The area was also known as Attakapa. Wilds, Dufour, and Cowan, *Louisiana Yesterday and Today*, 3, 253; Kniffen, *Indians of Louisiana*, 52–53. Hunter made no entries for February 14 and 15.

11. Andrew Hare (who died between October 1799 and February 1800) and "Mr. Roberts" are apparently two gentlemen with whom Hunter had conducted business during his time in the Illinois and Kentucky backcountry. McDermott identified Duncan as "a very prominent attorney in New Orleans." McDermott, "Western Journals of Hunter," 43, 118.

the richness of the soil, healthiness of the climate &cc.—He informed me that the goods generally saleable in that country are Handkerchiefs of all kinds, not the finest, or dearest, muslins, calicoes, blankets cloths not superfine, but second & coarse, Iron Mongery for house building & spades shovels plough irons &cc & for agriculture, axes, hoes &cc, nails, linnens, pistol muslins & for the Spaniards fine cloths blue & scarlet. Black striped Velvets He advised me not to purchase any place for a residence at present or until I had been some time in the country to look round & make choice after due deliberation—That it would be better to bring my family to that country & hire or purchase a small place for a short time than hazard a large purchase which I might afterwards repent upon seeing a more eligible situation.—[12]

20th. Feby.

HUNTER

Mr Clark has delivered to me a statement of what he owes Mr Roberts, by which the sum in reduced from 600, or 700 Dollars, as he acknowleged to Mr Clay, to 188 Dollars, & even this is attatched by a Patrick Morgan here, in the hands of Mr Clark as the property of Andw Hare deceased. to repel which I now write to Mr Peter January of Lexington to send down to Abner L. Duncan Attorney at Law Orleans, an official certificate of the judgment obtained there on the Marriage contract of Andw. Hare with Mrs Hare in favor of the trustees Geo & Wm Hunter.—which will prove that a Contract debt to it will come in before any other Creditor

22nd Feby

HUNTER

I have this day written to Mrs Hunter informing her of our sailing for New York this day, by the Brig Julian Capt Crooker—we accordingly went on board on Saturday 23rd. in the afternoon, but the vessel did not get under way untill Sunday forenoon on the 24th . Feby, when Mr Wm Donaldson a Merchant of Orleans[13]

12. William Wikoff (or Wykoff), a longtime resident of New Orleans, was appointed treasurer in the community of Opelousas, Louisiana. As inferred in Hunter's entry, Wikoff served in the Orleans Territory Legislative Council. It is apparent that Hunter is beginning to explore the prospects of starting business relations and of moving to New Orleans. Carter, *Territorial Papers of the United States,* 9:10, 28, 285, 601. On Attacapa, see Hunter's entry for February 13. "Iron Mongery" here means ironmongery, which referred to household hardware such as door knobs and knockers, hinges, pulls, bolts, latches, hooks, and so forth.

Hunter made no entries for February 17–19.

13. William Donaldson was a founder of a steam-engine sawmill in New Orleans and also later the founder of Donaldsonville, Louisiana. He opened a mill (possibly with one of the engines acquired

came on board with a letter for me, to procure for him of Oliver Evans in Philada. a Steam Engine & mill to saw timber with six saws at once. Upon enquiry, what sort of goods were the most proper to carry with me to Louisiana, with a veiw that if it should happen that I could more easily exchange some of my Lots for goods, then sell them for cash, Mr Donaldson informed me, that the following goods were always saleable viz

Britanias, Cheifly white, Platilla's, Estopillas, morlais, all these should be rather fine.—2½ & 3 point Blankets, with deep blue stripe & twilled, large sized.—Calicoes, handsom patterns wide & narrow rather fine.—as the Spaniards of late run upon white grounds & small figures—Nankeens white & yellow, cheifly yellow.—India Cottons such as Bastas[?] Cossas &ccc—Coarse Narrow cloths, such as can be sold from 5 to 9/—curry pr. ell,[14] dark mixtures & blues—Marseiles Quiltings, good quality & handsome patterns, white & coloured, cheifly white.—I paid 100 Dollars to Capt E. Crooker for the passages of myself & son to New York & laid in our own provisions & stores, which cost 54$. I paid Madam Chabau [blank in MS.] for our board for 14 days whil[s]t at Orleans, besides [blank in MS.] for washing & sundry small expences.

I left with Messrs. Chew & Relf by Mr Dunbar's desire all the stones, coal, clays, ores &cc which I had packed in a ½ barrel as samples of what were contained on the Ouachita.—except two peices of rough stones of the Silicious kind, which had been left in the boat & thrown on the Levey at Orleans, these I took with me to have tried at my leisure to see whether they would answer the same purpose as the Turkey Oil stones, which they somewhat resembled.—[15]

---

through Hunter) in Manchac, Louisiana, in 1807. Hunter later established a steam distillery in New Orleans, which may have used this same type of engine. Oliver Evans was the first steam-engine manufacturer in the United States. McDermott, "Western Journals of Hunter," 118.

14. Hunter oddly appears to be using the value in shillings.

15. Mystery surrounds the fate of the collected specimens that Hunter mentions in this entry. In the "Hunter Official Report," he wrote, "Previous to our sailing, I received a letter from Mr Dunbar intimating that he had not received enough of the Stones & samples, the coal, clays &c, that we collected in our tour, that the whitstones &c might be of use to him on his farm &c desiring me to pack up a bagfull more assorted, & leave it with Messrs Chew & Relf at New Orleans for him." Hunter therefore stated, "I thought it right under all these circumstances to leave with Chew & Relf agreeably to his order, the half barrel containing samples we had collected on our Tour; the Mr Dunbar might send to the President such parts thereof as he should judge proper." By March 16, however, the samples had not arrived in Natchez, and Dunbar wrote to Jefferson explaining,

> a number of Specimens were collected to be taken round by the Doctor in order that you might have the satisfaction of judging of their properties from your own view, the Doctor being arrived at New Orleans writes me that Gov. Claiborne had already sent you a number of speci-

Monday 25th

HUNTER

Having yesterday proceeded down the Mississipi till dark, & then made fast to the bank for the night, Therefore as soon as it was daylight, we carried out a small Anchor some distance in the river & hove upon it until we were sufficiently clear of the bank, when we set sail & continued our route till night, when we made fast to the banks as before[16]

Tuesday 26th

HUNTER

Set out as yesterday, & at night made fast to the bank as usual. This day passed the Fort at Plaquimin[.][17] it was not convenient to stop to go ashore to veiw the fort which appeared to be of a square form surrounded by a ditch with brick walls in the inside as high as to form ambrasures for the cannon, a glacies &cc

Wednesday 27th.

HUNTER

Set out at day light. At 9 a m passed the N E. Pass, one of the Mouths of the Mississipi. At10 a.m. passed the Balize & Pilot house, got a Spanish Pilot on board, & immediately proceeded for sea; When we came closs to the Bar, the wind which had been fair, now shifted a little, & we were obliged to drop Anchor. But previous

---

> mens from the Washita collected by a Richard King, from which circumstances the Doctor conceived it to be superfluous to carry you those Specimens & left them at N.O. to be sent to me: I am persuaded that Mr. King has never thought of collecting any Specimens from the Hot Springs & probably has only sought for metallic or chrysaline Specimens, or any thing possessing a showy appearance. I have therefore requested the Gentlemen at N.O. with whom the Doctor left the specimens to forward them to him at Philad. In order that the first intention of presenting them to you may be fulfilled. (Dunbar to Jefferson, Natchez, March 16, 1805, Jefferson Papers)

It is not clear whether Hunter ever received these specimens from New Orleans; if he did receive them, whether he forwarded them to the president; and if he sent them to Jefferson, what may have happened to them once they were received by the government. Jefferson did send some of the specimens that he received from Dunbar to other scientists for study. The fate of these specimens is also unknown. Jefferson to Dunbar, Washington, May 25, 1805, Dunbar Papers, Jackson; Rowland, *William Dunbar,* 147–148, 174–177.

16. Hunter made no entries for February 23 or 24.

17. Plaquimines Fort was located on the east bank of the Mississippi at a place called Plaquimines Bend in what is today Iberville Parish. The fortifications at Plaquimines Bend were built by the French in 1746. Carter, *Territorial Papers of the United States,* 9:53–54.

to this we as we passed the Balize we observed the Schooner 5 Brothers which had left Orleans bound for Philada. three days before us, & several other vessels which had been windbound who when they saw us pass them, they all hove up their anchors & followed us,—We were detained here about one hour, when the wind veered about favorably & we got up our anchor [&] stood out to sea touching the mud on the bar once or twice without sticking,[18] & then discharged the Pilot to whom Capt. Crooker paid 1½ dollar pr foot as Pilotage.—

From the 26th . feby till the 3rd March we had fair winds tho light breezes, & fine cloudy weather tollerably pleasant. The wind became now rather variable & continued so till Tuesday 5th March when we were boarded by the Brittish Frigate Francois Capt. Perkins, & Politely treated; They informed us that they detained Spanish Vessels when they had money aboard, & enquired when a certain Spanish Brig should sail from Orleans, which they had learned was taking the Cannon & Military stores on board belonging to the Spainish government on evacuating Louisiana.—We were now by Observation in N. Lat. 23°.3′.0″ & in sight of Cuba, & by estimation near Bayu Honda.—[19]

March 6th.

HUNTER

The wind is unfavorable, the current carrying us out of our way; saw a schooner to windward which kept in sight all this day. Lat. by Obs. 23°. 4′ 0″.

Mr Wright, a gentleman from Tennessee, & now a fellow passenger with me in the Brig Julian from Orleans to New York, informs me that Benjn. Grayson Orr. who owes me about 1000$ for Drugs furnished by his order & for which I have his note protested, now lives on the main road from Virginia to Kentucky & Tennessee, near Bean's old station, within three miles of where the road forks, one leading to Tennessee & the other to Kentucky. He keeps good public house in said road, & is supposed to be in good circumstances.

This road is on the waters of Holstein; he lives within 25, or 30 miles of Ross's Iron works on the north fork of Holstein,[20] not many miles from the Town of Ab-

18. The "Balize" was a 40-foot-high tower on the west bank of the Mississippi. It was utilized as a watchtower and not as a lighthouse or light tower. Ibid. Even after extensive channel work, the Mississippi River estuary region remains a precarious area to navigate because of constantly changing sandbars, submerged vegetation, and other washed-down material. U.S. Environmental Protection Agency, *Mississippi River Basin;* U.S. Army Corps of Engineers, *Mississippi River and Tributaries Project.*

19. Hunter probably means Bahia Honda Key (meaning deep bay). One of the islands of the Florida Keys, Bahia Honda is approximately 12 miles south of the town of Marathon on U.S. Highway 1. *National Geographic Atlas of the World,* 43.

20. Hunter probably means the Holston River in southwestern Virginia. The North Fork of the

bington near the borders of Virginia. Mr Rogers Innkeeper in Roger's Ville will convey a letter to him, Mr Wright will pass that way on his return & will convey a letter to him. He also informs me that Mr Wm Cumings lives in Martin's Ville[,] North Carolina, Guilford County[.] He no practises phisic there.

18th, March from the 6th. to this date the wind proved unfavorable & we were beating in the Gulf of Florida, sometimes in sight of Cuba & at other times in sight of the Islands forming the promontary of Cape of Florida.

This day the wind came from the south beginning with a gentle breeze which gradually encreased till the 23rd when it blew a brisk gale & changed again to the N West & continued aganst us till the 26th. when it again came favorable, & on the 27th. in the morning we arrived at New York after an absence of ten months, during which time I have gone over a distance of above 7000 miles including the land & water.[21]

Holston is near the town of Pearlsburg (not far from the Virginia-Kentucky border) and flows southwesterly and into Tennessee. Virginia Quadrangle Maps, U.S. Geological Survey, Denver, Colorado. According to McDermott, "Ross's Iron works" was located 15 miles east of Hawkins Courthouse in Kentucky. McDermott, "Western Journals of Hunter," 120.

21. In the "Hunter Official Report," Hunter included an April 1, 1805, entry that stated: "Found my family in health, tho my business had suffered by my absence. Immediately set about transcribing my rough notes but am considerably impeded by an inflamation that has taken place in my eyes, since arrival. In the interim my son has gone again to Mr Patterson at the Accademy & will begin calculating the Astronomical observations made during our tour, to compleat the journal." His discovery that his business interests were in a state of decline may have been the reason or one of the reasons he chose not to participate in any additional government-sanctioned explorations. The Mr. Patterson mentioned may be Robert Patterson (1743–1824) who became a member of the American Philosophical Society in 1783 and served as a professor of mathematics at the University of Pennsylvania. Patterson also became director of the U.S. Mint in 1805. McDermott, "Western Journals of Hunter," 120.

Explanation of Navigational Techniques

William Dunbar collected navigational and scientific instruments and books on these subjects much as modern-day sports fans amass baseball cards. He owned some of the finest telescopes, microscopes, thermometers, compasses, and chronometers, a reflecting circle (called by Hunter "Mr. Dunbar's circle of reflection"), an artificial horizon, and a sextant. As he began to experience ever-declining health following the Ouachita River expedition, he continued to accumulate this equipment. He joyfully reported in a letter to Jefferson in December 1805 about the arrival of a new telescope from London.[1]

His personal library of more than eight hundred books contained volumes such as *Marget's Longitudinal Tables, Vince's Astronomy,* and *Maritime Surveying* as well as a considerable collection of nautical almanacs.[2] Through this reading and the

1. On December 17, 1805, Dunbar wrote the president and explained, "I have been in the habit of receiving books and instruments free of duty . . . but Mr. Brown at New Orleans is so rigidly faithful as a public servant, that he admits of no exemption neither in favor of the Mississippi Society [Mississippi Society for the Acquirement and Dissemination of Useful Knowledge] for which I have lately imported a chest of books; nor in favor of this valuable instrument . . . I have sent off an order for . . . astronomical instruments & chronometers." Dunbar to Jefferson, December 17, 1805, in Rowland, *William Dunbar,* 186–187.

A chronometer is a very accurate timepiece usually set on Greenwich mean time (GMT) (longitude 0°) to aid in finding latitude and longitude. Dunbar described the new telescope as a "six foot gregorian reflecting Telescope with magnifying powers from 100 to 500." Dunbar to Jefferson, December 17, 1805, in Rowland, *William Dunbar,* 186–187.

2. Rev. Samuel Vince (1749–1821), who became a professor of astronomy and philosophy at Cambridge in 1796, published a three-volume work that included the latest astronomical tables or indexes for navigation. One of these tables may have been one of the three indexes that Dunbar used on the expedition. Rev. Samuel Vince, *A Complete System of Astronomy,* 3 vols. (Cambridge: J. Burgess, 1797, 1799, 1808), 1:581–582, 2:585–586. The royal observer Nevil Maskelyne (1732–1811) authorized the publishing of the *Nautical Almanac* beginning in 1766. In that same year, Maskelyne also compiled a table to be used with a nautical ephemeris. The reprint of his table and ephemeris between 1781 and 1802 may also have been used on the Ouachita River expedition. Dunbar probably carried the most recent issue of the *Nautical Almanac.* In July of 1805, he requested the 1806 *Nautical Almanac* for the Freeman-Custis Red River expedition. Dunbar to Henry Dearborn, July, 13, 1805, in Rowland, *William Dunbar,* 157; McDermott, "Philosophic Outpost on the Frontier," 7, 15–19.

more practical experience he obtained through his personal observatory at Union Hill near Natchez, and in the course of his service as a surveyor for the Spanish government, Dunbar perfected his abilities.[3]

The expedition was lucky to have a knowledgeable, skilled, and experienced surveyor-astronomer like William Dunbar, who had acquired the acumen to become a talented celestial observer. How Dunbar obtained these skills is uncertain, but like much of his scientific expertise, they may have been self-taught. On the joint Spanish-American surveying team that plotted the 31° latitude border, and in his surveying that helped lay out the early plans for Spanish Natchez, he gained valuable training and information concerning the latest procedures. He apparently became well known in the lower Mississippi Valley for his accomplishments in this area. His journal makes it clear that he was well versed in the methodology of celestial observation. Dunbar knew his craft. In addition, however, he was also constantly pursuing ways to prefect his techniques.[4]

George Hunter, a Philadelphia chemist, was not as experienced in astrological observation and navigation as his companion, but he did seem to have some basic knowledge of the procedures used in discerning longitude-latitude positions. He also seemed to have a good comprehension of navigational terminology and symbols. Unlike Dunbar, Hunter did not keep a separate journal to list his astronomical readings. For that reason, he often placed his readings in his daily entries. Hunter carried a sextant on the trip, but by his own admission the device was fairly useless to him. Using Dunbar's artificial horizon as a model, he apparently attempted and failed to construct his own.[5] Some of Hunter's readings are slightly different from those of his counterpart, indicating that Hunter was indeed making some of his own observations. It is logical to infer, however, that on most occasions Dunbar's readings may have been the more accurate, because of his training, his practical experience, and the variety of quality instruments he possessed.

Determining latitude and longitude by celestial observation during the early nineteenth century took special skill with various cumbersome instruments, a geographical understanding of the earth's rotation, a good grasp of geometry and as-

3. Rowland, *William Dunbar,* 10; McDermott, "Philosophic Outpost on the Frontier," 3–9.

4. For more on the survey and Dunbar's service to the Spanish government at Natchez, see the introduction. Dunbar to Jefferson, July 6, October 8, December 17, 1805, in Rowland, *William Dunbar,* 154–156, 182–188; Dunbar journal entry, December 16.

5. Hunter entry, December 13; Dunbar to Jefferson, November 9, 1804, Fort Miró, in Jefferson Papers. Dunbar wrote this letter from Fort Miró and described Hunter's difficulties in using the sextant. The obstructions made by trees, hills, and mountains made the sextant a troublesome piece of equipment for the chemist, though it was a perfect instrument for observations on the open ocean.

tronomy, and patience. The expertise needed to complete such a task as mapping a river, laying out a town, or surveying a border took months, if not years, to perfect.

It is clear that Dunbar had that expertise, but in today's world of handheld global positioning systems, calculators, digital timepieces, space telescopes, and the Internet, it may be difficult for most people to understand the basics of the navigational techniques of circa 1804–1805. A brief summary of the procedures used by Dunbar and Hunter may aid in the comprehension of some of their entries.

As the expedition team ascended the rivers, they immediately attempted to determine their position by latitude. In basic terms, latitude is found by measuring the altitude of a star or the Moon, or more commonly by solar observations. Dunbar used all of these celestial bodies in his determinations but seemed to use the Sun most often. During the day, and when weather permitted, they moved ashore to allow for these observations.

In finding latitude it is important to locate the horizon, a task that is almost impossible in terrestrial navigation because of obstructions. Therefore Dunbar used an instrument called an *artificial horizon.*[6] It was a metal box filled with mercury (Lewis and Clark used one that contained water instead) that had a glass covering. The mercury inside the box formed a level mirror for the observation. While one observes the Sun, as Dunbar usually did (rather than observing a star), the object is reflected in such a way that the angle between the reflected image of the Sun and the Sun itself is twice its angle above the horizon. Because of the double refraction, the observer must then divide the angle found by two to discover an artificial horizon that can be used to calculate the latitude. Most sextants could only measure 120°, so they were nearly useless in this type of calculations; however, Dunbar also carried an important instrument called a *reflecting circle* that could aid him in this process.

Usually supported on a brass pedestal with a wooden tripod, a reflecting circle was important because it was calibrated at 720° (360° × 2). Using it with an artificial horizon, one could get the necessary readings to find the horizon or an artificial horizon despite obstructions such as trees or hills (the artificial horizon that is found is also referred to as the *double apparent meridian altitude*).[7] After find-

6. This instrument was probably invented by a London mathematical instrument maker named George Adams (c. 1703–1773) in 1738. Cotter, *History of Nautical Astronomy,* 92.

7. Dunbar determined his latitude by discovering the double apparent meridian altitude, or as Hunter abbreviated it, "d. appt. mer. alt." This method called for an observation of the Sun's altitude at a certain time before and after noon. Again, after these two angles were determined, an ephemeris was consulted to determine the declination, the Sun's highest point or meridian altitude (also called *zenith* or *transit*). With the finding of the meridian altitude, an observer could determine latitude.

ing the artificial horizon, an observer would refer to a chart in an almanac to compute the *index error*, or the refraction of the light through the earth's atmosphere. The index error was then subtracted from the angle (by the use of an accurate chronometer, the exact time of the observation would be known).[8] At that point, the person making the observations would consult a chart or index called an *ephemeris* that listed the Sun's *declination* and right ascension for the year, month, day, hour, minute, and second. According to the season, the ephemeris would show that an observer needed to add or subtract from the reading to correct for declination. These calculations determined the person's latitude. Dunbar revealed that he carried "three indexes" with him; however, he did not state the type or the source of these tables.[9]

On several occasions, the explorers used the angle of a star to determine their latitude. The basic principle is that the altitude of the star is equal to the angle between the observer and the equator (or the observer's latitude). Dunbar used stars such as Aldebaran (Alpha Tauri), Aries, and the North Star to make these calculations. Using a chronometer and making these calculations at the same time each night was a more efficient means of determining latitude and longitude.[10] Many times Dunbar attempted to use this method.

In celestial observations it was common to use the symbol ☉ for the Sun and ☽ or ☾ for the Moon or the Moon's limbs or edges (Hunter often abbreviated ☉ d. appt. mer. alt.). Both Hunter and Dunbar used these similar symbols when recording some of their observations.[11]

The same procedure can be employed to calculate longitude. Once the meridian altitude is determined, the time is recorded from a chronometer set on GMT (0° longitude). An observer then subtracts 12 (for local 12:00 noon) from the GMT time recorded. Since the Sun travels at a rate of 15° per hour, the newly calculated hour must be multiplied by 15. The minutes and seconds are divided by the new

8. Smyth, "Thompson's Surveying Instruments and Methods."

9. An ephemeris listed the daily position of certain celestial bodies and their declination. The declination represented the difference between magnetic north and true north. A compass finds magnetic north. Because of the earth's slight movement on its axis as it rotates, magnetic north is not always true north. Indexes were available in the late eighteenth and early nineteenth centuries that listed the daily declination. Ibid.; Dunbar journal entry, November 17.

10. Dunbar journal entries, November 10, December 17. Smyth, "Thompson's Surveying Instruments and Methods."

11. Hunter journal entry, November 26; Dunbar, "Journal of a Geometrical survey," November 7, 1804.

hour (GMT − 12 [local noon]) and added to the degree and minute. The resulting number is the longitude.[12] Here is an example:

> GMT determined as 18 hrs. 12 min. 36 sec. − 12 (local time) = 6 hrs. 12 min. 36 sec.
> 6 hrs. × 15° (Sun's movement per hour) = 90°
> 12 min. ÷ 6 (GMT − local time) = 2 + 90° = 92°
> 36 sec. ÷ 6 (GMT − local time) = 6
> Longitude = 92°6′

Because of the calculations involved in determining longitude after an observation is taken, Dunbar rarely inserted these readings in his daily narrative entries; he kept them in a separate table.[13] Keeping a chronometer (timepiece) accurate was essential in making correct daily observations. However, even the finest timepiece of the early nineteenth century lost time each day. According to several of his daily entries, Dunbar attempted to reset his watch by examining the angle of celestial bodies.[14] By that time Dunbar would have known from his own calculations or by use of the three indexes he carried, what Greenwich mean time (GMT) was at certain angles. Knowing the difference between local time and GMT, he could reset his watch. Dunbar and Hunter could be even more precise by noting when the limbs (edges) of the reflecting circle passed a certain place on the instrument.[15] Doing so could allow them to calculate the correct time down to the second. Apparently Dunbar had kept accurate records in his many surveying ventures, and one of the indexes he carried may have consisted of his own calculations. Because determining longitude was time-consuming, Dunbar stated on several occasions that he planned to make the necessary mathematical calculations at a later date. Many of them he made after his return home to "the Forest."[16]

Dunbar and Hunter's diligence in making their daily observations as well as

12. Smyth, "Thompson's Surveying Instruments and Methods"; *Map Reading and Land Navigation* (Washington, DC: Department of the Army, 1993), 4-1–4-4.

13. "Dunbar Trip Journal, Vol. I"; "Dunbar Trip Journal, Vol. II"; Dunbar, "Journal of a Geometrical survey."

14. For example, Dunbar journal entry, December 17.

15. Dunbar allowed Hunter to make some of his observations using the reflecting circle. Hunter journal entry, December 17.

16. Dunbar, "Journal of a Geometrical survey"; Dunbar to Jefferson, February 23, 1805, in Rowland, *William Dunbar*, 145–146.

their directional river courses allowed the men to present data that aided in the drawing of the first accurate map of a river in the southern Louisiana Purchase. It is obvious from their journals that this was not only a trip of physical rigor; for William Dunbar and George Hunter it became an intellectual journey that constantly exercised their inquisitive minds.

Notes on Sources and Editorial Process

The journals of William Dunbar and George Hunter provide a rich, yet often puzzling, view of the natural world in 1804. Although Hunter's account of the journey up the Ouachita River has received historical editing, Dunbar's notes have never been edited. And the two explorers' journals have never been presented together in one volume. As stated in the introduction, Dunbar copied his journal after the expedition and sent it to Thomas Jefferson. Dunbar also sent some of his collected specimens to Jefferson, as well as to the pioneer botanists Benjamin Barton and Henry Muhlenberg for classification and study.[1] To date none of these specimens have been located.

Dunbar's copied journal, entitled "Journal of a voyage commencing at St. Catherines Landing on the East bank of the Mississippi, proceeding downwards to the mouth of the Red River, & from thence ascending that river, the black river and the Washita river, as high as the Hot Springs in the proximity of the last mentioned river," was given to the American Philosophical Society in Philadelphia in July of 1817. In this present version of the Dunbar-Hunter accounts, the copied Dunbar journal is referred to as the "Dunbar Report Journal" and is the Dunbar document edited in this work.

Dunbar's original journal (which he recorded en route) is less detailed than the "Dunbar Report Journal" and in some daily entries more abbreviated. The Mississippi Department of Archives and History in Jackson holds an original journal, but it begins on December 11, 1804, and constitutes only the second half of the notations Dunbar made during the trip. That journal is herein be called "Dunbar Trip Journal, Vol. II."[2]

Until the spring of 2002, the "Dunbar Report Journal" in the American Philosophical Society's library and the second half of the original in Jackson, the "Dun-

1. Actually, Dunbar hired a man who was engaged as a "Schoolmaster" to copy his journal. Dunbar to Jefferson, March 9, 1805, in Rowland, *William Dunbar,* 146. For more on Barton and Muhlenberg, see the introduction.

2. In a letter dated March 9, 1805, Dunbar revealed to Jefferson that his original journal was in two volumes. Dunbar to Jefferson, March 9, 1805, in Rowland, *William Dunbar,* 146.

bar Trip Journal, vol. II," were the only archived sources of the Dunbar references for the Ouachita River expedition. However, in March of that year a new journal came to light within a private collection in Natchez, Mississippi. This journal, which is undoubtedly the first half of the original journal (from October 16 to December 10, 1804), is now preserved in the Riley-Hickingbotham Library at Ouachita Baptist University. In this present work, the first volume of the original journal is called the "Dunbar Trip Journal, Vol. I."

Because the "Dunbar Report Journal" provides more detail and more analytical data than the "Dunbar Trip Journal," the "Report Journal" is a vital source of information and more thoroughly documents the expedition's experiences. Dunbar obviously remembered additional details and also refined his thoughts as he copied his daily entries. Where the content of the entries vary, the differences are noted. Simple variation of phrases or words, however, is not referenced.

The American Philosophical Society also preserves a Dunbar journal entitled "Journal of a Geometrical survey Commencing at St. Catherine's Landing on the East shore of the Mississippi descending to the mouth of the red river, and from thence ascending that river, the black river and river of the Washita as high as the Hot Springs in the proximity of the last named river." This document includes a meticulous listing of the Scottish explorer's astronomical observations and river course statistics.

In 1930 Eron Rowland published *Life, Letters, and Papers of William Dunbar of Elgin, Morayshire, Scotland, and Natchez, Mississippi: Pioneer Scientist of the Southern United States.* This book drew heavily on the Dunbar Papers at the Mississippi Department of Archives and History. Among the numerous items of correspondence between William Dunbar, Thomas Jefferson, and other contemporaries, Rowland included the first complete printed version of what we are calling the "Dunbar Trip Journal, Vols. I and II." Apparently, Rowland had access to the "Dunbar Trip Journal, Vol. II" through connections in her hometown of Natchez. This volume is a good resource for the "Trip Journal" and the Jefferson-Dunbar letters, but Rowland did little editing of the daily entries and simply presented them in printed form.

Three sources also give narrative summaries of Dunbar's notes. These include Jefferson's 1806 *Message from the President of the United States, Communicating Discoveries Made in Exploring the Missouri, Red River, and Washita, by Captains Lewis and Clark, Doctor Sibley and Mr. Dunbar with A Statistical Account of the Countries Adjacent.* Following a two-page introduction and overview of the boundaries of Louisiana and what had been accomplished in exploring the territory up to that time, Jefferson presented narrative descriptions of the various expeditions. That same year a Natchez printer, Andrew Marschall, published the same summary and included an "Appendix" by William Dunbar that listed his river course observations and included a section titled "Of the medical properties of the Hot Springs."[3] Both of these sources are merely brief synopses of the different explorations, lacking the detail found in the "Dunbar Report Journal."

George Hunter's four-volume journal was also deposited in the American Philosophical Society's collection. His detailed accounts are cited here as "Hunter Journals." In addition, Hunter completed a narrative form of his journal entitled "Manuscript Journal of Geo. Hunter up the Red & Washita Rivers with William Dunbar 1804 by Order US. & up to Hot Springs." There he expounded on various location name origins and made other observations as well. In this work, we refer to Hunter's narrative as the "Hunter Official Report." Because the "Hunter Journals" are divided into daily entries and correspond well to those of Dunbar, the expedition accounts in these journals are the source of most of the Hunter portions of this edited volume.

In 1962 the American Philosophical Society published John Francis McDermott's thorough historical edit of Hunter's journals, including the "Journal of an Excursion from Natchez on the Mississippi up the river Ouachita," in the *Transactions of the American Philosophical Society* under the title "The Western Journals of George Hunter, 1796–1805."[4] McDermott's work, now out of print, continues to be a rich source of biographical information on George Hunter; it also offers details about the identity of many of the individuals encountered by the expedition. Although McDermott did not pursue natural history sources or make many references to the scientific component of Hunter's entries, his work is a valuable source for understanding the expedition in a historical perspective.

The Philadelphia cartographer Nicholas King constructed an official map of the expedition in 1806 from the "Dunbar Report Journal," from Dunbar's observations, and from the sketches the Scottish explorer sent to the Jefferson administration. Called the "Map of the Washita River in Louisiana From the Hot Springs to the Confluence of the Red River with the Mississippi, Laid down from the Journal & Survey of Wm Dunbar, Esq. in the Year 1804," it remains one of the finest sources of early Louisiana and Arkansas geography and demography. The original maps may be found in the National Archives and the Library of Congress in Washington, D.C. A very fine version of the King map appeared in *Documents Relating to the Purchase and Exploration of Louisiana*. This book, published in 1904, pre-

3. *Discoveries Made in Exploring,* 113–164.
4. McDermott, "Western Journals of Hunter."

sented the map as an onionskin foldout.[5] We refer to the map as "Map of the Washita River in Louisiana."

Early-nineteenth-century spelling and punctuation were often idiosyncratic at best, even among the learned populace. But because the journal entries are not difficult to read, and for the sake of presenting authentic documents, we have made few attempts to correct the spelling of the two explorers. Punctuation and capitalization, however, are added when needed to preserve the continuity of thought and meaning.

In the same manner, much of the plant life noted in the journals is mentioned with such general descriptions that conclusive identifications would be impossible. In those instances, several species have been listed as possible *family, genus,* and/or *species* identities. Several entries contain nineteenth-century slang terms that have been forever lost, in referring to some plants; in these cases even speculation is improper.

Both Dunbar and Hunter made their journal entries at the end of the day or even the day after the one they were describing.[6] Because of this practice, some references are mistaken and some locations transposed. When these mistakes could be discerned, we added corrective or explanatory footnotes.

5. Dunbar's section in this book was titled "The Exploration of the Red, the Black, and the Washita Rivers." *Documents Relating to the Purchase and Exploration of Louisiana,* 159–167.

6. After Hunter's accident with his pistol (see Dunbar's and Hunter's entries for November 22), he was unable to write for some time. When he later wrote entries for the dates involved, he probably copied from Dunbar's journal and/or relied on his memory. Therefore the independence of Hunter's journal for a time following November 22 is an issue to be considered.

Bibliography

MANUSCRIPTS AND CORRESPONDENCE

Andrew Ellicott Papers, Library of Congress.

Bouligny, Francisco, to Bernardo de Gálvez, New Orleans, June 23, 1778. Archivo General de Indias, Seville, Spain.

D'Anemours, Charles Le Paulmier. "Mémoire sur le district de Ouachita dans le province de la Louisianne," 1803. American Philosophical Society, Philadelphia.

Dunbar Family Slave Ledger. Private Collection, Natchez, MS. Photocopy of scanned original in Special Collections, Riley-Hickingbotham Library, Ouachita Baptist University.

Dunbar, William. "The Exploration of the Red, the Black, and the Washita Rivers." In *Documents Relating to the Purchase and Exploration of Louisiana.* Boston: Houghton, Mifflin, 1904.

———. "Journal of a Geometrical survey Commencing at St. Catherine's Landing on the East shore of the Mississippi descending to the mouth of the red river, and from thence ascending that river, the black river and river of the Washita as high as the Hot Springs in the proximity of the last named river." Miscellaneous Manuscript Collection, American Philosophical Society, Philadelphia.

———. "Journal of a voyage commencing at St. Catherines Landing on the East bank of the Mississippi, proceeding downwards to the mouth of the Red River, & from thence ascending that river, the black river and the Washita river, as high as the Hot Springs in the proximity of the last mentioned river," October 16—December 10, 1804. Riley-Hickingbotham Library, Ouachita Baptist University. (Herein "Dunbar Trip Journal, Vol. I")

———. "Journal of a voyage commencing at St. Catherines Landing on the East bank of the Mississippi, proceeding downwards to the mouth of the Red River, & from thence ascending that river, the black river and the Washita river, as high as the Hot Springs in the proximity of the last mentioned river," December 11, 1804—January 26, 1805. Mississippi Department of Archives and History, Jackson, MS. (Herein "Dunbar Trip Journal, Vol. II")

———. "Journal of a voyage commencing at St. Catherines Landing on the East bank of the Mississippi, proceeding downwards to the mouth of the Red River, & from thence ascending that river, the black river and the Washita river, as high as the Hot Springs in the proximity of the last mentioned river." Miscellaneous Manuscript Collection, American Philosophical Society, Philadelphia. (Herein "Dunbar Report Journal")

George Hunter Papers. Miscellaneous Manuscript Collection, American Philosophical Society, Philadelphia.

Hunter, George. "Journal of an Excursion from Natchez on the Mississippi up the River Ouachita, 1804–1805." In "George Hunter Journals, 1796–1805," 4 vols. Miscellaneous Manuscript Collection, American Philosophical Society, Philadelphia. (Herein "Hunter Journals")

———. "Manuscript Journal of Geo. Hunter up the Red & Washita Rivers with William Dunbar 1804 by Order US. & up to Hot Springs." Miscellaneous Manuscript Collection, American Philosophical Society, Philadelphia. (Herein "Hunter Official Report")

Journal of Henry Muhlenberg. Miscellaneous Manuscript Collection, American Philosophical Society, Philadelphia.

Letters of Henry Muhlenberg. Miscellaneous Manuscript Collection, Historical Society of Pennsylvania, Philadelphia.

Letters Sent to the President. National Archives, Washington, DC.

"Relation of the Defeat of the Natchez by Mr. Perier, Commandant General of Louisiana. Attached to the letter of Mr. Perier, Governor of Louisiana, of March 25, 1731." Library of Congress, Washington, DC.

Thomas Jefferson Papers. Library of Congress, Washington, DC.

"Transactions on the Plantation of Wm Dunbar begun Monday the 27th May 1776, Mississippi, Richmond Settlement." Dunbar Papers, Mississippi Department of Archives and History. (Herein "Dunbar Diary")

William Dunbar Papers, Mississippi Department of Archives and History, Jackson.

William Dunbar Papers. Special Collections, Riley-Hickingbotham Library, Ouachita Baptist University.

William Dunbar Tomb Inscription, Family Cemetery, the Forest, Natchez, MS. Miscellaneous Manuscript Collection, American Philosophical Society, Philadelphia.

BOOKS, ARTICLES, AND OTHER SOURCES

Aber, James S. *Ouachita Mountains.* GO 568, Structural Geology, http://academic.emporia.edu/aberjame/struc_geo/ouachita/ouachita.htm, 1–5.

An Act authorizing the governor of the territory of Arkansas to lease the salt springs, in said territory, and for other purposes. 22nd Cong., 1st sess., ch. 70, 71, April 20, 1832.

Adams, Katherine J., and Lewis L. Gould. *Inside the Natchez Trace Collection: New Sources for Southern History.* Baton Rouge: Louisiana State University Press, 1999.

Adams, M. W. W., and R. M. Kelly. "Enzymes Isolated from Microorganisms That Grow in Extreme Environments." *Chemical and Engineering News,* December 18, 1995, 32–42.

Allen, John Logan, ed. *North American Exploration.* 3 vols. Lincoln: University of Nebraska Press, 1997.

Allen, Milford F. "Thomas Jefferson and the Louisiana-Arkansas Frontier." *Arkansas Historical Quarterly* 20, no. 1 (Spring 1961): 39–64.

Annals of the Congress of the United States, 1789–1824. 42 vols. Washington, DC: 1834–1856.

Arnold, Morris S. *Colonial Arkansas, 1686–1804: A Social and Cultural History.* Fayetteville: University of Arkansas Press, 1991.

———. *The Rumble of a Distant Drum: The Quapaws and Old World Newcomers, 1673–1804.* Fayetteville: University of Arkansas Press, 2000.

Arrowsmith, Aaron. "A Map exhibiting all new discoveries of the interior parts of North America," 1802. Map Collection, Sterling Library, Yale University.

Audubon, John James. *Journal of John James Audubon Made during His Trip to New Orleans in 1820–1821.* Ed. Howard Corning. Cambridge: Business Historical Society, 1929.

Baerg, W. J. *Birds of Arkansas.* Fayetteville: University of Arkansas, College of Agriculture, 1931.

Bailyn, Bernard. *Voyagers to the West: A Passage in the Peopling of America on the Eve of the Revolution.* New York: Alfred A. Knopf, 1986.

Baird, W. David. *The Quapaws.* New York: Chelsea House, 1989.

———. "The Reduction of a People: The Quapaw Removal, 1824–1834." *Red River Historical Review* 1 (Spring 1974): 21–36.

Bakeless, John, ed. *The Journals of Lewis and Clark.* New York: New American Library, 1964.

Baker, Charles. "A Study of Aboriginal Novaculite Exploitation in the Ouachita Mountains of South-Central Arkansas." Master's thesis, University of Arkansas, 1974.

Baker, T. Lindsay, ed. *The Texas River Country: The Official Surveys of the Headwaters, 1876.* College Station: Texas A&M Press, 1998.

Bates, Robert L., and Julia A. Jackson, eds. *Dictionary of Geological Terms.* 3rd ed. New York: Anchor Books, 1984.

Bedinger, M. S., F. J. Pearson Jr., J. E. Reed, R. T. Sniegocki, and G. G. Stone. *The Waters of Hot Springs National Park, Arkansas: Their Nature and Origin.* U.S. Geological Survey Professional Paper 1044-C. Washington, DC: U.S. Geological Survey, 1979.

Bergfelder, Bill. "The Origin of the Thermal Water of Hot Springs, Arkansas." Master's thesis, University of Missouri, Columbia, 1976.

Bolton, S. Charles. *Arkansas, 1800–1860: Remote and Restless.* Fayetteville: University of Arkansas Press, 1998.

———. *Territorial Ambition: Land and Society in Arkansas, 1800–1840.* Fayetteville: University of Arkansas Press, 1993.

Borror, Donald J., and Richard E. White. *A Field Guide to Insects: America North of Mexico.* Boston: Houghton Mifflin, 1970.

Branner, George C. *Cinnabar in Southwestern Arkansas.* Information Circular 2. Little Rock: Arkansas Geological Survey, 1932.

Brockman, C. Frank. *Trees of North America: A Field Guide to the Major Native and Introduced Species North of Mexico.* New York: Golden Press, 1986.

Brough, Charles H. "The History of Banking in Mississippi." *Journal of the Mississippi Historical Society* 3 (1901): 317–340.

Bruce, James. *Travels to Discover the Source of the Nile, in the Years 1768, 1769, 1770, 1771, 1772, and 1773.* 5 vols. Edinburgh: G. G. J. and J. Robinson, 1790.

Burt, William H., and Richard P. Grossenheider. *A Field Guide to the Mammals: Field Marks of All North American Species Found North of Mexico.* 3rd ed. Boston: Houghton Mifflin, 1976.

Calhoun, Robert Dabney. "A History of Concordia Parish." *Louisiana Historical Quarterly* 15, no. 1 (January 1932): 46–47.

Carpenter, Alma. "A Note on the History of the Forest Plantation, Natchez." *Journal of Mississippi History* 46 (May 1984): 130–137.

"Carte du Mississippi et ses embranchemens, 1802." Bibliothèque du Service Hydrographique de la Marine, Paris.

Carter, Cecile Elkins. *Caddo Indians: Where We Come From.* Norman: University of Oklahoma Press, 1995.

Carter, Clarence E., ed. *The Territorial Papers of the United States.* 28 vols. Washington, DC: Government Printing Office, continuing since 1933.

Chancery Records, 1810 (recorded in 1820). Adams County, Mississippi.

Chardon, Roland. "The Linear League in North America." *Annals of the Association of American Geographers* 70, no. 2 (June 1980): 143–147.

Chesterman, Charles W., and Kurt E. Lowe. *National Audubon Society Field Guide to Rocks and Minerals.* New York: Alfred A. Knopf, 2001.

Claiborne, John F. H. *Mississippi as a Province, Territory, and State.* Jackson, MS, 1880.

Clark, John G. *New Orleans, 1718–1812: An Economic History.* Baton Rouge: Louisiana State University Press, 1970.

Clark, William. "A Map of Part of the Continent of North America, 1805." Library of Congress, Washington, DC.

Clayman, Charles B., ed. *The American Medical Association Home Medical Encyclopedia.* 2 vols. New York: Random House, 1989.

Cohen, Paul E. *Mapping the West: America's Westward Movement 1524–1890.* New York: Rizzoli International Publications, 2002.

Coker, William S. "Research Possibilities and Resources for a Study of Spanish Mississippi." *Journal of Mississippi History* 34 (May 1972): 117–128.

Coker, William S., and Jack D. L. Holmes, eds. "Daniel Clark's Letter on the Mississippi Territory." *Journal of Mississippi History* 32 (May 1970): 153–169.

Cook, Phillip C. "The Pioneer Preachers of the North Louisiana Hill Country." *North Louisiana Historical Association* 14, no. 1 (Winter 1983): 1–12.

Cotter, Charles H. *A History of Nautical Astronomy.* New York: American Elsevier, 1968.

Cowdrey, Albert E. *This Land, This South: An Environmental History.* Rev. ed. Lexington: University Press of Kentucky, 1996.

Cox, Isaac. "The Exploration of the Louisiana Frontier, 1803–1806." *American Historical Association Annual Report* (1904): 151–174.

———. "The Louisiana-Texas Frontier." *Southwestern Historical Quarterly* 10 (July 1906): 1–75; 17 (July 1913): 1–42; 17 (October 1913): 140–187.

Cramer, Zadok. *The Ohio & Mississippi Navigator.* Pittsburgh, 1804. Available at the American Philosophical Society, Philadelphia.

Crouse, Nellis M. *Lemoyne d'Iberville: Soldier of New France.* Port Washington, NY: Kennikat Press, 1954.

Cushman, H. B. *History of the Choctaw, Chickasaw, and Natchez Indians.* Norman: University of Oklahoma Press, 1994.

Custis, Peter. "Observation relative to the Geography, Natural History, & etc., of the Country along the Red-River, in Louisiana." *Philadelphia Medical and Physical Journal* 2, pt. 2 (1806): 43–50.

Daniels, Jonathan. *The Devil's Backbone: The Story of the Natchez Trace.* Gretna: Pelican, 1998.

Davis, William C. *A Way through the Wilderness: The Natchez Trace and the Civilization of the Southern Frontier.* New York: HarperCollins, 1995.

Debeauvilliers. "Map of Louisiana, 1720." Bibliothèque du Service Hydrographique de la Marine, Paris.

De Ferrer, Jose Joaquin. "Astronomical Observations made by Jose Joaquin de Ferrer, chiefly for the Purpose of determining the Geographical Position of various Places in the United States, and other Parts of North America." *Transactions of the American Philosophical Society* 6 (1809): 1–6.

Department of Defense. *Map Reading and Land Navigation.* Washington, DC: Apple Pie Publishing, 1999.

DeRosier, A. H., Jr. "Natchez and the Formative Years of William Dunbar." *Journal of Mississippi History* 34 (February 1972): 29–47.

———. "William Dunbar, Explorer." *Journal of Mississippi History* 25 (July 1963): 165–185.

Dickinson, Mary B. *National Geographic Field Guide to the Birds of North America.* 3rd ed. Washington, DC: National Geographic Society, 1999.

Dickinson, Samuel D. "Don Juan Filhiol at *Écore à Fabri.*" *Arkansas Historical Quarterly* 46, no. 2 (Summer 1987): 133–155.

———. "An Early View of the Ouachita Region." *Old Time Chronicle: Folk History of Southwest Arkansas* 3, no. 7 (July 1990): 12–17.

———. "Historic Tribes of the Ouachita Drainage System in Arkansas." *Arkansas Archeologist* 21 (1980): 1–11

Din, Gilbert C. *Francisco Bouligny: A Bourbon Soldier in Spanish Louisiana.* Baton Rouge: Louisiana State University Press, 1993.

Din, Gilbert C., and Abraham P. Nasatir. *The Imperial Osages: Spanish Diplomacy in the Mississippi Valley.* Norman: University of Oklahoma Press, 1983.

Discoveries Made in Exploring the Missouri, Red River and Washita, by Captains Lewis and Clark, Doctor Sibley and William Dunbar, Esq. with a Statistical Account of the Countires Adjacent, with an Appendix by Mr. Dunbar. Natchez: Printed by Andrew Marschall, 1806.

Documents Relating to the Purchase and Exploration of Louisiana. Boston: Houghton, Mifflin, 1904.

Dorsey, George A. *Traditions of the Caddo.* Lincoln: University of Nebraska Press, 1997.

Dungan, James R. "Sir William Dunbar of Natchez: Planter, Explorer, and Scientist, 1792–1810." *Journal of Mississippi History* 23 (1961): 211–228.

DuPratz, M. LePage. *The History of Louisiana, or the Western Parts of Virginia, and Carolina.* Intro. by Henry Clay Dethloff. Baton Rouge: Claitor's Publishing Division, 1774. Reprint, 1972.

Early, Ann, ed. *Caddoan Saltmakers in the Ouachita Valley: The Hardman Site.* Arkansas Archeological Survey Research Series, no. 43. Fayetteville: Arkansas Archeological Survey, 1993.

Early, Ann, and Frederick Limp. "Fancy Hill: Part III, A Brief Study of the Arkansas Novaculite Quarries." In *Fancy Hill: Archeological Studies in the Southern Ouachita Mountains,* edited by Ann Early and Frederick Limp, 307–334. Arkansas Archeological Survey Research Series, no. 16. Fayetteville: Arkansas Archeological Survey, 1982.

"Early Roads of Hot Spring County." *Heritage* 2 (1971): 33–37.

Ellis, Joseph J. *American Sphinx: The Character of Thomas Jefferson.* New York: Alfred A. Knopf, 1997.

Etchieson, Meeks. "Prehistoric Novaculite Quarries in the Ouachita Mountains." 1998. Special Collections, Riley-Hickingbotham Library, Ouachita Baptist University.

———. "Prehistoric Use of Geological Resources in the Ouachita Mountains." 1989. Special Collections, Riley-Hickingbotham Library, Ouachita Baptist University.

"Extracts from the Travels of William Bartram." *Alabama Historical Quarterly* 17 (Fall 1955): 110–124.

"Extremophiles." *Microbiology Reviews* 18, nos. 2–3 (May 1996): 157–158.

Fay, R. O. "Geology of Paleozoic Strata in the Arbuckle Mountains of Southern Oklahoma." In *Geological Society of America, Centennial Field Guide, South-Central Section,* 183–188. Boulder, CO: Geological Society of America, 1988.

Featherstonhaugh, G. W. *Excursion through the Slave States, from Washington on the Potomac to the Frontier of Mexico; With Sketches of Popular Manners and Geological Notices.* 2 vols. London: John Murray, 1844.

Filhiol, Jean-Baptiste. "Description of the Ouachita in 1786." *Louisiana Historical Quarterly* 20 (1937): 476–478.

Flores, Dan, ed. *Jefferson and Southwestern Exploration: The Freeman and Custis Accounts of the Red River Expedition of 1806.* Norman: University of Oklahoma Press, 1984.

———. *Southern Counterparts to Lewis & Clark: The Freeman & Custis Expedition of 1806.* Norman: University of Oklahoma Press, 1984.

Ford, James A. *A Ceramic Decoration Sequence at an Old Indian Village Site Near Sicily Island, Louisiana.* New Orleans: Department of Conservation, Louisiana Geological Survey, 1935.

Ford, Paul L., ed. *Writings of Thomas Jefferson.* 10 vols. New York: G. P. Putnam, 1892–1899.

Fortier, Alcee, ed. *Louisiana: Comprising Sketches of Parishes, Towns, Events, Institutions, and Persons, Arranged in Cyclopedic Form.* 2 vols. New Orleans: Century Historical Association, 1914.

Foster, William C., ed. *The La Salle Expedition to Texas: The Journal of Henri Joutel, 1684–1687.* Austin: Texas State Historical Association, 1998.

Foti, Thomas, and Gerald Hanson. *Arkansas and the Land.* Fayetteville: University of Arkansas Press, 1992.

Galloway, Patricia, ed. *The Hernando de Soto Expedition: History, Historiography, and "Discovery" in the Southwest.* Lincoln: University of Nebraska Press, 1997.

Gardner, Juliet, ed. *Who's Who in British History.* London: Collins and Brown, 2000.

Garrett, Julia K. "Dr. John Sibley and the Louisiana-Texas Frontier, 1803–1814." *Southwestern Historical Quarterly* 45 (1942): 286–301. Also appears in five subsequent issues: 46 (1943), 47 (1944), 48 (1945), 49 (1946), 50 (1947).

Garvey, Joan B., and Mary Lou Widmer. *Beautiful Crescent: A History of New Orleans.* New Orleans: Garmer Press, 1984.

Gayarre, Charles. *History of Louisiana: The French Domination.* 4 vols. Gretna, LA: Pelican Publishing, 1903. Reprint, 1974. Page references are to the 1974 edition.

Gerstaecker, Frederick. *Wild Sports in the Far West.* Boston: Crosby, Nichols, 1859.

Giley, B. H., ed. *North Louisiana to 1865: Essays on the Region and Its History.* Ruston, LA: McGinty Trust Fund Publications, 1984.

Gingerich, Phillip D. *Pleistocene Extinctions in the Context of Origination-Extinction Equilibria in Cenozoic Mammals.* Tucson: University Arizona Press, 1984.

Green, John A. "Governor Perier's Expedition against the Natchez Indians: December, 1730—January, 1731." *Louisiana Historical Quarterly* 19, no. 3 (July 1936): 3–22.

Grimm, William Carey, and John T. Kartesz. *The Illustrated Book of Wildflowers and Shrubs: The Comprehensive Field Guide to More than 1,300 Plants of Eastern North America.* Mechanicsburg, PA: Stackpole Books, 1993.

Griswold, L. S. "Indian Quarries in Arkansas." *Proceedings of the Boston Society of Natural History* 26 (1895): 25–28.

———. "Whetstones and Novaculites of Arkansas." In *Arkansas Geological Surveys Annual Report for 1890,* vol. 3. Little Rock: Arkansas Geological Survey, 1892.

Haley, Boyd R., Ernest E. Glick, William V. Bush, Benjamin F. Clardy, Charles G. Stone, Mac B. Woodward, and Doy L. Zacry. "Geologic Map of Arkansas." Arkansas Geological Commission, Little Rock, AR, 1993.

Hall, James. "A Brief History of the Mississippi Territory." *Publications of the Mississippi Historical Society* 9 (1906): 539–576.

Halliday, E. M. *Understanding Thomas Jefferson.* New York: Harper Collins, 2001.

Hamilton, Peter J. "Running Mississippi's South Line." *Publications of the Mississippi Historical Society* 1 (1898): 157–168.

Hammon, Neal O., ed. *My Father, Daniel Boone: The Draper Interviews with Nathan Boone.* Lexington: University Press of Kentucky, 1999.

Hardin, J. Fair. "Don Juan Filhiol and the Founding of Fort Miró: The Modern Monroe, Louisiana." *Louisiana Historical Quarterly* 20 (April 1937): 463–465.

Heitman, Francis B. *Historical Register and Dictionary of the United States Army, 1789–1903.* 2 vols. Washington, DC: Government Printing Office, 1903.

———. *Historical Register of Officers of the Continental Army, 1775–1783.* Washington, DC: Government Printing Office, 1893.

Hilliard, Sam Bowers. *Hog Meat and Hoecake: Food Supply in the Old South.* Carbondale: Southern Illinois University Press, 1972.

"The History of Hot Spring County." *Heritage* 4 (1977): 3–7

Holmes, Jack D. L. "Stephen Minor: Natchez Pioneer." *Journal of Mississippi History* 42 (February 1980): 17–26.

Holmes, William H. "Aboriginal Novaculite Quarries in Garland County, Arkansas." *American Anthropologist* 4 (1891): 313–316.

Howell, Arthur H. *Birds of Arkansas.* U.S. Department of Agriculture, Biological Survey, Bulletin no. 38. Washington, DC: Government Printing Office, 1911.

Hudgins, Mary D. "William Dunbar, History Maker." *Arkansas Historical Quarterly* 1, no. 4 (December 1942): 331–341.

Hudson, Charles. *The Southeastern Indians.* Knoxville: University of Tennessee Press, 1976.

Jackson, Donald. *Thomas Jefferson and the Rocky Mountains: Exploring the West from Monticello.* Norman: University of Oklahoma Press, 1993. Reprint, 2002. Page references are to the 2002 edition.

———. *Thomas Jefferson and the Stony Mountains: Exploring the West from Monticello.* Urbana: University of Illinois Press, 1981.

———, ed. *Letters of the Lewis and Clark Expedition, with Related Documents, 1783–1854.* 2 vols. Urbana: University of Illinois Press, 1962. Reprint, Norman, OK: Red River Books, 1978.

Jackson, Jere L. *Nacogdoches: A Brief History.* Nacogdoches, TX: Stephen F. Austin University, 1995.

James, D. Clayton. *Antebellum Natchez.* Baton Rouge: Louisiana State University Press, 1968.

James, Douglas A., and Joseph C. Neal. *Arkansas Birds: Their Distribution and Abundance.* Fayetteville: University of Arkansas Press, 1986.

Jansma, Jerome, and Harriett H. Jansma. "George Engelmann in Arkansas Territory." *Arkansas Historical Quarterly* 50, no. 3 (Autumn 1991): 225–248.

Jefferson, Thomas. "The Limits and Bounds of Louisiana." In *Documents Relating to the Purchase and Exploration of Louisiana.* Boston: Houghton, Mifflin, 1904.

———. *Message from the President of the United States, Communicating Discoveries Made in Exploring the Missouri, Red River, and Washita, by Captains Lewis and Clark, Doctor Sibley and Mr. Dunbar with A Statistical Account of the Countries Adjacent.* New York: Hopkins and Seymour, 1806.

———. *Writings.* New York: Library of America, 1984.

Jones, R. W. "Some Facts concerning the Settlement and Early History of Mississippi." *Publications of the Mississippi Historical Society* 1 (1898): 85–89.

Jordan, Winthrop D. *Tumult and Silence at Second Creek: An Inquiry into a Civil War Slave Conspiracy.* Baton Rouge: Louisiana State University Press, 1993.

Kennedy, Roger G. *Mr. Jefferson's Lost Cause: Land, Farmers, Slavery, and the Louisiana Purchase.* Oxford: Oxford University Press, 2003.

Kennedy, W. James. "The Earliest Tissotiid Ammonite." *Paleontology* 20, pt. 4 (1977): 1–6.

King, Nicholas. "Map of the Red River in Louisiana. From the Spanish Camp where the exploring party of the U.S. was met by the Spanish Troops to where it enters the Mississippi," 1806. Library of Congress, Washington, DC.

———. "Map of the Washita River in Louisiana from the Hot Springs to the Confluence of the Red River with the Mississippi. Laid down from the Journal and Survey of Wm. Dunbar, Esq. in the Year 1804." In *Documents Relating to the Purchase and Exploration of Louisiana.* Boston: Houghton, Mifflin, 1904. Original map in the National Archives, Washington, DC.

Kircher, John, and Gordon Morrison. *A Field Guide to Eastern Forests: North America.* Boston: Houghton Mifflin, 1998.

Kniffen, Fred B. *The Indians of Louisiana.* Gretna, LA: Pelican, 1998.

Kukla, Jon. *A Wilderness So Immense: The Louisiana Purchase and the Destiny of America.* New York: Alfred A. Knopf, 2003.

Langenkamp, Robert D. *Handbook of Oil Industry Terms and Phrases.* New York: Pennwell Books, 1981. Reprint, 1984.

LaVere, David. *The Caddo Chiefdoms: Caddo Economics and Politics, 700–1835.* Lincoln: University of Nebraska Press, 1998.

Little, Elbert L. *National Audubon Society Field Guide to Trees, Eastern Region.* New York: Alfred A. Knopf, 2001.

"The Lower Ouachita Valley, ca. 1720–1725." Archives Nationales, Paris.

Lowery, Charles D. "The Great Migration to the Mississippi Territory, 1798–1819." *Journal of Mississippi History* 30 (August 1968): 173–192.

Lowrie, Walter, ed. *Early Settlers of Missouri: As Taken from Land Claims in the Missouri Territory.* Greenville, SC: Southern Historical Press, 1986. Originally published in *American State Papers, Public Lands.*

Macrery, Joseph. "Description of the Hot Springs and Volcanic Appearances in the Country adjoining the River Ouachitta, in Louisiana: Communicated in a Letter from Joseph Macrery, M.D. of Natchez, to Dr. Miller." In *New York Medical Repository,* second hexade, 47–50. 1806.

Madigan, Michael T., and Barry L. Marrs. "Extremophiles." *Scientific American,* April 1997, 82–87.

Maison Rouge Land Claims 7, Transcript of Proceedings in Case of *Landerneau vs. Coxe and Turner.* House of Representatives document 151, 27th Cong., 1841–1943.

Malone, Dumas. *Jefferson and His Time.* 6 vols. Boston: Little, Brown, 1952–1981.

Margry, Pierre, ed. *Découvertes et établissements des Français dans l'ouest et dans le sud de l'Amérique septentrionale, 1614–1754.* 6 vols. Paris: Jouast Printing, 1879–1888.

Martin, Edwin T. *Thomas Jefferson, Scientist.* New York: Schuman, 1952.

Martin, James C., and Robert Sidney Martin. *Maps of Texas and the Southwest, 1513–1900.* Austin: Texas State Historical Association, 1999.

Mathews, Catharine VanCortlandt. *Andrew Ellicott: His Life and Letters.* New York: Grafton Press, 1908.

McCain, William. "The Charter of Mississippi's First Bank." *Journal of Mississippi History* 1 (October 1939): 251–263.

McDermott, John Francis. "Philosophic Outpost on the Frontier: The Library of William

Dunbar of The Forest." Unpublished manuscript, copy at the Mississippi Department of Archives and History, Jackson, MS.

———. *The Spanish in the Mississippi Valley, 1762–1804*. Urbana: University of Illinois Press, 1974.

———. *Travelers on the Western Waters*. Urbana: University of Illinois Press, 1970.

———, ed. *Audubon in the West*. Norman: University of Oklahoma Press, 1965.

———, ed. "The Western Journals of Dr. George Hunter, 1796–1805." *Transactions of the American Philosophical Society* 53, pt. 4 (1962): 5–122.

McDonald, Forrest, and Grady McWhiney. "The Antebellum Southern Herdsmen: A Reinterpretation." *Journal of Southern History* 41 (May 1975): 155–157.

McFarland, J. D., III, and W. V. Bush, eds. *Contributions to the Geology of Arkansas*. Vol. 2. Arkansas Geological Commission Miscellaneous Publication no. 18-B. Little Rock: Arkansas Geological Commission, 1984.

McLemore, Richard Aubrey, ed. *A History of Mississippi*. 2 vols. Jackson: University and College Press of Mississippi, 1973.

McRaney, Joan Warren, and Carolyn Vance Smith. *Silhouettes of Settlers: Eight Sketches of Early Natchez Personalities*. Natchez, MS: Natchez Historical Society, 1974.

Metcalf, Bryce. *Original Members and Other Officers Eligible to the Society of Cincinnati, 1783–1938*. Strasbourg, VA: Shenandoah Public House, 1938. Reprint, Los Angeles: Eastwood, 1995. Page references are to the 1938 edition.

Miller, Mary Carol. *Lost Mansions of Mississippi*. Jackson: University Press of Mississippi, 1996.

Mills, William. *The Arkansas: An American River*. Fayetteville: University of Arkansas Press, 1988.

Mitchell, Jennie O'Kelly, and Robert Dabney Calhoun. "The Marquis de Maison Rouge, The Baron de Bastrop, and Colonel Abraham Morhouse: Three Ouachita Valley Soldiers of Fortune." *Louisiana Historical Quarterly* 20, no. 2 (April 1937): 289–462.

Moerman, Daniel E. *Native American Ethnobotany*. Portland, OR: Timber Press, 1998.

Moore, Dwight M. *Trees of Arkansas*. Little Rock: Arkansas Resources and Development Commission, Division of Forestry and Parks; Fayetteville: University of Arkansas, 1950.

Moore, Edith Wyatt. *Natchez Under-the-Hill*. Natchez, MS: Southern Historical Publications, 1958.

Moulton, Gary E., ed. *The Journals of the Lewis and Clark Expedition, August 25, 1804—April 6, 1805*. Vol. 3. Lincoln: University of Nebraska Press, 1987.

Murie, Olaus J. *A Field Guide to Animal Tracks*. 2nd ed. Boston: Houghton Mifflin, 1974.

National Geographic Atlas of the World. 7th ed. Washington, DC: National Geographic Society, 1999.

Neuman, Robert W. *An Introduction to Louisiana Archaeology*. Baton Rouge: Louisiana State University Press, 1984.

Norman, N. Phillip. "The Red River of the South." *Louisiana Historical Quarterly* 25 (April 1942): 397–535.

Nuttall, Thomas. *Journal of Travels into the Arkansas Territory during the Year 1819*. Ed. Savoie Lottinville. Norman: University of Oklahoma Press, 1979. Rev. ed., Fayetteville: University of Arkansas Press, 1999.O'Connell, Marjorie. "Phylogeny of the Ammonite Genus." *American Museum of Natural History Bulletin* 46, no. 5 (1922): 387- 412.

Opler, Paul A., and Vichai Malikul. *A Field Guide to Eastern Butterflies*. Boston: Houghton Mifflin, 1992.

Owsley, Frank Lawrence. *Plain Folk of the Old South*. Baton Rouge: Louisiana State University Press, 1949.

Parker, Sybil, ed. *McGraw-Hill Dictionary of Geology and Mineralogy*. New York: McGraw-Hill, 1997.

Petersen, James C., and David N. Mott. *Hot Springs National Park, Arkansas Water Resources Scoping Report*. Water Resources Division, National Parks Service, Department of the Interior. Washington, DC: Government Printing Office, 2002.

Peterson, Merrill D. *Thomas Jefferson and the New Nation*. New York: Oxford University Press, 1970.

———. *Thomas Jefferson: A Profile*. New York: Hill and Wang, 1967.

Peterson, Roger Tory. *A Field Guide to the Birds, Eastern and Central North America*. 4th ed. Boston: Houghton Mifflin, 1980.

Petrides, George A. *A Field Guide to Trees and Shrubs*. Boston: Houghton Mifflin, 1958.

Pough, Frederick H. *A Field Guide to Rocks and Minerals*. 3rd ed. Boston: Houghton Mifflin, 1960.

Randall, Henry S. *The Life of Thomas Jefferson*. 3 vols. New York, 1858.

Record of Pennsylvania Marriages, Prior to 1810. Vol. 2. Baltimore: Genealogical Publishing, 1968.

Reed, John C., and Francis G. Wells. *Geology and Ore Deposits of the Southwestern Arkansas Quicksilver District*. United States Geological Survey Bulletin 886-C. Washington, DC: Government Printing Office, 1938.

Richardson, Rupert N., Ernest Wallace, and Adrian Anderson. *Texas: The Lone Star State*. 4th ed. Englewood Cliffs, NJ: Prentice-Hall, 1981.

Riley, Franklin L. "Historical Adams County." *Publications of the Mississippi Historical Society* 1 (1898): 208–217.

———. "Sir William Dunbar—Pioneer Scientist of Mississippi." *Publications of the Mississippi Historical Society* 2 (1899): 85–111.

———. "Transition from Spanish to American Rule in Mississippi." *Publications of the Mississippi Historical Society* 3 (1900): 261–311.

Roberts, David C. *A Field Guide to Geology, Eastern North America*. Boston: Houghton Mifflin, 1996.

Robertson, James A., ed. *Louisiana under the Rule of the Spanish, French, and the United States, 1785–1807*. 2 vols. Cleveland: A. H. Clark, 1910–1911.

Robin, C. C. *Voyages dans l'intérieur de la Louisiane . . . pendant les années, 1802–1806*. 3 vols. Paris: Chez F. Buisson, 1807.

Robison, Henry W., and Robert T. Allen. *Only in Arkansas: A Study of the Endemic Plants and Animals of the State.* Fayetteville: University of Arkansas Press, 1995.

Roth, David. "Louisiana Hurricane History." www.srh.noaa.gov/lch/research/la18hu.htm.

Rowland, Dunbar, ed. *Mississippi Provincial Archives, French Dominion,* 3 vols. Jackson: Press of the Mississippi Department of Archives and History, 1927, 1929, 1932.

———, ed. *Official Letter Book of W.C.C. Claiborne, 1801–1816.* 6 vols. Jackson, MS: Printed for the State Department of Archives and History, 1917.

Rowland, Eron, comp. *Life, Letters, and Papers of William Dunbar of Elgin, Morayshire, Scotland, and Natchez, Mississippi: Pioneer Scientist of the Southern United States.* Jackson: Press of the Mississippi Historical Society, 1930.

Russell-Hunter, W. D. *The Life of Invertebrates.* New York: Macmillan, 1979.

Sansing, David, Sim C. Callon, and Carolyn Vance Smith. *Natchez: An Illustrated History.* Natchez, MS: Plantation, 1992.

Saunders, Charles Francis. *Edible and Useful Wild Plants of the United States and Canada.* New York: Dover, 1948. Reprint, 1976. Page references are to the reprint edition.

Schambach, Frank, and Leslie Newell. *Crossroads of the Past: 12,000 Years of Indian Life in Arkansas.* Little Rock: Arkansas Humanities Council; Fayetteville: Arkansas Archeological Survey, 1990.

Sealander, John A., and Gary A. Heidt. *Arkansas Mammals: Their Natural History, Classification, and Distribution.* Fayetteville: University of Arkansas Press, 1990.

Seeds, Michael A. *Foundations of Astronomy.* 3rd ed. Belmont, CA: Wadsworth, 1992.

Shinn, Josiah H. *Pioneers and Makers of Arkansas.* Little Rock, AR: Democrat Printing and Lithograph, 1908.

Shreve, Henry Miller. "Rough Sketch of that part of the Red Rivir [sic] in which the Great Raft is situated and the bayous, lakes, swamps, &c., belonging to or in the vicinity, 31.50′ to 32.20′ to 93.40′, 1833." Louisiana House Document 98, ser. 256.

Sibley, John. "Historical Sketches of the Several Tribes in Louisiana South of the Arkansas River and Between the Mississippi and the River Grand." In Thomas Jefferson, *Message from the President of the United States, Communicating Discoveries Made in Exploring the Missouri, Red River, and Washita, by Captains Lewis and Clark, Doctor Sibley and Mr. Dunbar with A Statistical Account of the Countries Adjacent.* New York: Hopkins and Seymour, 1806.

———. "A Report from Natchitoches in 1807." Ed. Annie Heloise Abel. Museum of the American Indian, New York, 1922.

Silverberg, Robert. *Mammoths, Mastodons, and Man.* New York: McGraw Hill, 1970.

Smith, Edwin B. *Keys to the Flora of Arkansas.* Fayetteville: University of Arkansas Press, 1994.

Smith, F. Todd. *The Caddo Indians: Tribes at the Convergence of Empires, 1542–1854.* College Station: Texas A&M University Press, 1995.

———. *The Caddos, the Wichitas, and the United States, 1846–1901.* College Station: Texas A&M University Press, 1996.

Smith, Ralph A., ed. and trans. "Exploration of the Arkansas River by Bénard de LaHarpe, 1721–1722." *Arkansas Historical Quarterly* 10, no. 4 (Winter 1951): 339–363.

Smyth, David. "David Thompson's Surveying Instruments and Methods in the Northwest, 1790–1812." *Cartographica* 18, no. 4 (1981): 1–17.

Stiles, Henry Reed, ed. *Joutel's Journal of LaSalle's Last Voyage, 1684–7.* Albany, NY: Josephy McDonough, 1906.

Stoddard, Major Amos. *Sketches, Historical and Descriptive, of Louisiana.* Philadelphia: Matthew Carey, 1812.

Stone, Charles G., Boyd R. Haley, and George W. Viale. *A Guidebook to the Geology of the Ouachita Mountains, Arkansas.* Little Rock: Arkansas Geological Commission, 1973.

Sutton, Keith, ed. *Arkansas Wildlife: A History.* Fayetteville: University of Arkansas Press, 1998.

Swanton, John R. *The Indians of the Southeastern United States.* Bureau of American Ethnology Bulletin 137. Washington, DC: Government Printing Office, 1946.

———. *Indian Tribes of the Lower Mississippi Valley and Adjacent Coast of the Gulf of Mexico.* Bureau of American Ethnology Bulletin 43. Washington, DC: Government Printing Office, 1911.

———. *The Indian Tribes of North America.* Bureau of American Ethnology Bulletin 145. Washington, DC: Government Printing Office, 1952.

———. *Source Material on the History and Ethnology of the Caddo Indians.* Norman: University of Oklahoma Press, 1996. Originally published as Bureau of American Ethnology Bulletin 132 (Washington, DC: Government Printing Office, 1942).

Tanner, H. S. "Map of Louisiana and Mississippi." 1820. Watson Library, Northwestern State University. Photocopy.

Thieret, John W., William A. Niering, and Nancy C. Olmstead. *National Audubon Society Field Guide to Wildflowers, Eastern Region.* New York: Alfred A. Knopf, 2001.

Thwaites, Reuben Gold. *Daniel Boone.* Williamstown, MA: Corner House Publishers, 1902.

Tonty, Henri de. "Memoir by the Sieur de la Tonty." Ed. Isaac J. Cox. *The Journeys of Rene Robert Cavelier Sieur de La Salle.* 2 vols. New York: Allerton Book, 1904–1905.

Troy, Alan A., T. Michael French, and John F. Ales. *Coal and Lignite Mining in Louisiana.* Baton Rouge: Louisiana Department of Natural Resources, 1993.

United States Army Corps of Engineers, New Orleans District. *The Mississippi River and Tributaries Project.* www.mvn.usace.army.mil/pao/bro/misstrib.htm.

United States Environmental Protection Agency. *Mississippi River Basin: Challenges.* www.epa.gov/msbasin/navigation.htm.

United States Geological Survey. Quadrangle maps of Mississippi, Arkansas, Colorado, Louisiana, New Mexico, and Texas. Denver, CO.

Van Doren, Mark, ed. *The Travels of William Bartram.* New York: Dover, 1928.

Venning, Frank D., and Manabu C. Saito. *Wildflowers of North America: A Guide to Field Identification.* New York: Golden Books, 1984.

Von Humboldt, Alexander. "General Map of the Kingdom of New Spain." In *Personal Travels to the Equinoctial Regions of America.* 3 vols. 1814–1826. Rev. ed., edited and translated by Thomasina Ross, 3 vols. London, 1907.

Walker, Winslow M. *The Troyville Mounds, Catahoula Parish, La.* Bureau of American Ethnology Bulletin 113. Washington, DC: Smithsonian Institution, 1936.

Wall, Bennett H., Light Townsend Cummins, Judith Kelleher Schafer, Edward F. Haas, and Michael Kurtz. *Louisiana: A History*. 3rd ed. Wheeling, IL: Harlan Davidson, 1997.

Wernert, Susan, ed. *North American Wildlife*. New York: Reader's Digest Association, 1982.

Whayne, Jeannie, comp. *Cultural Encounters in the Early South: Indians and Europeans in Arkansas*. Fayetteville: University of Arkansas Press, 1995.

Whayne, Jeannie, Thomas DeBlack, George Sabo III, and Morris S. Arnold. *Arkansas: A Narrative History*. Fayetteville: University of Arkansas Press, 2002.

White, J. M. "Territorial Growth of Mississippi." *Publications of the Mississippi Historical Society* 2 (1899): 125–132.

Whittington, G. P. "Doctor John Sibley of Natchitoches, 1757–1837." *Louisiana Historical Quarterly* 20 (October 1927): 467–512.

———. "Rapides Parish, Louisiana: A History." *Louisiana Historical Quarterly* 25 (October 1932): 573–580.

Wilds, John, Charles L. Dufour, and Walter G. Cowan. *Louisiana Yesterday and Today: A Historical Guide to the State*. Baton Rouge: Louisiana State University Press, 1996.

Williams, Ernest Russ. "Jean Baptiste Filhiol and the Founding of the Poste du Ouachita." In *North Louisiana to 1865: Essays on the Region and Its History*, ed. B. H. Giley, 1–10. Ruston, LA: McGinty Trust Fund Publications, 1984.

Young, Gloria A., and Michael P. Hoffman, eds. *The Expedition of Hernando de Soto West of the Mississippi, 1541–1543: Proceedings of the de Soto Symposia, 1988 and 1990*. Fayetteville: University of Arkansas Press, 1993.

Zim, Herbert S., Robert H. Baker, Mark R. Chartrand, and James Gordon Irving. *Stars: A Guide to the Constellations, Sun, Moon, Planets, and Other Features of the Heavens*. New York: St. Martin's Press, 2001.

Index